Wireless Networks

Series Editor

Xuemin Sherman Shen, University of Waterloo, Waterloo, ON, Canada

The purpose of Springer's Wireless Networks book series is to establish the state of the art and set the course for future research and development in wireless communication networks. The scope of this series includes not only all aspects of wireless networks (including cellular networks, WiFi, sensor networks, and vehicular networks), but related areas such as cloud computing and big data. The series serves as a central source of references for wireless networks research and development. It aims to publish thorough and cohesive overviews on specific topics in wireless networks, as well as works that are larger in scope than survey articles and that contain more detailed background information. The series also provides coverage of advanced and timely topics worthy of monographs, contributed volumes, textbooks and handbooks.

** Indexing: Wireless Networks is indexed in EBSCO databases and DPLB **

More information about this series at http://www.springer.com/series/14180

Hongliang Zhang • Boya Di • Lingyang Song
Zhu Han

Reconfigurable Intelligent Surface-Empowered 6G

Springer

Hongliang Zhang
Princeton University
Princeton, NJ, USA

Lingyang Song
Peking University
Beijing, China

Boya Di
Imperial College London
London, UK

Zhu Han
University of Houston
Houston, TX, USA

ISSN 2366-1186 ISSN 2366-1445 (electronic)
Wireless Networks
ISBN 978-3-030-73501-2 ISBN 978-3-030-73499-2 (eBook)
https://doi.org/10.1007/978-3-030-73499-2

This Springer imprint is published by the registered company Springer Nature Switzerland AG
The registered company address is: Gewerbestrasse 11, 6330 Cham, Switzerland

Preface

Launching from fifth-generation communications, which provide a single platform enabling a variety of data services, evolution towards sixth-generation (6G) has been kicked off, envisioning future wireless networks to become distributed intelligent communications, sensing, and localization systems. Though such a demand has gained support from existing techniques such as massive MIMO and small cells, they heavily depend on the quality of the uncontrollable wireless environments. Differently, reconfigurable intelligent surface (RIS) as a new type of ultra-thin meta material inlaid with multiple sub-wavelength scatters can create favorable propagation conditions by controlling the phase shifts of the reflected waves at the surface such that the received signals are directly reflected towards the receivers without any extra cost of power sources or hardware. It provides a revolutionarily new approach to actively improve the link quality and coverage, which sheds light into the future 6G. Due to the coupling between radio propagation and discrete phase shifts of RIS, existing protocols and design methodologies cannot be directly applied any more. Therefore, it is essential to develop new communication and signal processing techniques and to explore various RIS-based applications such as intelligent sensing and localization.

In this book, novel RIS-based smart radio techniques are discussed, targeting at achieving high-quality channel links in cellular communications via design and optimization of RIS construction. Unlike traditional antenna arrays, three unique characteristics of RIS will be revealed. First, the built-in programmable configuration of RIS enables analog beamforming inherently without extra hardware or signal processing. Second, the incident signals can be controlled to partly reflect and partly transmit through the RIS simultaneously, adding more flexibility to signal transmission. Third, RIS has no digital processing capability to actively send signals nor any radio frequency (RF) components. As such, it is necessary to develop novel channel estimation and communication protocols, design joint digital and RIS-based analog beamforming schemes, and perform interference control via mixed reflection and transmission.

Benefited from its ability to actively shape the propagation environment, the RIS technique is further investigated to achieve two types of wireless applications, that is, RF sensing and localization.

- In RF sensing, the influence of the sensing objectives on the wireless signal propagation can be potentially recognized by the receivers, which is then utilized to identify the objectives. Unlike traditional sensing techniques, RIS-aided sensing can actively customize the wireless channels and generate a favorable massive number of independent paths interacting with the sensing objectives. It is desirable to design RIS-based image recovery algorithms, optimize RIS configurations, and study efficient tracking methods.
- For the second application, that is, RIS aided localization, RIS is deployed between the access point (AP) and users such that the AP can analyze the reflected signals from users via different RIS configurations to obtain the accurate locations of users. However, this is a challenging task due to the dynamic user topology as well as the mutual influence between multiple users and the RIS. Therefore, the operations of RIS, AP, and multiple users need to be carefully coordinated. A new RIS-based localization protocol for device cooperation and an RIS configuration optimization algorithm are also required. Implementations with respect to different real-world scenarios will be delivered to illustrate the proposed approaches separately in the above applications.

The aim of this book is to educate control and signal processing engineers, computer and information scientists, applied mathematicians and statisticians, as well as systems engineers to carve out the role that analytical and experimental engineering has to play in RIS research and development. This book will emphasize on RIS technologies and applications for future communications.

This book is organized as below. Chapter 1 provides an overview of RISs and introduces fundamentals of RIS aided wireless communications. In Chap. 2, we present some study cases on RIS-aided multi-input multi-output (MIMO) communications. In Chap. 3, we show how to integrate RISs into existing wireless technologies. Finally, in Chap. 4, we give study cases to show the possible applications in RF sensing and localization.

Princeton, NJ, USA Hongliang Zhang

London, UK Boya Di

Beijing, China Lingyang Song

Houston, TX, USA Zhu Han

Contents

Acronyms

2D	Two-Dimensional
3D	Three-Dimensional
5G	Fifth-Generation
6G	Sixth-Generation
ADC	Analog-Digital Converter
AF	Amplify-and-Forward
AoA	Angle-of-Arrival
AP	Access Point
BS	Base Station
CDF	Cumulative Distribution Function
CPU	Central Processing Unit
CSI	Channel State Information
DAS	Distributed Antenna System
DC	Difference of Concave/Convex Functions
DF	Decode-and-Forward
D2D	Device-to-Device
EM	Electromagnetic
eMBB	Enhanced Mobile Broadband
FGPA	Field-Programmable Gate Array
GPS	Global Positioning System
HBF	Hybrid Beamforming
IRS	Intelligent Reflecting Surface
ISI	Inter-symbol Interference
IOS	Intelligent Omni-surface
IoT	Internet-of-Things
LNA	Low-Noise Amplifiers
LoS	Line-of-Sight
MDP	Markov Decision Process
MIMO	Multiple-Input and Multiple-Output
MLP	Multi-Layer Perceptrons
mMTC	Massive Machine Type Communications

MU	Mobile User
NLoS	Non-Line-of-Sight
NN	Neural Network
PIN	Positive Intrinsic Negative
PPS	Pulses-per-Second
QoS	Quality of Services
RF	Radio Frequency
RFID	Radio Frequency Identification
RIS	Reconfigurable Intelligent Surface
RSS	Received Signal Strength
SA	Simulated Annealing
SBS	Small Base Station
SDP	Semi-definite Program
SINR	Signal-to-Interference-plus-Noise Ratio
SNR	Signal-to-Noise Ratio
SOI	Space of Interest
TOA	Time-of-Arrival
TDM	Time Division Multiplex
UAV	Unmanned Aerial Vehicle
UE	User Equipment
URLLC	Ultra-reliable and Low Latency Communications
USRP	Universal Software Radio Peripheral
WLAN	Wireless Local Area Network
ZF	Zero-Forcing

Chapter 1
Introductions and Basics

In this chapter, we will first introduce the background and requirements in Sect. 1.1, and then overview the basics of the RIS in Sect. 1.2. Finally, we will present some fundamentals of RIS-aided wireless communications in Sect. 1.3.

1.1 Background and Requirements

The unprecedented demands for high-quality and seamless wireless services impose continuous challenges to existing cellular networks. Applications like enhanced mobile broadband (eMBB), ultra-reliable and low latency communications (URLLC), and massive machine type communications (mMTC) services are pushing the evolution of cellular systems towards the fifth-generation (5G). However, 5G cannot meet all requirements of future cellular systems in 2030, and researchers from academia and industry now start to focus on the six generation (6G) wireless communication networks.

In comparison with 5G networks, 6G wireless communication networks are expected to provide much higher spectral/energy/cost efficiency, higher data rate, 10 times lower latency, 100 times higher connection density, more intelligence for full automation, sub-centimeter geo-location accuracy, near 100% coverage, and sub-millisecond time synchronization [1]. These requirements have brought three main challenges:

- **Conflicts between low hardware cost and high spatial resolution:** In 5G communications, there are three types of techniques to achieve high spatial resolution. The first one is *high frequency communications* since it is easier to generate narrow beams in the high-frequency band [2]. However, high-frequency communications require dedicated radio frequency (RF) chains, whose cost increases rapidly as the number of users grows. The second technique is *massive Multi-Input Multi-Output (MIMO) systems* [3], while a huge number of antennas

H. Zhang et al., *Reconfigurable Intelligent Surface-Empowered 6G*, Wireless Networks, https://doi.org/10.1007/978-3-030-73499-2_1

each with a phase shifter imposes a significant cost in network deployment. Tlhe third method is *ultra-dense networking* [4]. With the coordination among densely deployed base stations (BSs), the ultra-dense networking can provide a high spatial resolution but it requires extremely high cost of deployment and coordination.

- **Conflicts between flexible network deployment and low energy consumption:** On the one hand, if access points (APs) are fixed, such as terrestrial infrastructure, there is no guarantee to adapt to dynamic user traffics since it is not cost-effective to deploy abundant access points to support the extreme situations. On the other hand, if APs are moving such as unmanned aerial vehicles (UAVs) [5], extra propulsion energy consumption is brought for these mobile APs [17].
- **Conflicts between convenience and high sensing accuracy:** Sensing and localization are important use cases in the future 6G. One possible solution is to deploy Internet-of-Things (IoT) devices to obtain sensing results. However, the size of these devices is typically negatively related to the sensing accuracy. Therefore, the non-negligible size and weights of these IoT devices will bring inconvenience in practice especially for health applications. Another possible scheme is MIMO assisted RF sensing where objects are detected by recognizing the impact of the objects on wireless signals. However, undesirable and uncontrollable signal propagation environments lead to low sensing accuracy.

Therefore, innovative technologies are expected to be low-cost in manufacture, easy and flexible deployment, and compatible with 6G demands on communications and sensing services.

Fortunately, recent developments in meta-materials have provided such an opportunity. Reconfigurable intelligent surfaces (RISs), also known as large intelligent surfaces [6], consisting an array of meta-materials have exhibited their capability to address above challenges for the following advantages [7]:

- **RISs are cost efficient in manufacture and deployment:** RISs are made of meta-materials, which are almost passive devices. In other words, RISs do not require extra circuits or power to process signals. On the other hand, RISs can be made in different shapes, including but not limited to building facades, indoor walls, and roadside billboards, which can facilitate the deployment.
- **RISs can control and customize favorable radio environments:** RISs can customize the radio propagation environments by reconfiguring the phase and amplitude of the incident signals. Through introducing virtual line-of-sight (LoS) links, RISs can control the power of received signals according to service requirements.
- **RISs are able to provide high accuracy contact/contactless sensing:** Due to the capability to provide the favorable propagation environment, RISs can increase the differences between two neighboring objects on the wireless signals, thereby improving the RF sensing resolution.

1.2 Overview of RISs

Due to attractive features introduced in the previous section, RISs have been recognized as an emerging technique in various applications and services. In this section, we will first briefly overview the development of meta-materials, which are major components of RISs. In what follows, we will introduce two types of RISs and their working principles. Finally, we will present the major applications of RISs.

1.2.1 History of Meta-Materials

The dielectric permittivity ϵ and magnetic conductivity μ of materials determine the capability of controlling electromagnetic (EM) waves, such as reflection and refraction. However, limited possibility of atom arrangement in natural materials leads to limited available values of ϵ and μ. As illustrated in Fig. 1.1, it is rare to have negative values of both ϵ and μ in natural materials, which motivates us to develop artificial materials with any values of ϵ and μ. Explorations of such artificial materials can be dated back to the nineteenth century. Some milestones of meta-materials are listed as follows:

- **Concept proposal of left-handed materials:** Left-handed materials were first described by Victor Veselago in 1968 [8], where ϵ and μ are negative. He proved that such materials could transmit light, but the phase velocity is contrary to wave propagation in natural materials.
- **First implementation of left-handed materials:** The first successful implementation of artificial materials with negative value of ϵ and μ was realized by John Pendry in 1996 and 1999, respectively [8]. He demonstrated that metallic wires aligned along the direction of a wave could provide $\epsilon < 0$, and a split ring with its axis placed along the direction of wave propagation could achieve $\mu < 0$.
- **Proposal of meta-surfaces:** In 1999, Dan Sievenpiper designed a 2D meta-material, i.e., meta-surface [9], which was fabricated as a printed circuit board. The structure consists of a lattice of metal plates connected to a solid metal sheet by vertical conducting vias, which simplified the design and manufacture compared to 3D meta-materials.

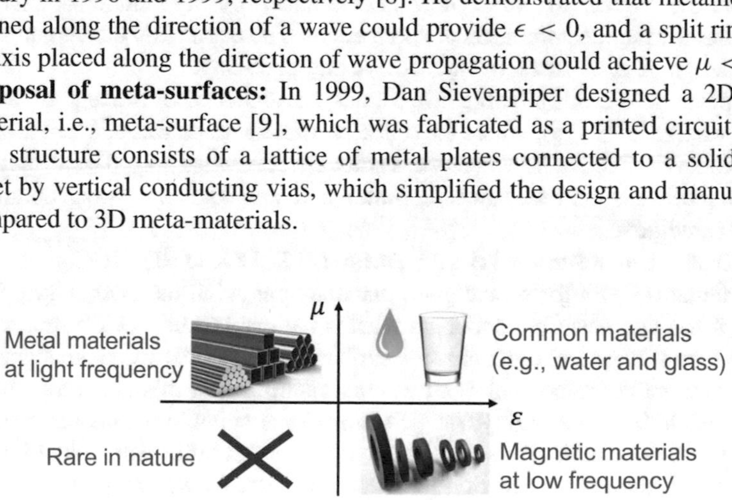

Fig. 1.1 ϵ and μ for natural materials

- **Proposal of programmable meta-surfaces:** The property of traditional meta-surfaces is fixed after they are manufactured, which cannot satisfy the changing requirements in practical applications. This motivates the development of programmable meta-surfaces. One possible solution is to incorporate adjustable materials into the meta-surfaces, such as liquid crystal and semiconductor. In 2002, varactor-based programmable meta-surfaces were developed [10]. By changing the biased voltage of the varactors, a near 360° continuous reflection phase tuning is achieved.
- **Transformation optics:** In 2006, a theory called transformation optics was proposed [11], which provides a theory to construct a meta-material with an arbitrary value of ϵ and μ. This is a powerful tool to enable a flexible control of EM.
- **Generalized Snell's law:** In 2011, the generalized Snell's law was proposed [12]. This law reveals that we can control the refraction direction by changing the phase gradient of the meta-surface.

From 2010s, researchers attempt to utilize meta-materials in wireless communication networks. Some related milestones are listed as follows:

- **Reconfigurable large reflectarray with PIN diodes:** The design and manufacture of varactor-based meta-surfaces are still difficult since continuous phase shifts are required to be achieved. However, in some practical applications, we only need finite phase shifts and it becomes promising to reduce the design complexity by decreasing the number of phase shifts. In 2011, PIN diodes were utilized as phase shifters in a large reflectarray [13]. As a result, the reflecting element structure can be simple and easily controlled.
- **Programmable meta-surfaces with PIN diodes:** Programmable meta-surfaces enabled by positive intrinsic negative (PIN) diodes were developed in 2014 [14], where the number of phase shifts is determined by the number of PIN diodes according to the requirements of applications. The use of PIN diodes can simplify the design of meta-surfaces and achieve digital coding.
- **Proposal of reconfigurable intelligent surfaces:** The concept of RIS was proposed in 2019 [15], where the reflective meta-surface was adopted to wireless networks to shape radio waves via a software controlled way. The availability of RISs motivates the redesign of common and well-known network communication paradigms.
- **Protype of meta-material reflectors:** NTT DOCOMO, INC., working in collaboration with the global glass manufacturer AGC Inc., announced in 2020 that it has successfully conducted what is believed to be world's first trial of a prototype transparent dynamic metasurface using 28 GHz 5G radio signals [16]. The new metasurface achieves dynamic manipulation of radio-wave reflection and penetration in a highly transparent package suitable for unobtrusive use in the windows of buildings and vehicles as well as on billboards. According to the test result, the communication data rate improves by 500 Mbps for the vehicle equipped with the 5G mobile station.

- **Proposal of intelligent omni-surfaces:** In 2020, our group proposed a type of RIS, which is referred to intelligent omni-surface (IOS) [18]. This type of RIS enables dual function of reflection and transmission, which can further extend the coverage of wireless networks.

1.2.2 Working Principles

To enable the functional of controlling the propagation environment, an RIS is composed of multiple layers, as shown in Fig. 1.2:

- **Outer layer** is a two dimensional (2D)-array of RIS elements, which will directly interact with incident signals;
- **Middle layer** is a copper plate which can prevent the signal energy leakage;
- **Inner layer** is a printed circuit connecting to the RIS controller which can control the phase shifts of the RIS elements.

Each RIS element is a low-cost sub-wavelength programmable meta-material particle, whose working frequency can vary from sub-6 GHz to THz [19]. When an EM wave impinges into the RIS element, a current will be induced by the EM wave, and this induced current will emit another EM radiation based on permittivity ϵ and permeability μ of the RIS. This is how the RIS element controls the wireless signals. An example of the meta-material particle is given in Fig. 1.3. As illustrated in this figure, PIN diodes are embedded in each element. By controlling the biasing voltage through the via hole, the PIN diode can be switched between "ON" and "OFF" states. The "ON" and "OFF" states of the PIN diodes lead to different values of ϵ and μ. As a result, this element will have different responses to incident signals by imposing different phase shifts and amplitudes.

It is worthwhile to point out that the amplitudes with different phase shifts will be fixed. In other words, once the phase shift is given, the corresponding amplitude

Fig. 1.2 Components of an RIS

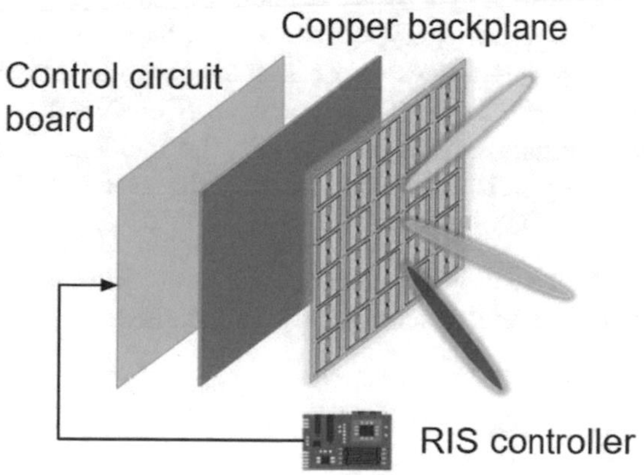

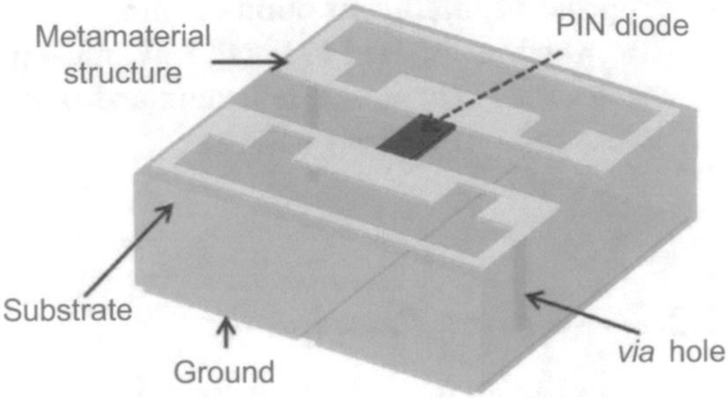

Fig. 1.3 An example of a programmable meta-material particle

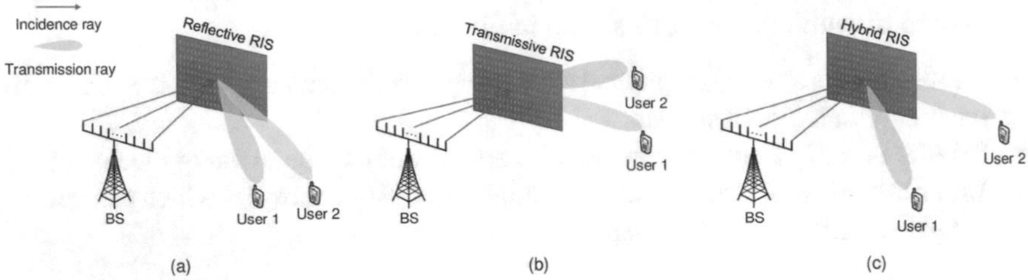

Fig. 1.4 Different types of RISs: (**a**) Reflective type; (**b**) Transmissive type; (**c**) Hybrid type

will also be determined. On the other hand, the amplitudes for different phase shifts can be the same or different, which is determined by the meta-material structure. In practice, we will make the amplitude almost the same for the ease of the control. For the phase shifts, the intervals between any two neighboring phase shifts are uniform in practical implementations.

According to the implementation of the RIS, RISs can be categorized into three types: reflective, transmissive, and hybrid [20]. In the following, we will elaborate on these types. These three types of RISs are also illustrated in Fig. 1.4.

- **Reflective Type:** In this type, the RIS only reflects incident signals towards the users on the same side of the BS. In the literature, this type of RIS is also referred to as intelligent reflecting surfaces (IRS) [21].
- **Transmissive Type:** In this type, incident signals will penetrate the RIS and be transmitted towards users on the opposite side of the BS.
- **Hybrid Type:** This type of RIS enables the dual function of reflection and transmission. In other words, the incident signals will be split into two parts: one part is transmitted and the other is reflected. This type of RIS is also referred to as IOS [18].

In the following chapters, the term "RIS" typically refers to the reflective type RIS for brevity unless otherwise specified.

1.2.3 *Applications of RISs*

Due to the capability of controlling the propagation environment, RISs have shown their potentials in a variety of applications. Typical applications of RISs can be broadly categorized into two types: wireless communications and RF sensing.

1.2.3.1 Wireless Communications

As shown in Fig. 1.5, the RIS can have but not limit to the following applications in wireless communications and networks:

- **Spectrum efficiency enhancement:** Since each RIS element can provide an extra communication link, the RIS can provide extra spatial diversity gain by exploiting these channels.
- **Coverage extension:** By deploying the RIS at the cell-edge or near the dead zone, i.e., the quality-of-services (QoS) of direct links between users and the BS are not satisfactory, the RIS can serve as passive relay to forward signals to these users.
- **Energy efficiency improvement:** Since the deployment of the RISs does not require extra energy-consuming hardware, the RISs are regarded as a promising solution to improve the energy efficiency by increasing the data rate while keeping the same amount of energy consumption.

On the other hand, when the phase shifts of the RIS are fixed, the RIS will be transparent to the BS and users. In other words, an RIS will be compatible with existing communication techniques or systems, and thus, it is possible to integrate the RIS with existing communication techniques.

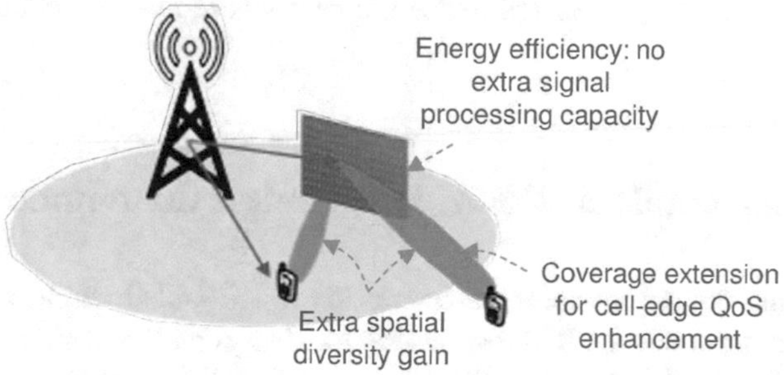

Fig. 1.5 Applications of the RIS for wireless communications

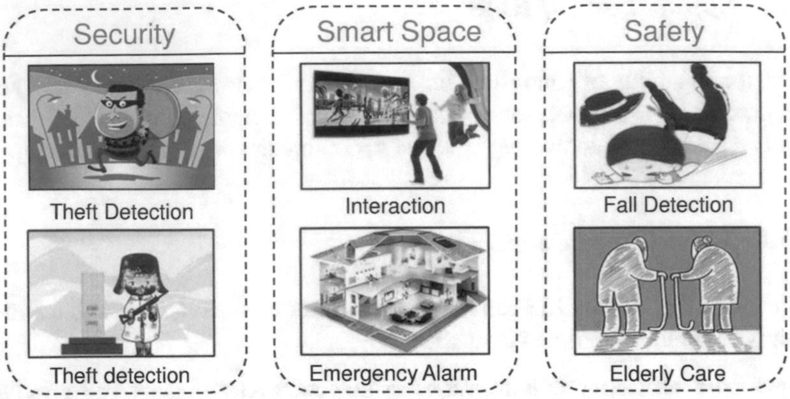

Fig. 1.6 Applications of the RIS for RF sensing

1.2.3.2 RF Sensing

Our current environment is covered by wireless signals. These ubiquitous signals provide the opportunities for RF sensing, where the contact or line-of-sight view of sensing targets is not necessary. As shown in Fig. 1.6, RF sensing has a wide range of applications in security, smart space, and safety.

Some traditional methods in RF sensing, such as WiFi sensing and mmWave Radar, utilize the impact of the target on WiFi signals or mmWave beams, where the sensing accuracy is limited by channel conditions. Fortunately, RISs have shown their capability to customize the propagation environment which can further improve the sensing accuracy. By changing the phase shifts of the RIS, the differences between the impact of two neighboring objects or blocks on wireless signals can be enhanced, and thus, the sensing accuracy can be improved [22]. The same idea can be easily extended to the use case where utilize wireless signals for localization.

1.3 Fundamentals of RIS-Aided Wireless Communications

In this section, we present some basics related to RIS-aided wireless communications. We first introduce the RIS response to incident signals and channel models. In what follows, we will discuss the differences between RISs and existing techniques. Finally, we will present some key challenges in RIS-aided wireless communications and RF sensing.

1.3.1 Response Model

Without loss of generality, we first consider the response of one RIS element. We denote the phase shift of the RIS by θ. As we have introduced in the previous section, if the RIS is incorporated with varactors, θ can be continuous, while θ has finite values if the RIS is implemented by PIN diodes. Mathematically, if K PIN diodes are implemented in the RIS, we have N possible phase shifts with $N \le 2^K$, which can be expressed by $\mathcal{N} = \{0, \ldots, 2n\pi/N, \ldots 2(N-1)\pi/N\}, 1 \le n \le N-1$. Since the RIS can be divided into three types, we will show how these three types of RISs response to the incident signals in the following.

1.3.1.1 Reflective Type

As shown in Fig. 1.7, let x be the incident signal and y be the reflected signal through the RIS element. Therefore, we have

$$y = \Gamma e^{-j\theta} x, \tag{1.1}$$

where j is the imaginary unit, i.e., $j^2 = 1$. Here, $\Gamma \in [0, 1]$ is the amplitude of the RIS response where $\Gamma = 1$ indicates that the incident signals are fully reflected while $\Gamma = 0$ implies that the incident signals are fully absorbed. In other words, the response of a reflective RIS element can be written by

$$\eta = \Gamma e^{-j\theta}. \tag{1.2}$$

Typically, we can know the response of a reflective RIS element by having the information of (Γ, θ).

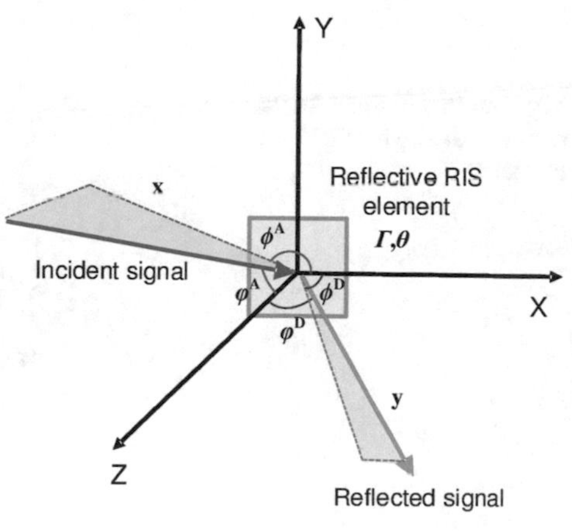

Fig. 1.7 Illustrations for the response of a reflective RIS element

1.3.1.2 Transmissive Type

As shown in Fig. 1.8, similar to the reflective type, the response of a transmissive RIS element can also be expressed in the form of (1.2). The only difference is that the incident signals fully penetrate the RIS element when $\Gamma = 1$.

1.3.1.3 Hybrid Type

The hybrid RIS element has the function of both reflective and transmissive types [23], as shown in Fig. 1.9. Therefore, the RIS will first split the energy of incident signals into two parts: one for transmissive signals and the other for reflective signals. To quantify the energy separation, we introduce a metric $\beta \in [0, +\infty)$, which is the power ratio of reflected signals to transmitted signals [24]. Therefore, assume that the no energy leakage for signal retransmission, and the response of the RIS elements to reflected and transmitted signals can be expressed by

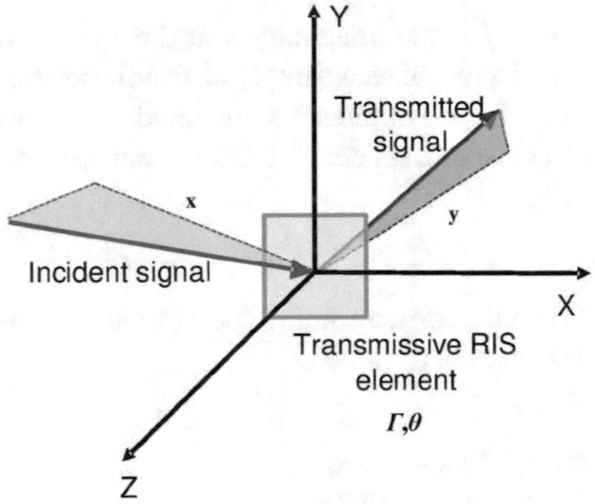

Fig. 1.8 Illustrations for the response of a transmissive RIS element

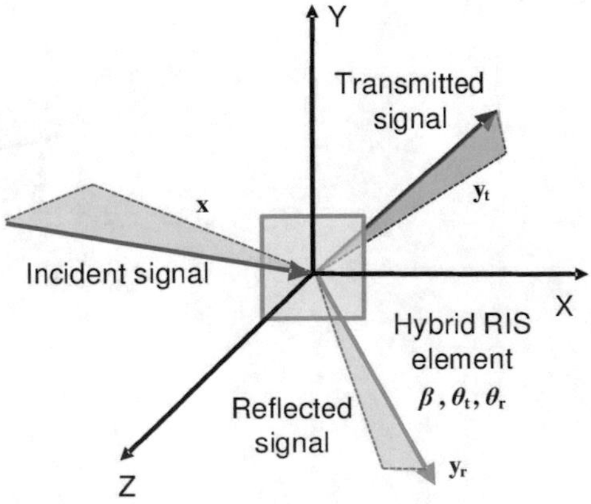

Fig. 1.9 Illustrations for the response of a hybrid RIS element

$$\eta_r = \sqrt{\frac{\beta}{1+\beta}} e^{-j\theta_r}; \tag{1.3}$$

$$\eta_t = \sqrt{\frac{1}{1+\beta}} e^{-j\theta_t}, \tag{1.4}$$

respectively. From (1.3), we can infer that the hybrid type will be reduced to the reflective type with $\beta = +\infty$, and the transmissive type with $\beta = 0$. Here, it is worthwhile to point out that the phase shift between incident and reflected signals θ_r could be different from the phase shift between incident and transmitted signals θ_t. Thus, we can know the response of a hybrid RIS element by having the information of $(\beta, \theta_r, \theta_t)$.

1.3.2 Channel Model

In this part, we discuss how to incorporate the existence of an RIS in the propagation environments into the channel model. This is a paramount in the performance evaluation of the RIS-aided communication network. In the following, we only focus on the reflective type RIS, the channel model of other two types can be set up in the same approach.

In an RIS-aided communication network, the receiver generally can receive signals from two paths: the direct link from the transmitter and the reflection link through the RIS. Assume a narrowband flat-fading scenario, and thus, the received signal can be written as

$$\boldsymbol{y} = \boldsymbol{H}_{\text{direct}}\boldsymbol{x} + \boldsymbol{H}_{\text{RIS}}\boldsymbol{x} + \boldsymbol{n}, \tag{1.5}$$

where $\boldsymbol{x}$ is the transmitted vector, $\boldsymbol{n}$ is the white Gaussian noise vector at the receiver. Here, $\boldsymbol{H}_{\text{direct}}$ represents the direct channel between the transmitter and the receiver and $\boldsymbol{H}_{\text{RIS}}$ represents the channel between the transmitter and the receiver through the RIS. The modeling of the channel through the RIS, $\boldsymbol{H}_{\text{RIS}}$, will be discussed below.

1.3.2.1 Large-Scale Path Loss Model

An important issue related to channel modeling is how to model the large scale propagation path loss from a transmitter to a receiver through the RIS. Consider one RIS element as shown in Fig. 1.10. In the case where the element is electrically

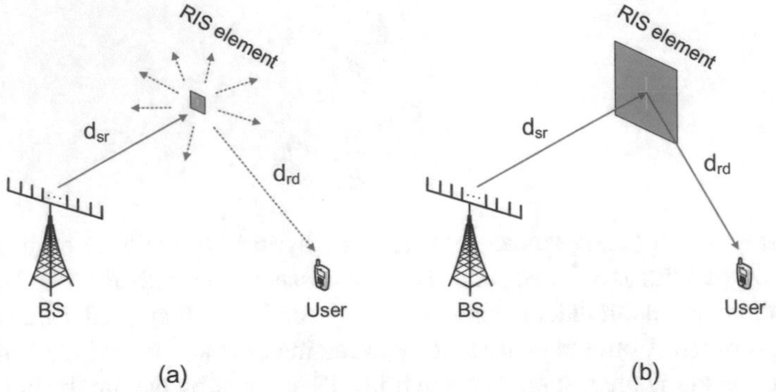

Fig. 1.10 Implementation effects on large scale path loss: (**a**) Electrically small size; (**b**) Electrically large size

small,[1] i.e., in Fig. 1.10a, the element acts as a diffuse scatter. In particular, it only receives *a point on the incident wavefront* and then diffusely scatter it in all directions around the element resulting in further power loss toward the receiver. On the other hand, in the case where the element is electrically large,[2] i.e., in Fig. 1.10b, each element acts as a smooth reflector. It receives *a section of the incident wavefront* and redirects it according to a programmable reflection angle. There is no further spreading of the wavefront at the RIS. It is usually referred to as anomalous reflection [25].

Hence, in case (b), the path loss through the RIS element will be proportional to the overall distance [26], $d_{sr} + d_{rd}$, i.e.,

$$\mathrm{PL}_{\mathrm{reflected}} \propto \frac{1}{(d_{sr} + d_{rd})^{\alpha}}, \tag{1.6}$$

where α is the path loss exponent. In case (a), the path loss will be proportional to the product of the distances, d_{sr} and d_{rd}, i.e.,

$$\mathrm{PL}_{\mathrm{scattered}} \propto \frac{1}{(d_{sr} \times d_{rd})^{\alpha}}. \tag{1.7}$$

It is evident that the difference in path loss between the two cases can be immense. Although case (b) can provide a lower pathloss in the far-field communications, case (a) can provide more freedom and higher spatial resolution due to more elements

[1]The element is called electrically small when the dimension of the element is on the same order of the wavelength.

[2]The element is called electrically large when its dimensions are orders of magnitudes larger than the wavelength.

can be implemented in a fixed space, and thus, is widely adopted in practical implementations.

1.3.2.2 Dyadic Backscatter Channel Model

As discussed above, each element in the RIS typically can be regarded as a regular omnidirectional scatter subject to the effects of fading and we can use the so-called dyadic backscatter channel [27] to model the channel through the RIS. Using this model, the channel through the RIS, $\boldsymbol{H}_{\text{RIS}}$, can be written as

$$\boldsymbol{H}_{\text{RIS}} = \boldsymbol{F}\boldsymbol{Q}\boldsymbol{G}, \tag{1.8}$$

where $\boldsymbol{F}$ is the channel from the RIS to the receiver, $\boldsymbol{G}$ is the channel from the transmitter to the RIS. The matrix, $\boldsymbol{Q}$, represents the interaction of the RIS with the transmitted waveform. Assuming no coupling between the RIS elements, $\boldsymbol{Q}$ should be diagonal matrix whose diagonal element should be the response of each element, as η introduced in Sect. 1.3.1.

From (1.8), for any assumed statistical distribution for the entries of $\mathbf{G}$ and $\mathbf{F}$, the overall channel distribution will be given by their product distribution. In general, this kind of cascaded fading is known to have more detrimental effects on performance compared to regular fading. However, increasing the number of RIS elements, does improve the fading characteristics [27].

1.3.2.3 Spatial Scattering Channel Model

Another possible method for channel modeling is to create a distinct propagation path. By this characterization, we can use the spatial model to write the received signal as

$$\boldsymbol{H}_{\text{RIS}} = \sum_{\ell=1}^{L} \alpha_\ell\, \eta_\ell\, \mathbf{a}_R\left(\theta_{R,\ell}, \phi_{R,\ell}\right)\, \mathbf{a}_T^*\left(\theta_{T,\ell}, \phi_{T,\ell}\right), \tag{1.9}$$

where L is the number of RIS elements, α_ℓ is a complex scalar representing the ℓ-th path gain excluding the effects of the RIS element, η_ℓ is a complex scalar representing the response of the ℓ-th RIS element, and $\mathbf{a}_R$ and $\mathbf{a}_T$ represent the array steering vectors at the receiver and the transmitter, respectively, with θ representing the azimuth angle and ϕ representing the elevation angle.

In general, the amplitude/phase parameter β_ℓ will be controllable by the RIS, regardless of its implementation technology. Furthermore, a metasurface-based RIS element may also control the angle of the reflection, and hence the angles of arrival at the receiver, $\theta_{R,\ell}$ and $\phi_{R,\ell}$, which will change the receiving array response accordingly. However, optimizing the reflection angles may be too complex to be practical. Note that the parameter β_ℓ is deterministic based on the current configuration of the RIS while the parameters α_ℓ can be stochastic to model the fading resulting from scattering around the receiver [28]. By this formulation, statistical characterization is possible without forcing the assumption of cascaded fading.

1.3.3 Comparisons with Existing Techniques

In this part, we will discuss the differences between the RIS and some similar existing techniques, such as massive MIMO, relay, and backscatter communications.

1.3.3.1 RIS vs. Massive MIMO

Similar to the RIS, massive MIMO can also consist of a large number of antennas which can be used for improving the spatial resolution. However, unlike the RIS working in a passive manner, the massive MIMO needs to process these signals in an active way. In general, the massive MIMO antennas only transmit or receive signals instead of forwarding the signals from the source to the destination. In other words, the RISs are full-duplex while the massive MIMO can be half-duplex. In addition, as shown in Fig. 1.11, the number of RF chains is equal to that of antennas [29]. Therefore, the cost and energy consumption is ineluctably extremely high due to the massive number of antennas compared with the RIS.

1.3.3.2 RIS vs. Relay

Similar to the RIS, a relay will receive the signals and forward the received signals to the destination [30]. Different from the RISs which do not require any signal processing, the relay nodes need some modules to process these received signals. According to the operating protocols, the relay nodes could be either half-duplex or full-duplex. Since the relay nodes need to amplify (amplify-and-forward (AF)

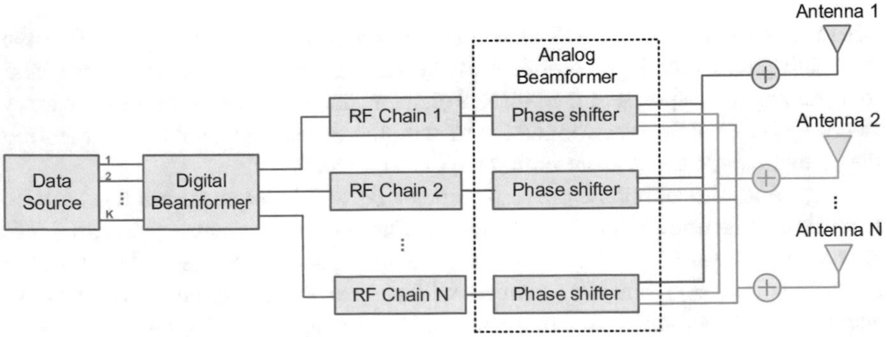

Fig. 1.11 Hybrid beamforming framework in massive MIMO systems

relays) or decode (decode-and-forward (DF) relays) the signals, these relay nodes need extra processing circuits which have a higher hardware cost and higher energy consumption compared to the RISs.

1.3.3.3 RIS vs. Backscatter

Similar to the RIS, the Radio Frequency Identification (RFID) tag will also reflect signals to the reader receiver in backscatter communications passively. However, as shown in Fig. 1.12, the controller will control the equivalent impedance according to the status of the sensor, and the reflected signals will be different [31]. In other words, the tag has information to transmit through reflection in backscatter communications, which makes it different from the RIS. Besides, the communication range should be short in order to distinguish the data transmitted by the tag. Due to the passive manner, the energy consumption is very low which can be supported by a button battery. In addition, the hardware cost is also low as the RFID tag is quite cheap.

In summary, the differences can be shown in Table 1.1.

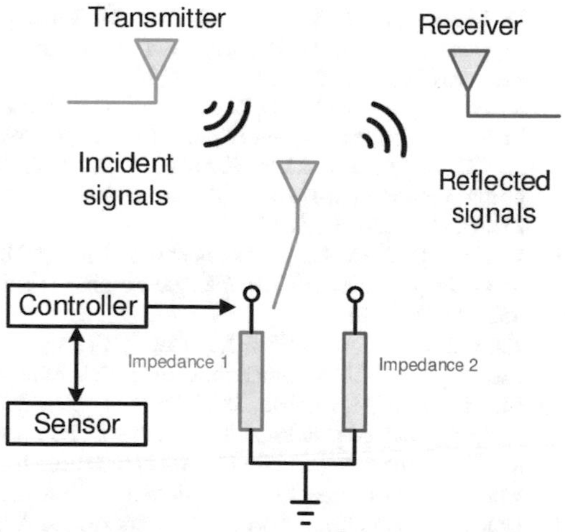

Fig. 1.12 System level diagram of backscatter communications

Table 1.1 RIS vs. existing techniques

Technology	Operating mechanism	Duplex	Hardware cost	Energy consumption
RIS	Almost passive, reflection/refraction	Full	Low	Low
Massive MIMO	Active, transmission/reception	Half/full	Very high	Very high
Relay	Active reception & transmission	Half/full	High	High
Backscatter	Passive, reflection with own data	Full	Very low	Very low

References

1. X. You, et al., Towards 6G wireless communication networks: vision, enabling technologies, and new paradigm shifts. Sci. China Inf. Sci. **64**(1), 110301 (2021)
2. W. Roh, J.-Y. Seol, J. Park, B. Lee, J. Lee, Y. Kim, J. Cho, K. Cheun, F. Aryanfar, Millimeter-wave beamforming as an enabling technology for 5G cellular communications: theoretical feasibility and prototype results. IEEE Commun. Mag. **52**(2), 106–113 (2014)
3. E.G. Larsson, O. Edfors, F. Tufvesson, T.L. Marzetta, Massive MIMO for next generation wireless systems. IEEE Commun. Mag. **52**(2), 186–195 (2014)
4. H. Zhang, L. Song, Y. Li, G.Y. Li, Hypergraph theory: applications in 5G heterogeneous ultra-dense networks. IEEE Commun. Mag. **55**(12), 70–76 (2017)
5. H. Zhang, L. Song, Z. Han, *Unmanned Aerial Vehicle Applications over Cellular Networks for 5G and Beyond* (Springer, Berlin, 2020)
6. Y. Liang, R. Long, Q. Zhang, J. Chen, H. Cheng, H. Guo, Large intelligent surface/antennas (LISA): making reflective radios smart. J. Commun. Inf. Netw. **4**(2), 4050 (2019)
7. M.A. Elmossallamy, H. Zhang, L. Song, K. Seddik, Z. Han, G.Y. Li, Reconfigurable intelligent surfaces for wireless communications: Principles, challenges, and opportunities. IEEE Trans. Cognitive Commun. Netw. **6**(3), 990–1002 (2020)
8. V.I. Slyusar, Metamaterials on antenna solutions, in *International Conference Antenna Theory and Techniques*, Lviv (2009)
9. D. Sievenpiper, L. Zhang, R.F.J. Broas, N.G. Alexópolous, E. Yablonovitch, High-impedance electromagnetic surfaces with a forbidden frequency band. IEEE Trans. Microw. Theory Tech. **47**(11), 2059–2074 (1999)
10. D. Sievenpiper, J. Schaffner, R. Loo, G. Tangonan, S. Ontinveros, R. Harold, A tunable impedance surface performing as a reconfigurable beam steering reflector. IEEE Trans. Antennas Propag. **50**(3), 384–390 (2002)
11. D. Schurig, J. Pendry, D.R. Smith, Calculation of material properties and ray tracing in transformation media. Opt. Express **14**(21), 9794–9804 (2006)
12. N. Yu, P. Genevet, M.A. Kats, F. Aieta, J.-P. Tetienne, F. Capasso, Z. Gaburro, Light propagation with phase discontinuities: generalized laws of reflection and refraction. Science **334**(6054), 333–337 (2011)
13. H. Kamoda, T. Iwasaki, J. Tsumochi, T. Kuki, O. Hashimoto, 60-GHz electronically reconfigurable large reflectarray using single-bit phase shifters. IEEE Trans. Antennas Propag. **59**(7), 2524–2531 (2011)
14. T.J. Cui, M.Q. Qi, X. Wan, J. Zhao, Q. Cheng, Coding metamaterials, digital metamaterials and programmable metamaterials. Light Sci. Appl. **3**(e208), 1–9 (2014)
15. M. Di Renzo, M. Debbah, D.-T. Phan-Huy, A. Zappone, M.-S. Alouini, C. Yuen, V. Sciancalepore, G.C. Alexandropoulos, J. Hoydis, H. Gacanin, J.D. Rosny, A. Bounceu, G. Lerosey, M. Fink, Smart radio environments empowered by AI reconfigurable meta-surfaces: an idea whose time has come. EURASIP J. Wirel. Commun. Netw. **2019**(1), 120 (2019)
16. DOCOMO Conducts World's First Successful Trial of Transparent Dynamic Metasurface. https://www.nttdocomo.co.jp/english/info/media_center/pr/2020/0117_00.html
17. S. Zhang, H. Zhang, B. Di, L. Song, "Cellular UAV-to-X Communications: Design and Optimization for Multi-UAV Networks". IEEE Trans. Wireless Commun. **18**(2), 1346–1359 (2019)
18. S. Zhang, H. Zhang, B. Di, Y. Tan, Z. Han, L. Song, Reflective-transmissive metasurface aided communications for full-dimensional coverage extension. IEEE Trans. Veh. Technol. **69**(11), 13905–13909 (2020)
19. H.-T. Chen, W.J. Padilla, J.M.O. Zide, A.C. Gossard, A.J. Taylor, R.D. Averitt, Active terahertz metamaterial devices. Nature **444**, 597–600 (2006)
20. S. Zeng, H. Zhang, B. Di, Y. Tan, Z. Han, H.V. Poor, L. Song, Reconfigurable intelligent surfaces in 6G: reflective, transmissive, or both?, IEEE Commun. Lett. arxiv: https://arxiv.org/abs/2102.06910

21. X. Yu, D. Xu, Y. Sun, D.W.K. Ng, R. Schober, Robust and secure wireless communications via intelligent reflecting surfaces. IEEE J. Sel. Areas Commun. **38**(11), 2637–2652 (2020)
22. J. Hu, H. Zhang, B. Di, L. Li, L. Song, Y. Li, Z. Han, H.V. Poor, Reconfigurable intelligent surfaces based radio-frequency sensing: design, optimization, and implementation. IEEE J. Sel. Areas Commun. **38**(11), 2700–2716 (2020)
23. H. Zhang, S. Zeng, B. Di, Y. Tan, M.D. Renzo, M. Debbah, L. Song, Z. Han, H.V. Poor, Intelligent reflective-transmissive metasurfaces for full-dimensional communications: principles, technologies, and implementation. arxiv: https://arxiv.org/pdf/2104.12313.pdf
24. S. Zhang, H. Zhang, B. Di, Y. Tan, M.D. Renzo, Z. Han, H.V. Poor, L. Song, Intelligent Omni-surface: ubiquitous wireless transmission by reflective-transmissive metasurface. arxiv: https://arxiv.org/abs/2011.00765
25. A.D. Rubio, V.S. Asadchy, A. Elsakka, S.A. Tretyakov, From the generalized reflection law to the realization of perfect anomalous reflectors. Sci. Adv. **3**(8), e1602714 (2017)
26. M.R. Akdeniz, Y. Liu, M.K. Samimi, S. Sun, S. Rangan, T.S. Rappaport, E. Erkip, Millimeter wave channel modeling and cellular capacity evaluation. IEEE J. Sel. Area Commun. **32**(6), 1164–1179 (2014)
27. J.D. Griffin, G.D. Durgin, Gains for RF tags using multiple antennas. IEEE Trans. Antennas Propag. **56**(2), 563–570 (2008)
28. O.E. Ayach, S. Rajagopal, S.A. Surra, Z. Pi, R.W. Heath, Spatially sparse precoding in millimeter wave MIMO systems. IEEE Trans. Wirel. Commun. **13**(3), 1499–1513 (2014)
29. T.E. Bogale, L.B. Le, A. Haghighat, L. Vandendorpe, On the number of RF chains and phase shifters and scheduling design with hybrid analog-digital beamforming. IEEE Trans. Wirel. Commun. **15**(5), 3311–3326 (2016)
30. E. Bjornson, Ö. Özdogan, E.G. Larsson, Intelligent reflecting surface versus decode-and-forward: How large surfaces are needed to beat relaying? IEEE Wirel. Commun. Lett. **9**(2), 244248 (2020)
31. C. Boyer, S. Roy, Backscatter communication and RFID: coding, energy, and MIMO analysis. IEEE Trans. Commun. **62**(3), 770–785 (2014)

Chapter 2
RIS Aided MIMO Communications

The explosive growth of the number of mobile devices has brought new user requirements and applications, and innovative networking characteristics for future communications [1], which necessitate radically novel communication paradigms [2]. During the past few years, there has been a growing interest in developing new transmission technologies for exploiting the implicit randomness of the propagation environment, so as to provide high-speed and seamless data services [3], such as spatial modulation [4] and massive multiple-input and multiple-output (MIMO) technologies [5]. However, the implementation of massive MIMO is still constrained by implementation bottlenecks, which include the hardware cost, the total energy consumption, and the high complexity for signal processing [6]. Due to these complexity constraints, therefore, the quality of service (QoS) is not always guaranteed in harsh propagation environments [7].

The recent development of metasurfaces has motivated the introduction of a new hardware technology for application to wireless communications, i.e., the reconfigurable intelligent surface (RIS) [8, 9], which can improve the spectral efficiency, the energy efficiency, the security, and the communication reliability of wireless networks [10]. An RIS is an ultra-thin surface containing multiple sub-wavelength nearly-passive scattering elements [11]. The sub-wavelength separation between adjacent elements of the RIS enables exotic manipulations of the signals impinging upon the surface [12]. A typical implementation of an RIS consists of many passive elements that can control the electromagnetic responses of the signals through the appropriate configuration of positive intrinsic negative (PIN) diodes distributed throughout the surface [13]. Depending on the ON/OFF status of the PIN diodes, several signal transformations can be applied [14]. The programmable characteristics of the RIS enables it to shape the propagation environment as desired [15], and allows for the re-transmission of signals to the receiver at a reduced cost, size, weight, and power [16].

In this chapter, we will discuss how the performance will be in RIS-aided MIMO communications. In Sect. 2.1, we consider the physical constraint of RIS, where

H. Zhang et al., *Reconfigurable Intelligent Surface-Empowered 6G*, Wireless Networks, https://doi.org/10.1007/978-3-030-73499-2_2

the number of the phase shift for each element is limited. In Sect. 2.2, we consider how many reflective elements of the RIS can offer an acceptable data rate. These two sections focus on the point-to-point communications, followed by multi-user communications. In Sect. 2.3, we optimize the placement and orientation of the RIS to extend the coverage. In Sect. 2.4, a hybrid beamforming scheme is introduced in the RIS-aided MIMO communications. Finally, in Sect. 2.5, the hybrid type RIS is utilize to achieve a full-dimensional coverage extension.

2.1 Limited Phase Shifts: How Many Phase Shifts Are Enough?

2.1.1 Motivations

In the literature, some initial works have studied the phase shifts optimizations in the RIS assisted wireless communications. Since the phase shifts on the RIS will significantly influence the received energy, it provides another dimension to optimize for further QoS improvement. In [17], beamforming and continuous phase shifts of the RIS were optimized jointly to maximize the sum rate for an RIS assisted point-to-point communication system. For multi-user cases, in [18], a joint power allocation and continuous phase shift design is developed to maximize the system energy efficiency.

However, most works assume continuous phase shifts, which are hard to be implemented in practical systems [19]. Therefore, it is worthwhile to study the impact of the limited phase shifts on the achievable data rate. In this section, we consider an uplink cellular network where the direct link between the base station (BS) and the user suffer from deep fading. To improve the QoS at the BS, we utilize a practical RIS with limited phase shifts to reflect signal from the user to the BS. To evaluate the performance limits of the RIS assisted communications, we provide an analysis on the achievable data rate with continuous phase shifts of the RIS, and then discuss how the limited phase shifts influence the data rate based on the derived achievable data rate.

The rest of this section is organized as follows. In Sect. 2.1.2, we introduce the system model for the RIS assisted communications. In Sect. 2.1.3, the achievable data rate is derived. The impact of limited phase shifts is discussed in Sect. 2.1.4. Numerical results in Sect. 2.1.5 validate our analysis. Finally, conclusions are drawn in Sect. 2.1.6.

2.1.2 System Model

Consider a narrow-band uplink cellular network[1] consisting of one BS and one cellular user [20]. Due to the dynamic wireless environment involving unexpected fading and potential obstacles, the Line-of-Sight (LoS) link between the cellular user and the BS may not be stable or even falls into a complete outage. To tackle this problem, we adopt an RIS to reflect the signal from the cellular user towards the BS to enhance the QoS. In the following, we first introduce the RIS assisted communication model, and then present the reflection-dominant channel model.

2.1.2.1 RIS Assisted Communication Model

The RIS is composed of $M \times N$ electrically controlled RIS elements. Each element can adjust the phase shift by leveraging PIN diode. In Fig. 2.1, we give an example of the element's structure. The PIN diode can be switched between "ON" and "OFF" states by controlling its biasing voltage, based on which the metal plate can add a different phase shift to the reflected signal. It is worthwhile to point out that the phase shifts are limited rather than continuous in practical systems [19].

In this section, we assume that the RIS is K bit coded, that is, we can control the PIN diodes to generate 2^K patterns of phase shifts with a uniform interval $\Delta\theta = \frac{2\pi}{2^K}$. Therefore, the possible phase shift value can be given by $s_{m,n}\Delta\theta$, where $s_{m,n}$ is an integer satisfying $0 \leq s_{m,n} \leq 2^K - 1$. Without loss of generality, the reflection factor of RIS element (m, n) at the m-th row and the n-th column is denoted by $\Gamma_{m,n}$, where

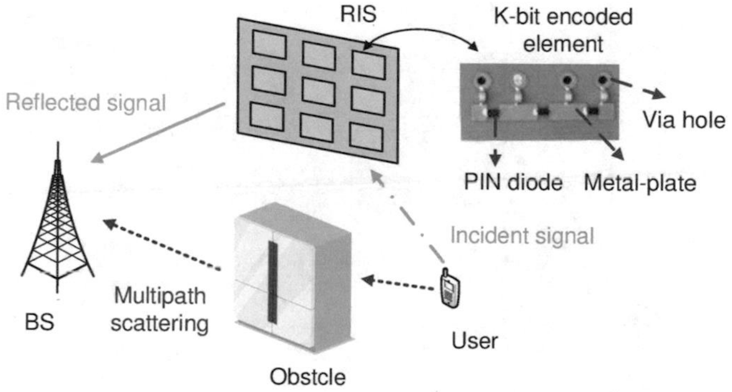

Fig. 2.1 System model for the RIS assisted uplink cellular network

[1] In the downlink case, although the working frequency and transmission power might be different, a similar method can be adopted since the channel model is the same due to the channel reciprocity.

$$\Gamma_{m,n} = \Gamma e^{-j\theta_{m,n}}, \tag{2.1}$$

where the reflection amplitude $\Gamma \in [0, 1]$ is a constant.

2.1.2.2 Reflection Dominant Channel Model

In this part, we introduce the channel modeling between the BS and the user. Benefited from the directional reflections of the RIS, the received power from the BS-RIS-user links actually is much stronger than the multipath effect as well as the degraded direct link between the BS and the user. For this reason, we use the Rician model to model the channel. Here, the BS-RIS-user links act as the dominant LoS component and all the other paths contributes to the non-LoS (NLoS) component. Therefore, the channel model between the BS and the user via RIS element (m, n) can be written by

$$\tilde{h}_{m,n} = \sqrt{\frac{\kappa}{\kappa+1}} h_{m,n} + \sqrt{\frac{1}{\kappa+1}} \hat{h}_{m,n}, \tag{2.2}$$

where $h_{m,n}$ is the LOS component, $\hat{h}_{m,n}$ is the NLOS component, and κ is the Rician factor indicating the ratio of the LoS component to the NLoS one.

As shown in Fig. 2.2, let $D_{m,n}$ and $d_{m,n}$ be the distance between the BS and RIS element (m, n), and that between element (m, n) and the user, respectively. Define the transmission distance through element (m, n) as $L_{m,n}$, where $L_{m,n} = D_{m,n} + d_{m,n}$. Therefore, the reflected LoS component of the channel between the

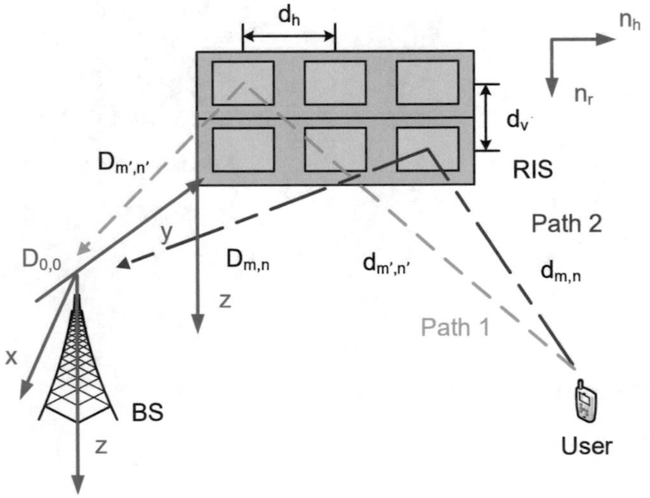

Fig. 2.2 Channel model for the RIS-based uplink cellular network

BS and the user via RIS element (m, n) can then be given by

$$\begin{aligned} h_{m,n} &= \sqrt{G D_{m,n}^{-\alpha} d_{m,n}^{-\alpha}} e^{-j\frac{2\pi}{\lambda} L_{m,n}} \\ &= \sqrt{G}\left[\sqrt{D_{m,n}^{-\alpha}} e^{-j\frac{2\pi}{\lambda} D_{m,n}}\right] \cdot \left[\sqrt{d_{m,n}^{-\alpha}} e^{-j\frac{2\pi}{\lambda} d_{m,n}}\right] \\ &= \sqrt{G} h_{m,n}^{t} h_{m,n}^{r}, \end{aligned} \tag{2.3}$$

where α is the channel gain parameter, G is the antenna gain, and λ is the wave length of the signal. Here, $h_{m,n}^{t}$ and $h_{m,n}^{r}$ are the channel between the BS and RIS element (m, n), as well as that between RIS element (m, n) and the user, respectively. Similarly, the NLoS component can be written by

$$\hat{h}_{m,n} = PL(D_{m,n}) PL(d_{m,n}) g_{m,n}, \tag{2.4}$$

where $PL(\cdot)$ is the channel gain for the NLoS component and $g_{m,n} \sim \mathcal{CN}(0, 1)$ denotes the small-scale NLoS components.

Using the geometry information, we can rewritten the channel gain by the following proposition.

Proposition 2.1 *When the distance $D_{m,n}$ between RIS element (m, n) and the BS and the distance $d_{m,n}$ between element (m, n) and the user are much larger than the horizontal and vertical distances between two adjacent elements, d_h and d_v, i.e., $D_{m,n}, d_{m,n} \gg d_h, d_v$, for $\forall m, n$, we have*

$$G D_{m,n}^{-\alpha} d_{m,n}^{-\alpha} \triangleq PL_{LoS}, \; PL(D_{m,n}) PL(d_{m,n}) \triangleq PL_{NLoS}, \tag{2.5}$$

where PL_{LoS} and PL_{NLoS} are constants.

Proof Assume that the BS is located at the local origin and the angle between the direction of the RIS and the x-y plane as θ_R. Define the principle directions of the RIS as $\boldsymbol{n}_h$ and $\boldsymbol{n}_v$, we have $\boldsymbol{n}_h = \boldsymbol{n}_x \cos\theta_R + \boldsymbol{n}_y \sin\theta_R$ and $\boldsymbol{n}_v = \boldsymbol{n}_z$, where $\boldsymbol{n}_x$, $\boldsymbol{n}_y$, and $\boldsymbol{n}_z$ are directions of x, y, and z axis, as shown in Fig. 2.3.

Denote $\boldsymbol{c}_{m,n}$ as the position of RIS element (m, n), where

$$\boldsymbol{c}_{m,n} = m d_h \boldsymbol{n}_h + n d_v \boldsymbol{n}_v + D_{0,0} \boldsymbol{n}_y. \tag{2.6}$$

Here, $D_{0,0}$ is the projected distance on y-asix between the BS and the RIS. Therefore, we have

$$\begin{aligned} D_{m,n} &= \left[(m d_h \cos\theta_R)^2 + (m d_h \sin\theta_R + D_{0,0})^2 + (n d_v)^2\right]^{\frac{1}{2}} \\ &\approx (m d_h \sin\theta_R + D_{0,0}) + \frac{(m d_h \cos\theta_R)^2 + (n d_v)^2}{2 D_{0,0}}, \end{aligned} \tag{2.7}$$

which can be achieved by $\sqrt{1+a} \approx 1 + a/2$ when $a \ll 1$. We can obtain the expression of the distance between the RIS and the user using the similar method.

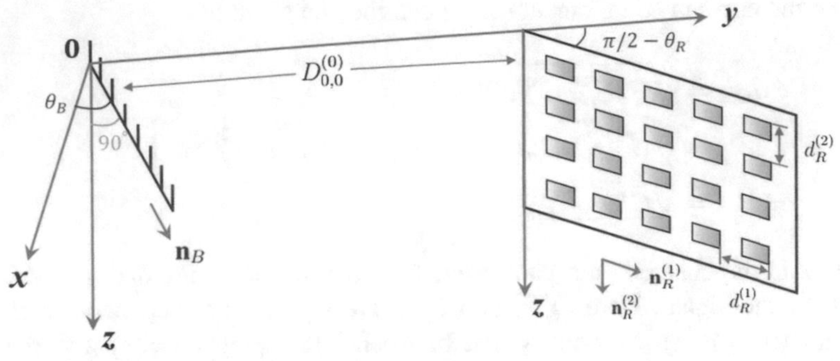

Fig. 2.3 Placement of the antenna arrays at the BS and the RIS

Since the distance between two RIS elements is much smaller than that between the BS and user, the pathloss of the BS and user via different RIS element can be regarded a constant. □

Besides, we can have the following remark to show how the location of the RIS influences the channel gain.

Remark 2.1 Given the transmission distance, i.e., $D_{m,n} + d_{m,n} = L$, where L is a constant, the channel gain will decrease first and then increase when the RIS is further to the BS.

Proof For the LoS component, according to (2.5), we have

$$\begin{aligned} PL_{LoS} &= G((L - D_{m,n})D_{m,n})^{-\alpha} \\ &= G(-(D_{m,n} - L/2)^2 + L^2/4)^{-\alpha}. \end{aligned} \tag{2.8}$$

Therefore, when $D_{m,n}$ increases, the LoS channel gain will decrease first and then increase. This trend will be the same for the NLoS component but the LoS component is typically dominant, and thus, the NLoS one can be neglected. This ends the proof. □

2.1.3 Achievable Data Rate Analysis

After traveling through the reflection dominated channel, the received signal at the user can be expressed by

$$y = \sum_{m,n} \Gamma_{m,n} \tilde{h}_{m,n} \sqrt{P} s + w, \tag{2.9}$$

where $w \sim \mathcal{CN}(0, \sigma^2)$ is the additive white Gaussian noise, P is the transmit power, and s is the transmitted signal with $|s|^2 = 1$. Therefore, the received Signal-to-Noise Ratio (SNR) can be expressed by

$$\gamma = \frac{P}{\sigma^2}\left(\sum_{m,n} \Gamma_{m,n}\tilde{h}_{m,n} \sum_{m',n'} \Gamma^*_{m',n'}\tilde{h}^*_{m,n}\right), \tag{2.10}$$

where s^* is the conjugate of a complex number s.

$$K_{req} = \log_2 \pi - \log_2 \arccos$$
$$\sqrt{\frac{\kappa+1}{\kappa\eta_{LoS}M^2N^2}\left(\left(1+\frac{\eta_{NLoS}}{\kappa+1}MN+\frac{\kappa\eta_{LoS}}{\kappa+1}M^2N^2\right)^{\epsilon_0} - 1 - \frac{\eta_{NLoS}}{\kappa+1}MN\right)}, \tag{2.11}$$

The received SNR can be maximized by optimizing the response of each RIS element, and thus the achievable data rate can be expressed by

$$R = \max_{\{\theta_{m,n}\}} \mathbb{E}\left[\log_2(1+\gamma)\right], \tag{2.12}$$

where

$$\mathbb{E}\left[\log_2(1+\gamma)\right] \approx \log_2$$
$$\left(1+\frac{\eta_{LoS}}{\kappa+1}MN+\frac{\kappa\eta_{NLoS}}{\kappa+1}\sum_{m,m',n,n'} e^{-j[\phi_{m,n}-\phi_{m',n'}+\theta_{m,n}-\theta_{m',n'}]}\right). \tag{2.13}$$

Here, $\eta_{NLoS} = \frac{P\Gamma^2}{\sigma^2}PL_{NLOS}$, $\eta_{NLoS} = \frac{P\Gamma^2}{\sigma^2}PL_{LOS}$, and $\phi_{m,n} = \frac{2\pi}{\lambda}L_{m,n}$. Derivations are given in Appendix 2.1.7.

To maximize the data rate, we need to let $\phi_{m,n} - \phi_{m',n'} + \theta_{m,n} - \theta_{m',n'} = 0$ for any (m, n) and (m', n'). Thus, we have the following proposition.

Proposition 2.2 *The optimal phase shifts with continuous value $\theta^*_{m,n}$ should satisfy the following equation:*

$$\theta^*_{m,n} + \phi_{m,n} = C, \tag{2.14}$$

where C is an arbitrary constant. The achievable data rate is

$$R = \log_2\left(1+\frac{\eta_{NLoS}}{\kappa+1}MN+\frac{\kappa\eta_{LoS}}{\kappa+1}M^2N^2\right). \tag{2.15}$$

Based on the expression of the achievable data rate, we can have the following remarks to show the upper and lower bounds of the data rate.

Remark 2.2 The upper bound of the data rate is achieved when we consider the pure LoS channel, i.e., $\kappa \to \infty$, where an asymptotic received power gain of $O(M^2N^2)$ can be obtained.

Remark 2.3 The lower bound of the data rate is achieved when we consider the Rayleigh channel, i.e., $\kappa \to 0$, where an asymptotic squared received power gain of $O(MN)$ can be obtained.

These two Remarks show that the data rate grows with κ, since the received SNR increases from the order of $O(MN)$ to that of $O(M^2N^2)$.

2.1.4 Analysis on the Number of Phase Shifts

In this subsection, we will discuss the influence of limited phase shifts on the data rate. Since the number of phase shifts is finite in practice, we will select the one which is the closest to the optimal one $\theta^*_{m,n}$ as given in (2.14), and denote it by $\hat{\theta}_{m,n}$. Define the phase shift errors caused by limited phase shifts as

$$\delta_{m,n} = \theta^*_{m,n} - \hat{\theta}_{m,n}. \tag{2.16}$$

With K coding bits, we have $-\frac{2\pi}{2^{K+1}} \leq \delta_{m,n} < \frac{2\pi}{2^{K+1}}$.

The SNR expectation $\hat{\gamma}$ with limited phase shifts can be written by

$$\begin{aligned}\mathbb{E}[\hat{\gamma}] &= \tfrac{\eta_{NLoS}}{\kappa+1} MN + \tfrac{\kappa\eta_{LoS}}{\kappa+1} \sum_{m,n,m',n'} e^{-j(C+\delta_{m,n}-C-\delta_{m'n'})} \\ &= \tfrac{\eta_{NLoS}}{\kappa+1} MN + \tfrac{\kappa\eta_{LoS}}{\kappa+1} \left| \sum_{m,n} e^{-j\delta_{m,n}} \right|^2 .\end{aligned} \tag{2.17}$$

Since $K \geq 1$, we have $\frac{2\pi}{2^{K+1}} \leq \frac{\pi}{2}$. Thus, the following inequality holds:

$$\left| \sum_{m,n} e^{-j\delta_{m,n}} \right|^2 \geq \left| MN\mathrm{Re}(e^{-j\frac{2\pi}{2^{K+1}}}) \right|^2 = M^2N^2 \cos^2\left(\frac{2\pi}{2^{K+1}}\right), \tag{2.18}$$

where $\mathrm{Re}(a)$ refers to the real part of a complex number a.

To quantify the data rate degradation, we define the error ϵ brought by limited phase shifts as the ratio of the data rate with limited phase shifts to that with continuous ones. To guarantee the system performance, ϵ should be larger than ϵ_0 with $\epsilon_0 < 1$, i.e.,

$$\epsilon = \log_2(1 + \mathbb{E}[\hat{\gamma}]) / \log_2(1 + \mathbb{E}[\gamma]) \geq \epsilon_0. \tag{2.19}$$

Recalling (2.17), we can obtain the requirements on the coding bits as

$$\frac{1+\frac{\eta_{NLoS}}{\kappa+1}MN+\frac{\kappa\eta_{LoS}}{\kappa+1}M^2N^2\cos^2\left(\frac{2\pi}{2^{K+1}}\right)}{\left(1+\frac{\eta_{NLoS}}{\kappa+1}MN+\frac{\kappa\eta_{LoS}}{\kappa+1}M^2N^2\right)^{\epsilon_0}}\geq 1. \tag{2.20}$$

Therefore, the required number of coding bits can be written as in (2.11).

In the following, we will provide a proposition to discuss impact of the RIS size on the required number of coding bits, i.e., the number of phase shifts.

Proposition 2.3 *Given the performance threshold ϵ_0, the required coding bits is a decreasing function of the RIS size. Specially, when the RIS size is sufficiently large, i.e., $MN\to\infty$, 1 bit is sufficient to satisfy the performance threshold.*

Proof We first discuss how the number of phase shifts influences the required coding bits. Define the RIS size $x=MN\geq 1$, and

$$f(x)=\frac{\kappa+1}{\kappa\eta_{LoS}x^2}\left(\left(1+\frac{\eta_{NLoS}}{\kappa+1}x+\frac{\kappa\eta_{LoS}}{\kappa+1}x^2\right)^{\epsilon_0}-1-\frac{\eta_{NLoS}}{\kappa+1}x\right). \tag{2.21}$$

According to (2.11), the required coding bits K has the same monotonicity as $f(x)$. Therefore, we only need to investigate how $f(x)$ changes when x increases.

Let $a=\frac{\kappa\eta_{LoS}}{\kappa+1}$ and $b=\frac{\eta_{NLoS}}{\kappa+1}$. We have

$$f'(x)=\frac{1}{ax^3}\left(2+bx+\frac{(\epsilon_0-1)2ax^2+(\epsilon_0-2)bx-2}{(1+bx+ax^2)^{1-\epsilon_0}}\right). \tag{2.22}$$

Note that $x\geq 1$ and $\epsilon_0\leq 1$, we have

$$\begin{aligned} f'(x)&\leq\frac{x^{-3}}{a}\left(2+bx+\left((\epsilon_0-1)2ax^2+(\epsilon_0-2)bx-2\right)\right)\\ &=\frac{x^{-3}}{a}\left((\epsilon_0-1)2ax^2+(\epsilon_0-1)bx\right)\leq 0. \end{aligned} \tag{2.23}$$

This implies that $f(x)$ decreases as x grows, i.e., the required number of coding bits decreases as the RIS size grows.

When the size of the RIS is sufficiently large, i.e., $MN\to\infty$, since $\epsilon<1$, we have $\frac{1}{M^2N^2}\left(\left(1+\frac{\eta_{NLoS}}{\kappa+1}MN+\frac{\kappa\eta_{LoS}}{\kappa+1}M^2N^2\right)^{\epsilon_0}-1-\frac{\eta_{NLoS}}{\kappa+1}MN\right)=0$. Therefore, the required number of coding bits should be $\log_2\pi-\log_2(\pi/2)=1$. □

2.1.5 Simulation Results

In this part, we verify the derivation of the achievable data rate and evaluate the optimal phase shift design for the RIS-assisted cellular network, the layout of which is given in Fig. 2.4. The parameters are selected according to 3GPP standard [21]

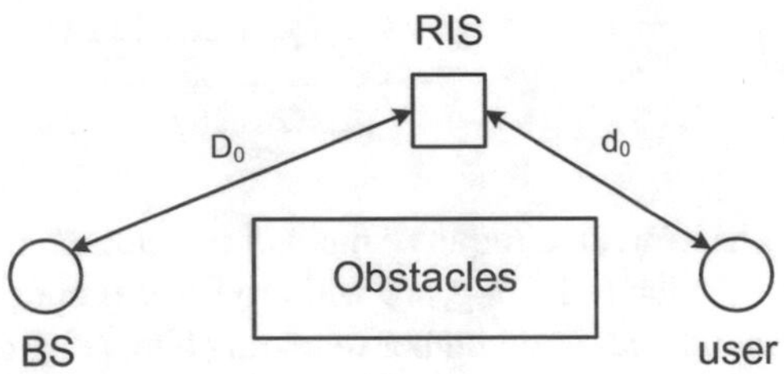

Fig. 2.4 Simulation layout for the RIS-based cellular network (top view)

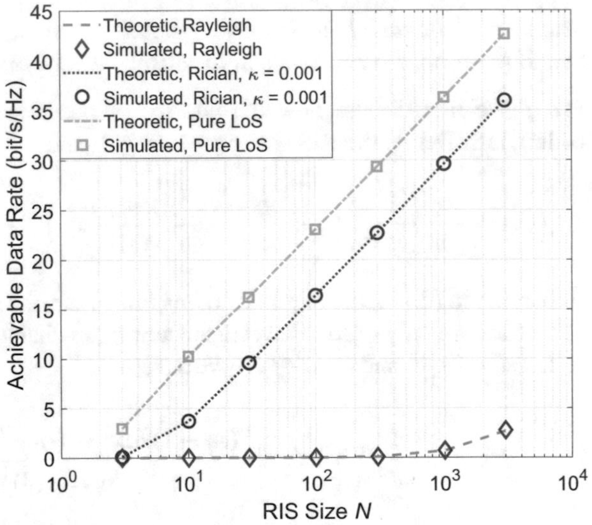

Fig. 2.5 Achievable data rate vs. RIS size N with continuous phase shifts

and existing works [19]. The distances between the BS and the user to the center of the RIS are given by $D_0 = 95$ m and $d_0 = 65$ m, and the heights of the BS and the RIS are 25 m and 10 m, respectively. The RIS separation is set as $d_h = d_v = 0.03$ m and the reflection amplitude is assumed to be ideal, i.e., $\Gamma = 1$. The transmit power is $P = 20$ dBm, the noise power is $\sigma^2 = -96$ dBm, and the carrier frequency f is 5.9 GHz. The UMa model in [21] is utilized to describe the path loss for both LoS and NLoS components. For simplicity, we assume that $M = N$ in this simulation. All numeral results are obtained by 3000 Monte Carlo simulations.

In Fig. 2.5, we plot the achievable data rate vs. the RIS size N with continuous phase shifts. From this figure, we can observe that our theoretic results are tightly close to the simulated ones. We can also observe that the data rate increases with the RIS size N, as more energy is reflected. In addition, when the RIS size is sufficiently large, the slope of the curve with the pure LoS channel is 4, which implies that the received SNR is proportional to the squared number of RIS elements.[2] This

[2]From (2.15), when N is sufficiently large in the pure LoS scenario, the data rate can be written by $R = 4\log_2(N) + z$, where z is a constant. Since we use the logarithmic coordinates for the RIS size N, the slope of the curve being 4 is equal to the received SNR being an order of $O(N^4)$.

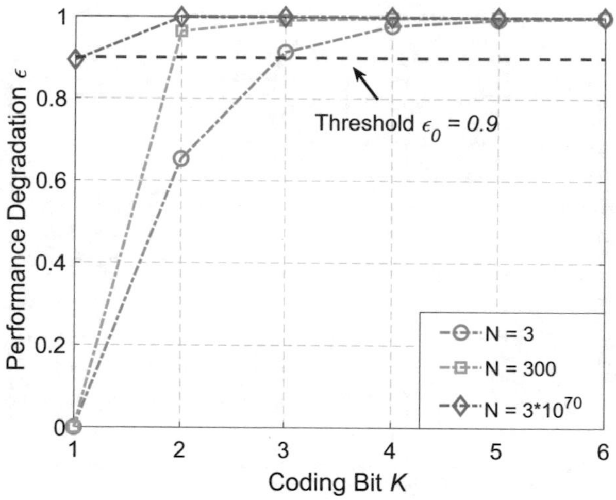

Fig. 2.6 Data rate degradation ϵ vs. coding bit K with $\kappa = 4$

result is consistent with Remark 2.2. Similarly, when the RIS size is sufficiently large, the slope of the curve with the Rayleigh channel is 2, which corresponds to Remark 2.3. In addition, we can observe that the data rate increases with the Rician factor, i.e., κ.

In Fig. 2.6, we plot the performance degradation ϵ vs. coding bit K with $\kappa = 4$ in the Rician channel model and threshold $\epsilon_0 = 0.9$. From this figure, we can find that the required number of phase shifts with the Rician channel model is: (1) 3 bit when the RIS size is small, e.g., $N = 3$; (2) 2 bit when the RIS size is moderate, e.g., $N = 300$; (3) 1 bit when the RIS size approaches to infinity, e.g., $N = 3 * 10^{70}$. These observations imply that the required coding bits decrease as the number of RIS elements grows, and 1 bit is enough when the RIS size goes to infinity, which verify Proposition 2.3.

In Fig. 2.7, we plot the performance degradation ϵ vs. distance between the BS and the RIS D_0 with $\kappa = 4$ to show how the location of the RIS influences the data rate degradation. For fairness, we assume that the transmission distance remains the same, i.e., $D_0 + d_0 = 160$. We can easily observe that the data rate degradation will decrease first and then increase as D_0 increases given RIS size N and coding bit K. This is because the channel gain will decrease and then increase as D_0 increases, which has been proved in Remark 2.1, and the performance degradation ϵ is a increasing function of the channel gain.[3] In addition, we can also learn that the variance caused by the location change of the RIS will be smaller when the RIS size

---

[3]When coding bit K and RIS size is given, the performance degradation ϵ can be written by $\epsilon = \frac{\log_2(1+g'x)}{\log_2(1+gx)}$, where $g' < g$ due to the limited phase shifts and x is the channel gain. It is easy to check that this function is increasing as x grows by calculating its first order derivative.

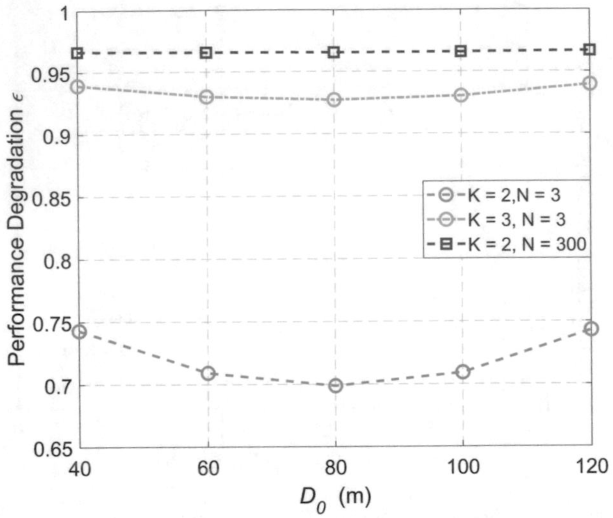

Fig. 2.7 Data rate degradation ϵ vs. distance between the BS and the RIS D_0 with $\kappa = 4$ and $D_0 + d_0 = 160$

becomes larger. This implies that the location change of the RIS might influence the number of required coding bits when the RIS size is small, while the number of required coding bits keeps unchanged when the RIS size is sufficiently large.

2.1.6 Summary

In this section, we have derived the achievable data rate of the RIS assisted uplink cellular network and have discussed the impact of limited phase shifts design based on this expression. Particularly, we have proposed an optimal phase shift design scheme to maximize the data rate and have obtained the requirement on the coding bits to ensure that the data rate degradation is lower than the predefined threshold.

From the analysis and simulation, given the location of the RIS, we can have the following conclusions: (1) We can achieve an asymptotic SNR of the squared number of RIS elements with a pure LoS channel model, and an asymptotic SNR of the number of RIS elements can be obtained when the channel is Rayleigh faded; (2) The required number of phase shifts will decrease as the RIS size grows given the data rate degradation threshold; (3) A number of phase shifts are necessary when the RIS size is small, while 2 phase shifts are enough when the RIS size is infinite.

2.1.7 Appendix

Due to the property of the logarithmic function [22], we have

$$\mathbb{E}[\log_2(1+\gamma)] \approx \log_2(1+\mathbb{E}[\gamma]). \tag{2.24}$$

Since $\frac{P}{\sigma^2}$ is constant, we will derive $\mathbb{E}[\gamma]$ in the following.

$$\begin{aligned}\mathbb{E}[\gamma] &= \frac{P\Gamma^2}{\sigma^2}\mathbb{E}\left[\sum_{m,n} e^{-j\theta_{m,n}}\tilde{h}_{m,n}\sum_{m',n'} e^{j\theta_{m',n'}}\tilde{h}^*_{m',n'}\right]\\ &= \frac{P\Gamma^2}{\sigma^2}\sum_{m,n,m',n'} e^{-j(\theta_{m,n}-\theta_{m',n'})}\left(\frac{1}{\kappa+1}\mathbb{E}[\hat{h}_{m,n}\hat{h}^*_{m',n}] + \frac{\kappa PL_{LoS}}{\kappa+1}e^{-j[\phi_{m,n}-\phi_{m',n'}]}+\right.\\ &\left. 2\frac{\sqrt{\kappa PL_{LOS}}}{\kappa+1}\mathrm{Re}\left\{e^{j\phi_{m,n}}\mathbb{E}[\hat{h}_{m,n}]\right\}\right).\end{aligned} \tag{2.25}$$

Since $\hat{h}_{m,n}$ has a zero mean, the final term in (2.25) equals to 0. Moreover, since $\hat{h}_{m,n}$ is independent for different elements (m, n) and (m', n'), the following equation holds:

$$\mathbb{E}[\hat{h}_{m,n}\hat{h}^*_{m',n'}] = \begin{cases} PL_{NLoS}, \text{if } m = m', n = n', \\ 0, \text{otherwise.}\end{cases} \tag{2.26}$$

Therefore, we have

$$\mathbb{E}[\gamma] = AMN + B\sum_{m,m',n,n'} e^{-j[\phi_{m,n}-\phi_{m',n'}+\theta_{m,n}-\theta_{m',n'}]}. \tag{2.27}$$

This ends the proof. □

2.2 Size Effect: How Many Reflective Elements Do We Need?

2.2.1 Motivation

In the previous section, we focus on the phase shift optimization/analysis in RIS assisted wireless communications. According to the results, the size of the RIS will also influence the system sum-rate. However, *how many RIS reflective elements are sufficient to provide an acceptable system sum-rate* remains an open problem in the literature. In this section, we consider an RIS assisted downlink multi-user cellular network. To quantify the impact of the number of RIS reflective elements on the system sum-rate, we first provide an asymptotic analysis of the system capacity for the RIS-assisted downlink multi-user MISO communications with zero-forcing (ZF) precoding which is easy to implement. Based on this capacity analysis, we further discuss how many RIS reflective elements are sufficient to provide an acceptable system sum-rate.

The rest of this section is organized as follows. In Sect. 2.2.2, we introduce a system model for RIS assisted multi-user communications. In Sect. 2.2.3, the asymptotic achievable data rate is derived. The relation between the number of RIS reflective elements and the system sum-rate is discussed in Sect. 2.2.4. Numerical results in Sect. 2.2.5 validate our analysis. Finally, conclusions are drawn in Sect. 2.2.6.

2.2.2 System Model

2.2.2.1 Scenario Description

Consider a narrow-band downlink multi-user MISO network as shown in Fig. 2.8, where one BS with M antennas serves K single-antenna users, where $M \geq K$ [23]. Due to the dynamic wireless environment involving unexpected fading and potential obstacles, the LoS link between the cellular users and the BS may not be stable or even falls into a complete outage. To enhance the QoS of the communication link, an RIS is adopted to reflect the signal from the BS and directly project the signals to the users by actively shaping the propagation environment into a desirable form.

The RIS is composed of N electrically controlled RIS reflective elements with the size of $a \times b$. Each reflective element can adjust the phase shift by leveraging PIN diodes. A PIN diode can be switched between "ON" and "OFF" states, based on which the metal plate can add a different phase shift to the reflected signal. Define θ_n as the phase shift for reflective element n, and the reflection factor of

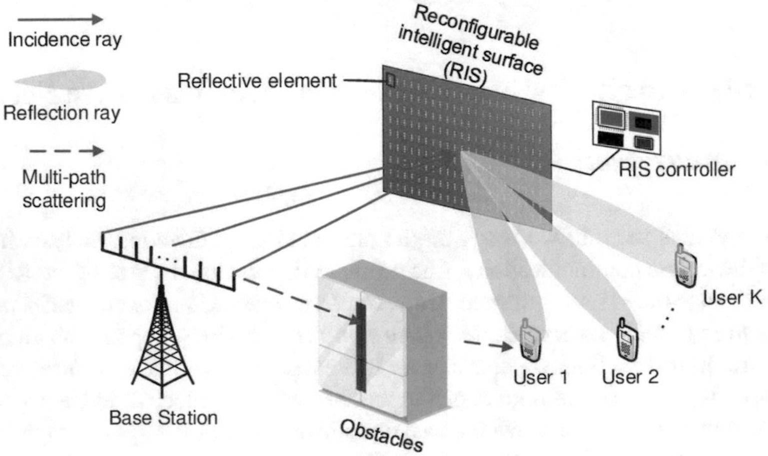

Fig. 2.8 System model of an RIS assisted downlink multi-user MISO network

reflective element n can be written by $\Gamma_n = \Gamma e^{-j\theta_n}$, where the reflection amplitude $\Gamma \in [0, 1]$ is a constant.

2.2.2.2 Channel Model

Let $\boldsymbol{G} \in \mathbb{C}^{K \times M}$ be the channel matrix between the BS and users, where $g^{km} \triangleq [\boldsymbol{G}]_{km}$ is the channel coefficient between user k and antenna m at the BS. Since each reflective element will reflect the signals from the BS to the users, the channel g_{km} consists of N paths, and we denote the channel gain from antenna m to user k through the n-th reflective element as g_n^{km}.

The channel matrix $\boldsymbol{G}$ models independent fast fading, path loss, and RIS response. To be specific, g_n^{km} can be written by

$$g_n^{km} = (\beta_n^k)^{-1/2} h_n^{km} \Gamma_n. \tag{2.28}$$

Here, h_n^{km} is independent fast fading coefficient with $\mathbb{E}[h_n^{km}] = 0$ and $\mathbb{E}[(h_n^{km})^2] = 1$. Based on the results in [24], the path loss is related with the incident angle α_n^k, i.e., $\beta_n^k = \frac{64\pi^3}{Aab\lambda^2} \frac{(l_n d_n^k)^2}{\cos^3(\alpha_n^k)}$, where A is the antenna gain, λ is the wavelength, l_n is the distance between the BS and the n-th reflective element,[4] and d_n^k is the distance between user k and the n-the reflective element. Therefore, we have

$$g^{km} = \sum_{n=1}^{N} g_n^{km} = (\bar{\beta}^k)^{-1/2} \sum_{n=1}^{N} h_n^{km} \Gamma_n, \tag{2.29}$$

where $\bar{\beta}^k$ is the equivalent path loss between the BS and user k through the RIS. Therefore, channel matrix $\boldsymbol{G}$ can be written by

$$\boldsymbol{G} = \boldsymbol{B}^{-1/2} \boldsymbol{H}, \tag{2.30}$$

where $\boldsymbol{B}$ is a diagonal matrix whose k-th diagonal reflective element being $\bar{\beta}^k$ and the reflective element at the k-th column and the m-th row of $\boldsymbol{H}$ is equal to $\sum\limits_{n=1}^{N} h_n^{km} \Gamma_n$.

[4]We assume that the distance between the BS and the RIS is much larger than the margin of two antennas, and thus the distances between the RIS and different antennas are assumed to be the same.

2.2.2.3 Achievable Rate with Zero-Forcing Precoding

To serve K users simultaneously, the BS first encodes the data symbols for different users with a normalized precoding matrix $\boldsymbol{W} \in \mathbb{C}^{M \times K}$. In this letter, we adopt the ZF precoder as it can obtain a near-optimal solution with a low complexity [25]. At the BS, the signals are assumed to be transmitted under a normalized power allocation matrix $\boldsymbol{\Lambda}$ which satisfies $\mathrm{Tr}\left\{\boldsymbol{\Lambda}\boldsymbol{\Lambda}^H\right\} = 1$. Denote the intended signal vector for K users as $\boldsymbol{s}$ satisfying $\mathbb{E}[\boldsymbol{s}\boldsymbol{s}^H] = \frac{P}{K}\boldsymbol{I}_K$, where P is the transmitted power. Therefore, the transmitted signals at the BS can be given by $\boldsymbol{x} = \boldsymbol{W}\boldsymbol{\Lambda}\boldsymbol{s}$. Due to ZF precoding, after the reflection by the RIS, the received signal at the users can be written by

$$\boldsymbol{y} = \boldsymbol{G}\boldsymbol{W}\boldsymbol{\Lambda}\boldsymbol{s} + \boldsymbol{n}, \tag{2.31}$$

where the k-th reflective element of $\boldsymbol{y}$ is the received signal for user k and $\boldsymbol{n} \sim \mathcal{CN}(\boldsymbol{0}, \sigma^2)$ is the additive white Gaussian noise. Assume that the channel information is available for the transmitter. For ZF precoding, the k-th column of $\boldsymbol{W}$ can be written by

$$\boldsymbol{w}_k = \frac{\boldsymbol{v}_k}{\|\boldsymbol{v}_k\|_F}, \tag{2.32}$$

where $\boldsymbol{v}_k$ is the k-th column of matrix $\boldsymbol{V}$ where

$$\boldsymbol{V} = \boldsymbol{G}^H\left(\boldsymbol{G}\boldsymbol{G}^H\right)^{-1}. \tag{2.33}$$

Consequently, with the ZF procoding, the signal-to-noise ratio (SNR) for user k can be expressed as

$$\gamma_k = \frac{P\Lambda_k}{K\sigma^2}\left(\boldsymbol{g}_k\boldsymbol{w}_k\right)\left(\boldsymbol{g}_k\boldsymbol{w}_k\right)^H, \tag{2.34}$$

where $\boldsymbol{g}_k$ is the k-th row of $\boldsymbol{G}$ and $\Lambda_k = \left[\boldsymbol{\Lambda}\boldsymbol{\Lambda}^H\right]_{k,k}$. Here, we assume that the channel state information is perfectly obtained by the BS, and thus $\boldsymbol{\Lambda}$ is assumed to be a constant in the following which can be obtained by water-filling algorithm [25].

Use above notations, we can express the data rate for user k as

$$r_k = \mathbb{E}\left[\log_2\left(1 + \gamma_k\right)\right]. \tag{2.35}$$

2.2.3 Analysis on Asymptotic Capacity

According to [26], the system capacity can be achieved by dirty paper coding, where

$$C = \mathbb{E}\left[\sum_k \log_2\left(1 + \frac{P\Lambda_k}{K\sigma^2}[\boldsymbol{G}\boldsymbol{G}^H]_{k,k}\right)\right]. \tag{2.36}$$

In the following, we will provide a proposition to show the upper bound of the system capacity. Before this position, we first give a proposition on the channel matrix.

Proposition 2.4 *When the number of reflective element N is large and the phase shifts of the RIS are given, i.e., θ_n is constant, we have*[5]

$$\frac{\sum_{n=1}^{N} h_n^{km}\Gamma_n}{\Gamma\sqrt{N}} \sim \mathscr{CN}(0, 1). \tag{2.37}$$

Proof Note that the fast fading for these channels is assumed to be independent. When the response of each reflective element is given, we have $\mathbb{E}[\Gamma_n h_n^{km}] = 0$ since $\mathbb{E}[h_n^{km}] = 0$. Besides, $\mathbb{E}[(\Gamma_n h_n^{km})^2] = \Gamma^2$ holds as $\mathbb{E}[(h_n^{km})^2] = 1$ and $|\Gamma_n| = \Gamma$. Therefore, according to the central limit theorem [27], we have (2.37) when N is a sufficiently large number.

Remark 2.4 Proposition 2.4 implies that phase shifts of the RIS will not influence the distribution of the SNR when the number of reflective elements N is sufficiently large. This observation shows that the expectation of the SNR will keep unchanged even when we select phase shifts of the RIS randomly.

From this proposition, we can learn that the columns of $\boldsymbol{H}$ are zero-mean independent Gaussian vectors. Based on this, we can have the following proposition.

Proposition 2.5 *With the optimal power allocation, when the number of reflective elements is large but finite, the system capacity can be upper bounded by*

$$C \leq \sum_k \log_2\left(1 + \frac{P\Lambda_k}{K\sigma^2}\bar{\beta}^k \Gamma^2 MN\right). \tag{2.38}$$

Proof According to Jensen's inequality, we have

$$\begin{aligned} C &\leq \sum_k \log_2\left(1 + \frac{P\Lambda_k}{K\sigma^2}\mathbb{E}\left[[\boldsymbol{G}\boldsymbol{G}^H]_{k,k}\right]\right) \\ &= \sum_k \log_2\left(1 + \frac{P\Lambda_k}{K\sigma^2}\frac{\bar{\beta}^k}{K}\mathbb{E}[\mathrm{Tr}\{\boldsymbol{H}\boldsymbol{H}^H\}]\right). \end{aligned} \tag{2.39}$$

Based on Proposition 2.4, $\frac{1}{N\Gamma^2}\boldsymbol{H}\boldsymbol{H}^H \sim \mathscr{W}_K(M, \boldsymbol{I}_M)$ is a central Wishart matrix with M degrees of freedom where $M \geq K$. According to the results in [28], we have

[5] It is worthwhile to point out that this proposition holds with any distribution of h_n^{km}.

$$\mathbb{E}[\mathrm{Tr}\{\boldsymbol{H}\boldsymbol{H}^H\}] = MNK\Gamma^2. \tag{2.40}$$

Therefore, we can have (2.38).

Remark 2.5 The upper bound can be achieved when the eigenvalues of $\boldsymbol{G}\boldsymbol{G}^H$ are the same. This can be obtained by optimizing the phase shifts of the RIS to maximize the effective rank of $\boldsymbol{G}$ if the problem has a feasible solution [29].

It is worthwhile to point out that the system capacity will not increase infinitely. When the number of the reflective elements grows, the equivalent channel coefficient will decrease as well. In the following, we will give a proposition on the path loss $\bar{\beta}^k$ at the extreme case, i.e., $N \to \infty$.

Proposition 2.6 *Without loss of generality, we assume that the RIS is on the plane with* $z = 0$*, and we assume that the coordinates of the BS and user k are* $(-x_0, 0, z_0)$ *and* $(x_0, 0, z_k)$*, respectively, as illustrated in Fig. 2.9. When* $N \to \infty$*, we have* $\bar{\beta}_k N \approx \frac{2Az_0^3\lambda^2}{5\pi^2(z_0+z_k)^5}$.

Proof Based on the coordinates given in Fig. 2.9, we have

$$\beta_n^k = \frac{Aabz_0^3\lambda^2}{64\pi^3((x-x_0)^2+y^2+z_k^2)((x+x_0)^2+y^2+z_0^2)^{5/2}}. \tag{2.41}$$

Therefore, we have $\bar{\beta}^k N = \sum_{n=1}^{N} \beta_n^k$.

When $N \to \infty$, in the most area of the RIS, the distance from the reflective element and the center between the BS and user k is much longer than that between the BS and user k. Therefore, we can assume that the distances from the reflective element to the BS and the user are almost the same, and thus have the approximations below:

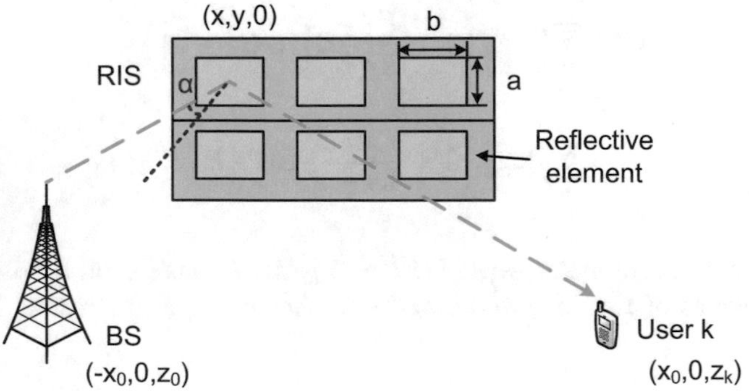

Fig. 2.9 Channel model on the RIS assisted downlink communications

$$((x - x_0)^2 + y^2 + z_k^2) \approx ((x + x_0)^2 + y^2 + z_0^2) \\ \approx (x^2 + y^2 + (z_0 + z_k)^2/4). \tag{2.42}$$

Define the area of the RIS as Φ. The received energy of the signals reflected by reflective elements in this area can be expressed by

$$\sum_{(x,y)\in\Phi} \beta_n^k \approx \int_{(x,y)\in\Phi} \frac{A z_0^3 \lambda^2}{64\pi^3 (x^2 + y^2 + (z_0 + z_k)^2/4)^{7/2}} dxdy \\ = \frac{A z_0^3 \lambda^2}{32\pi^2} \int_0^{+\infty} \frac{1}{(\rho^2 + (z_0 + z_k)^2/4)^{7/2}} \rho d\rho = \frac{2 A z_0^3 \lambda^2}{5\pi^2 (z_0 + z_k)^5}, \tag{2.43}$$

where $\rho^2 = x^2 + y^2$.

Based on the above two propositions, we have the following proposition.

Proposition 2.7 *When the number of reflective elements goes infinity, i.e., $N \to \infty$, the system capacity cannot exceed*

$$\tilde{C} = \sum_k \log_2 \left(1 + \frac{P \Lambda_k \Gamma^2 M}{K \sigma^2} \frac{2 A z_0^3 \lambda^2}{5\pi^2 (z_0 + z_k)^5}\right). \tag{2.44}$$

2.2.4 Analysis on the Number of Reflective Elements

In the following, we will discuss the effect of the number of reflective elements on the system sum-rate. We first give a proposition on the data rate for each user, and based on this, we will try to investigate how many reflective elements will be needed so that the sum-rate reach η of the system capacity.

Proposition 2.8 *The data rate for user k can be rewritten by*

$$r_k = \mathbb{E}\left[\log_2\left(1 + \frac{P \Lambda_k}{K \sigma^2 \left[\left(\boldsymbol{G}\boldsymbol{G}^H\right)^{-1}\right]_{k,k}}\right)\right]. \tag{2.45}$$

Proof According to the definition in (2.34), since $\frac{P\Lambda_k}{K}$ is a constant, we can derive $(\boldsymbol{g}_k \boldsymbol{w}_k)(\boldsymbol{g}_k \boldsymbol{w}_k)^H$ as follows:

$$(\boldsymbol{g}_k \boldsymbol{w}_k)(\boldsymbol{g}_k \boldsymbol{w}_k)^H = \boldsymbol{g}_k \frac{\boldsymbol{v}_k}{\|\boldsymbol{v}_k\|_F} \frac{\boldsymbol{v}_k^H}{\|\boldsymbol{v}_k\|_F} \boldsymbol{g}_k^H \stackrel{(a)}{=} \frac{1}{\|\boldsymbol{v}_k\|_F^2}. \tag{2.46}$$

Here, (a) can be achieved by

$$\boldsymbol{g}_k \boldsymbol{v}_k \boldsymbol{v}_k^H \boldsymbol{g}_k^H = \left[\boldsymbol{G}\boldsymbol{G}^H(\boldsymbol{G}\boldsymbol{G}^H)^{-1}\right]_{k,k} \left[\boldsymbol{G}\boldsymbol{G}^H(\boldsymbol{G}\boldsymbol{G}^H)^{-1}\right]_{k,k}^H = 1. \tag{2.47}$$

Note that $\|\boldsymbol{v}_k\|_F^2 = \boldsymbol{v}_k^H \boldsymbol{v}_k$, we have

$$\boldsymbol{v}_k^H \boldsymbol{v}_k = \left[\boldsymbol{V}^H \boldsymbol{V}\right]_{k,k} = \left[(\boldsymbol{G}\boldsymbol{G}^H)^{-1}\boldsymbol{G}\boldsymbol{G}^H(\boldsymbol{G}\boldsymbol{G}^H)^{-1}\right]_{k,k} = \left[(\boldsymbol{G}\boldsymbol{G}^H)^{-1}\right]_{k,k}. \tag{2.48}$$

With this proposition, the sum rate can be written by $R = \sum_k r_k$. Define ϵ as the ratio of the sum-rate to the system capacity with infinite reflective elements, i.e.,

$$\epsilon = R/\tilde{C}. \tag{2.49}$$

Therefore, the problem can be written by

$$\min_N N, \quad s.t. \ \epsilon \geq \eta. \tag{2.50}$$

However, the aforementioned problem is hard to solve since the phase shifts of the reflective elements are not determined. In what follows, we first derive a lower bound of ϵ, i.e., $\hat{\epsilon}$, to guarantee $\epsilon \geq \eta$, and then transform the original problem into

$$\min_N N, \quad s.t. \ \hat{\epsilon} \geq \eta. \tag{2.51}$$

2.2.4.1 Lower Bound of ϵ

The lower bound of the ratio of the sum-rate to the system capacity ϵ can be given by the proposition below.

Proposition 2.9 *ϵ can be lower bounded by*

$$\hat{\epsilon} = \frac{\sum_k \log_2\left(1 + \frac{P}{\sigma^2}\Lambda_k \Gamma^2 \bar{\beta}^k N(\mu - 1)\right)}{\sum_k \log_2\left(1 + \frac{P}{\sigma^2}\Lambda_k \Gamma^2 \mu \frac{2Az_0^3\lambda^2}{5\pi^2(z_0+z_k)^5}\right)}, \tag{2.52}$$

where $\mu = M/K$.

Proof According to the Jensen's inequality, we have

$$R \geq \sum_k \log_2 \left(1 + \frac{P\Lambda_k}{K\sigma^2 \mathbb{E}\left[\left[(\boldsymbol{G}\boldsymbol{G}^H)^{-1}\right]_{k,k}\right]}\right). \tag{2.53}$$

Here,

$$\left[(\boldsymbol{G}\boldsymbol{G}^H)^{-1}\right]_{k,k} = \frac{\mathbb{E}[\mathrm{Tr}\{(\boldsymbol{H}\boldsymbol{H}^H)^{-1}\}]}{K\bar{\beta}^k}. \tag{2.54}$$

Note that $\frac{1}{N\Gamma^2}\boldsymbol{H}\boldsymbol{H}^H$ is a central Wishart matrix with M degrees of freedom. Therefore, according to the results in [28], we have

$$\mathbb{E}[\mathrm{Tr}\{(\boldsymbol{H}\boldsymbol{H}^H)^{-1}\}] = \frac{K}{N\Gamma^2(M-K)}. \tag{2.55}$$

Based on these, we have

$$\begin{aligned} \epsilon &\geq \frac{\sum_k \log_2\left(1 + \frac{P\Lambda_k}{K\sigma^2}\Gamma^2\bar{\beta}^k N(M-K)\right)}{\sum_k \log_2\left(1 + \frac{P\Lambda_k\Gamma^2 M}{K\sigma^2}\frac{2Az_0^3\lambda^2}{5\pi^2(z_0+z_k)^5}\right)}, \\ &= \frac{\sum_k \log_2\left(1 + \frac{P}{\sigma^2}\Lambda_k\Gamma^2\bar{\beta}^k N(\mu-1)\right)}{\sum_k \log_2\left(1 + \frac{P}{\sigma^2}\Lambda_k\Gamma^2\mu\frac{2Az_0^3\lambda^2}{5\pi^2(z_0+z_k)^5}\right)} \end{aligned} \tag{2.56}$$

which ends the proof.

Remark 2.6 From the lower bound of ϵ, we can infer that a moderate number of antennas are required at the BS side to achieve an acceptable performance. Especially, when the number of antennas at the BS is equal to the number of users, we have $\mu = 1$, which leads to the lower bound of ϵ being 0.

2.2.4.2 Solution of Problem (2.51)

To satisfy the constraint in (2.51), we have

$$\prod_k \frac{K + \frac{P}{\sigma^2}\Lambda_k\Gamma^2\bar{\beta}^k N(\mu-1)}{\left(K + \frac{P}{\sigma^2}\Lambda_k\Gamma^2\mu\frac{2Az_0^3\lambda^2}{5\pi^2(z_0+z_k)^5}\right)^\eta} \geq 1. \tag{2.57}$$

We assume that the received SNR is high, i.e., $\frac{P}{\sigma^2}\Lambda_k\Gamma^2\bar{\beta}^k N(\mu-1) \gg K$ and $\frac{P}{\sigma^2}\Lambda_k\Gamma^2\mu \gg K$. Thus, we have

$$\prod_k \frac{\frac{P}{\sigma^2}\Lambda_k\Gamma^2\bar{\beta}^k N(\mu-1)}{\left(\frac{P}{\sigma^2}\Lambda_k\Gamma^2\mu\frac{2Az_0^3\lambda^2}{5\pi^2(z_0+z_k)^5}\right)^\eta} \geq 1. \tag{2.58}$$

In other words,

$$N \geq \left(\prod_k \frac{\left(\frac{P}{\sigma^2}\Lambda_k\Gamma^2\mu\frac{2Az_0^3\lambda^2}{5\pi^2(z_0+z_k)^5}\right)^\eta}{\frac{P}{\sigma^2}\Lambda_k\Gamma^2\bar{\beta}^k(\mu-1)}\right)^{1/K}. \tag{2.59}$$

2.2.5 Simulation Results

In this section, we verify the derivation of the achievable asymptotic capacity and evaluate the impact of the number of reflective elements on the data rate in the RIS assisted multi-user communications, the layout of which is given in Fig. 2.10. The parameters are selected according to 3GPP standard [21] and existing works [20]. The height of the BS is 25 m and the distance between the RIS and the BS $D_B = 100$ m. We set the number of users $K = 5$. The users are uniformly located in a square area whose side length is set as $L = 100$ m. The distance between the RIS and the closest side of the square area is $D_u = 10$ m, and the horizontal distance between the BS and the center of the square area is $D = 100$ m. The center of the RIS is located at the middle between the BS and the square area with the height being 25 m. The working frequency of the RIS is set as $f = 5.9$ GHz and reflection amplitude is assumed to be $\Gamma = 1$. The length and width of a reflective element are set the same, i.e., $a = b = 0.02$ m. Transmit power is set as $P = 46$ dBm, noise

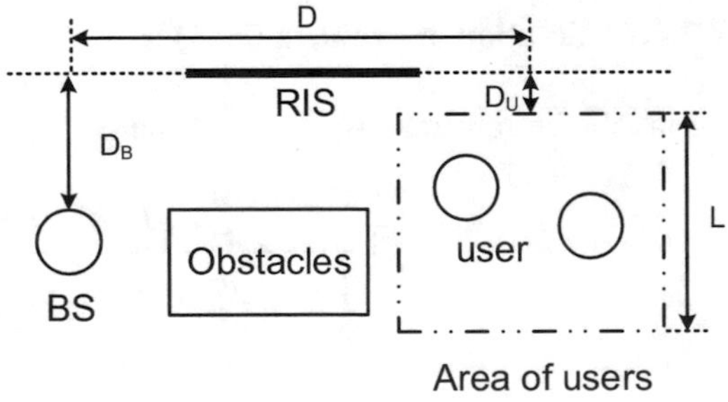

Fig. 2.10 Simulation layout for the RIS-based multi-user cellular network (top view)

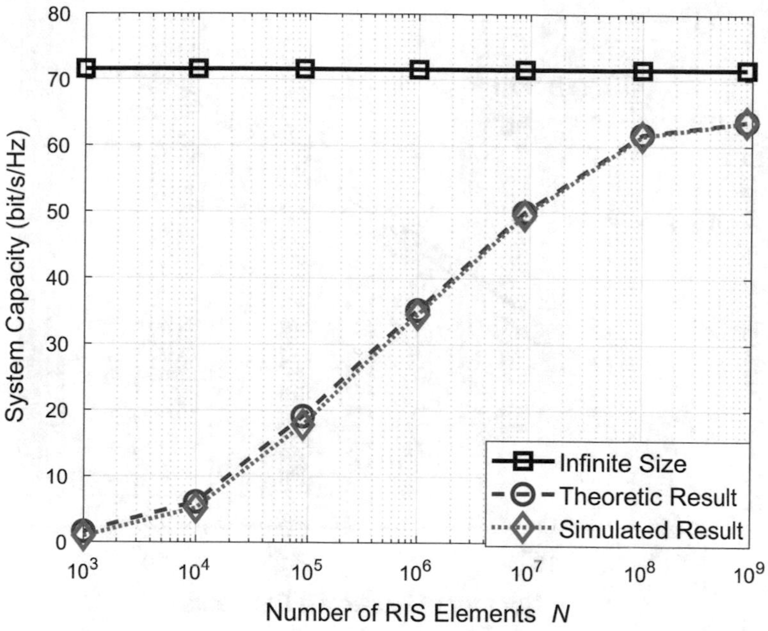

Fig. 2.11 System capacity vs. Number of reflective elements N with the number of antennas $M = 10$

power is set as $\sigma^2 = -96\,\text{dBm}$, and antenna gain is $A = 0\,\text{dB}$. All numeral results are obtained by 100 Monte Carlo simulations.

In Fig. 2.11, we plot the system capacity vs. the number of reflective elements N with the number of antennas $M = 10$. From this figure, we can observe that the theoretic result is a tight approximation of the simulated one, and the gap will decrease as the number of reflective elements increases because it can make the singular values of the channel matrix more equal. Moreover, we can observe that the capacity per user will increase as the number of reflective elements N increases first and then become saturated, as suggested by Proposition 2.7. We can observe that the system capacity with an infinite size of the RIS can reach 71.5 bit/s/Hz for 5 users.

In Fig. 2.12, we plot the ratio of sum-rate to system capacity $\hat{\epsilon}$ vs. the number of reflective elements N with threshold $\eta = 0.75$ for different values of μ. From this figure, we can observe that the ratio $\hat{\epsilon}$ first increases linearly with $\log N$ and then saturates, which suggests that it is more cost-effective to operate the RIS on the linear growth zone, i.e., $N < 10^7$. We can also observe that it requires the number of reflective elements $N = 8 \times 10^6$ to achieve 75% of the system capacity with $\mu = 20$. In other words, the side length of the RIS should be 56.7 m. Besides, we can observe that the required number of reflective elements will decrease with a higher value of μ. This implies that the size of the RIS can be reduced with more antennas at the BS. Therefore, there exists an trade-off between the number of reflective elements and

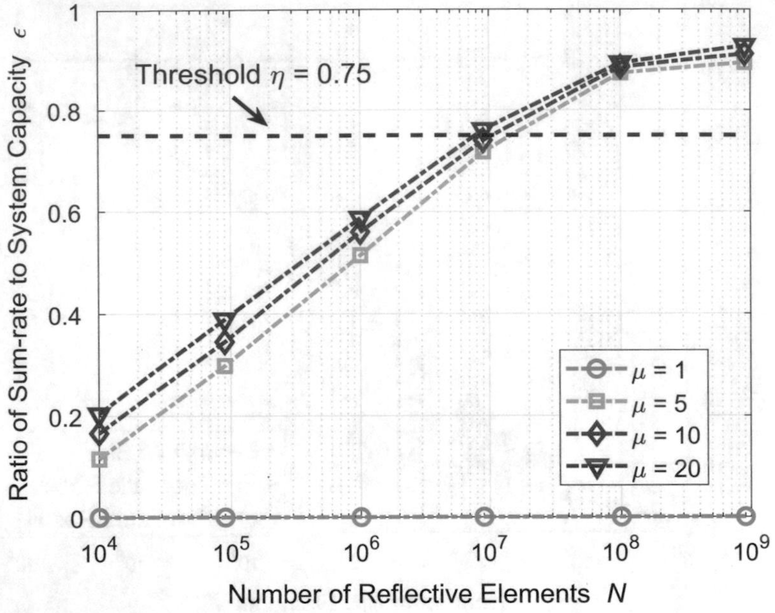

Fig. 2.12 Ratio of sum-rate to system capacity $\hat{\epsilon}$ vs. Number of reflective Elements N

the number of antennas at the BS, and we can reduce the total cost by optimizing the size of the RIS. Moreover, we can find that the ratio remains 0 when $\mu = 1$, i.e., the number of antennas at the BS is the same as the number of users. This implies that the antennas at the BS should be larger than the number of users, which is consistent with Remark 2.6.

2.2.6 Summary

In this section, we have investigated RIS assisted downlink multi-user communications and have derived the corresponding asymptotic capacity. Based on the derived capacity, we also have discussed the impact of the number of the reflective elements on the system sum-rate. Particularly, we have obtained a requirement on the RIS size to achieve an acceptable data rate. From the analysis and simulation, given the location of the RIS, we can draw the following conclusions: (1) The system capacity cannot increase infinitely as the RIS size grows, but rather is upper bounded; (2) To achieve a data rate threshold, the required number of reflective elements will decrease with more antennas at the BS.

2.3 Coverage Extension: RIS Orientation and Location Optimization

2.3.1 *Motivations*

In the literature, RIS-assisted wireless networks have been studied to increase coverage and improve link quality. However, existing works only utilized the RIS for coverage extension given the RIS location, but how to deploy the RIS to further maximize the cell coverage has not been studied yet. In this section, we consider an RIS-assisted downlink cellular network with one BS and one user equipment (UE). The cell coverage of this network is analyzed first. Then, since the RIS orientation and the horizontal distance between the RIS and the BS will significantly influence the cell coverage, we maximize the cell coverage by optimizing the RIS orientation and horizontal distance. To solve the RIS placement optimization problem, we propose a coverage maximization algorithm (CMA), where a closed-form optimal RIS orientation is derived first, and the horizontal distance is then optimized using the interior point method.

The rest of this section is organized as follows. In Sect. 2.3.2, the system model for the RIS-assisted network is introduced. The cell coverage is analyzed in Sect. 2.3.3. In Sect. 2.3.4, we first formulate a coverage maximization problem, and then propose the CMA to solve the problem. In Sect. 2.3.5, simulation results validate our analysis. Finally, we conclude our work in Sect. 2.3.6.

2.3.2 *System Model*

In this section, we first introduce the RIS-assisted network model. The channel model is then constructed.

2.3.2.1 Scenario Description

As shown in Fig. 2.13, we consider a narrow-band downlink network with one UE and one BS [30]. Due to the dynamic wireless environment involving unexpected fading, the link between the UE and the BS can be unstable or even fall into a complete outage [20]. To solve this problem, we introduce an RIS to assist the communication. To describe the topology of the system, a Cartesian coordinate is adopted, where the x-y plane coincides with the RIS surface, and the z-axis is vertical to the RIS. Based on the x-y plane, the space can be divided into two sides. To ensure that the RIS can reflect the signal from the BS towards the UE, we assume that the UE and the BS are in the same side of the RIS, i.e., $z > 0$ in the coordinate system.

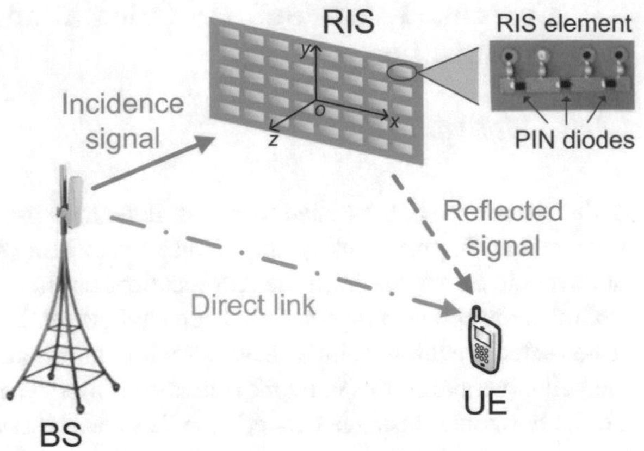

Fig. 2.13 System model of the RIS-assisted cellular network

The RIS is composed of $M \times N$ sub-wavelength elements, each with the size of $s_M \times s_N$. As shown in Fig. 2.13, the RIS element contains several PIN diodes [31]. When the biased voltages applied to the PIN diodes change, the RIS element will change the reflection phase shift accordingly. Define Γ and $\varphi^{m,n}$ as the reflection amplitude change and phase shift of the (m, n)-th element, respectively.[6] The reflection coefficient of the (m, n)-th element can thus be written as $\Gamma_{m,n} = \Gamma e^{-j\varphi_{m,n}}$. Assume that the BS is in the far field of the RIS. Amplitude change Γ can be modeled by $\cos\theta_i$ [32], where θ_i is the incidence angle from the BS to the RIS. Besides, we assume that phase shift $\varphi_{m,n}$ is not influenced by the incidence and reflection angles.

2.3.2.2 Channel Model

The channel from the BS to the UE is composed of MN RIS-based channels and a direct link, where the (m, n)-th RIS-based channel represents the channel from the BS to the UE via the (m, n)-th RIS element.[7]

The channel gain for the (m, n)-th RIS-based channel can be expressed as [20],

$$h_{m,n} = \frac{\lambda\sqrt{G s_M s_N}}{(4\pi)^{\frac{3}{2}}\sqrt{D_{m,n}^{\alpha} d_{m,n}^{\alpha}}} e^{-j\frac{2\pi}{\lambda}(D_{m,n}+d_{m,n})}, \tag{2.60}$$

---

[6] Without loss of generality, we assume that the states of the PIN diodes in an element do not influence reflection amplitude change Γ of the element.

[7] Since we consider average performance here, the small scale fading of the RIS-based channel is averaged. Besides, it is assumed that the small-scale fading corresponding to different RIS elements are independently distributed.

where λ is the wavelength corresponding to the carrier frequency, G is antenna gain, α represents the pathloss exponent, $D_{m,n}$ and $d_{m,n}$ are the distance between the BS and the (m, n)-th RIS element, and the distance between the (m, n)-th RIS element and the UE, respectively. Since the BS is far away from the RIS, its distances from different RIS elements are approximately the same, i.e., $D_{m,n} = D$, where D is the distance between the BS and the center of the RIS. Besides, in order to derive the area of the cell coverage, we focus on finding the cell edge, and thus, it can be assumed that the UE is in the far field of the RIS. Therefore, we have $d_{m,n} = d$, where d is the distance between the UE and the center of the RIS. As a result, the pathloss is common to all RIS elements, i.e., $GD_{m,n}^{-\alpha}d_{m,n}^{-\alpha} = GD^{-\alpha}d^{-\alpha} \triangleq PL_R$.

For the direct link, the channel model can be written as

$$h_D = \frac{\lambda\sqrt{G}}{4\pi d_{BU}} e^{-j\frac{2\pi}{\lambda}d_{BU}}, \tag{2.61}$$

where d_{BU} represents the distance between the BS and the UE. Based on (2.60) and (2.61), the channel from the BS to the UE is given by

$$h = \sum_{m,n} \Gamma_{m,n} h_{m,n} + h_D, \tag{2.62}$$

Therefore, the received signal-to-noise ratio (SNR) can be expressed as $\gamma = \frac{P|h|^2}{\sigma^2}$, where P denotes the transmit power of the BS, and σ^2 is the variance of the additive white Gaussian noise (AWGN) received at the UE.

2.3.3 Cell Coverage Analysis

Before analyzing the cell coverage, we first introduce its definition.

Definition 2.1 The cell coverage is defined as an area where the received SNR at the UE is larger than a certain threshold γ_{th} [33, 34], i.e.,

$$\gamma \geq \gamma_{th} = \gamma_s L_{mar}, \tag{2.63}$$

where γ_s represents UE sensitivity, and L_{mar} is margin for penetration loss.

In the following, we first find the optimal phase shifts of the RIS that maximizes the SNR. Then, the cell coverage is analyzed.

2.3.3.1 Optimal Phase Shifts of RIS

To improve the system performance, the phase shifts are optimized to maximize the SNR. Define $\eta_R = \frac{P}{\sigma^2}\frac{\lambda^2}{(4\pi)^3}\big(\cos^2(\theta_i)s_M s_N\big)$, $\eta_D = \frac{P}{\sigma^2}\frac{\lambda^2 G}{(4\pi)^2}$, and $\eta_X = 2\frac{P}{\sigma^2}\frac{\lambda^2\sqrt{G}}{(4\pi)^{\frac{5}{2}}}\big(\cos(\theta_i)\sqrt{s_M s_N}\big)$. The maximum SNR can be derived, given in the following theorem.

Theorem 2.1 *When the phase shifts are set as*

$$\varphi_{m,n} = \Big(\frac{2\pi}{\lambda}d_{BU} - \frac{2\pi}{\lambda}(D_{m,n} + d_{m,n})\Big) \mod 2\pi, \tag{2.64}$$

the SNR is maximized, and its value can be given by

$$\gamma = \eta_R M^2 N^2 P L_R + \eta_D d_{BU}^{-2} + \eta_X MN\frac{\sqrt{PL_R}}{d_{BU}}. \tag{2.65}$$

Proof Based on (2.62), the SNR can be rewritten as

$$\begin{aligned}\gamma =& \eta_R P L_R \sum_{m,m',n,n'} e^{j(\varphi_{m,n}+\theta_{m,n}-\varphi_{m',n'}-\theta_{m',n'})} + \eta_D d_{BU}^{-2}\\ &+ \eta_X \frac{\sqrt{PL_R}}{d_{BU}}\sum_{m,n}\cos(\varphi_{m,n}+\theta_{m,n}-\frac{2\pi}{\lambda}d_{BU}),\end{aligned} \tag{2.66}$$

where $\theta_{m,n} = \frac{2\pi}{\lambda}(D_{m,n} + d_{m,n})$. We can find that when the phase shifts are set as shown in (2.64), $\sum_{m,m',n,n'} e^{j(\phi_{m,n}+\theta_{m,n}-\phi_{m',n'}-\theta_{m',n'})}$ and $\sum_{m,n}\cos(\varphi_{m,n}+\theta_{m,n}-\frac{2\pi}{\lambda}d_{BU})$ are simultaneously maximized, with the values being M^2N^2 and MN, respectively [20]. Therefore, Theorem 2.1 holds. □

2.3.3.2 Cell Coverage Analysis

In this section, the cell coverage in a given direction is investigated first. The area of the cell coverage is then derived.

1. Cell coverage in a given Direction: As shown in Fig. 2.14, we use ϕ to represent the angle between the given direction and the direction from the BS to the RIS, where $\phi \in [0, 2\pi)$. To derive the cell coverage in direction ϕ, we first analyze how SNR γ changes with the horizontal distance between the BS and the UE, denoted by d_{BU}^h.

When d_{BU}^h is shorter than the horizontal distance between the BS and the RIS, denoted by D^h, it is hard to find out the influence of d_{BU}^h on SNR γ. This is because when d_{BU}^h becomes larger, the horizontal distance between the UE and the RIS, denoted by d^h, first decreases and then increases. To avoid discussing the trend of

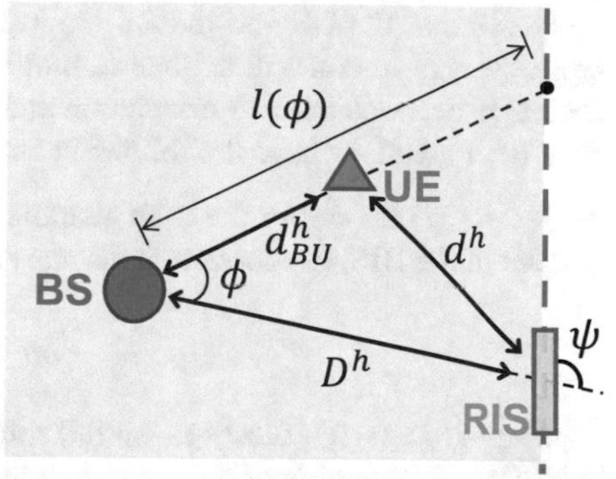

Fig. 2.14 Top view of the RIS-assisted network

γ with respect to d_{BU}^h, we require that SNR γ is always above threshold γ_{th} when $d_{BU}^h < D^h$. Note that γ is lower bounded by $\eta_D d_{BU}^{-2}$, which decreases as the UE moves away from the BS. Therefore, by assuming that when $d_{BU}^h = D^h$, $\eta_D d_{BU}^{-2}$ does not fall below γ_{th}, i.e.,

$$\eta_D\left(\sqrt{(D^h)^2 + (H_B - H_U)^2}\right)^{-2} \geq \gamma_{th}, \tag{2.67}$$

where H_B and H_U are the height of the BS and the UE, respectively, the SNR is guaranteed to exceed γ_{th}.

When d_{BU}^h exceeds D^h, it is obvious that γ strictly decreases as the UE moves away from the BS. Besides, γ approaches to 0 when the UE is sufficiently far away from the BS.

Based on above discussions, it can be found that the horizontal distance d_{BU}^h with $\gamma = \gamma_{th}$ is unique, which is denoted by $d_{th}(\phi)$. Define g as the function of γ with respect to the angle ϕ and the horizontal distance d_{BU}^h, i.e., $\gamma = g(\phi, d_{BU}^h)$. We then have

$$g\big(\phi, d_{th}(\phi)\big) = \gamma_{th}, \forall \phi. \tag{2.68}$$

The expression of g is given by

$$\begin{aligned} g(\phi, d_{BU}^h) &= \eta_R M^2 N^2 PL_R(\phi, d_{BU}^h) + \eta_D\left(d_{BU}(\phi, d_{BU}^h)\right)^{-2} \\ &\quad + \eta_X MN\sqrt{PL_R(\phi, d_{BU}^h)}/d_{BU}(\phi, d_{BU}^h), \end{aligned} \tag{2.69}$$

where $PL_R(\phi, d_{BU}^h) = GD^{-\alpha}\left((D^h - d_{BU}^h \cos\phi)^2 + (d_{BU}^h \sin\phi)^2 + (H_R - H_U)^2\right)^{-\frac{\alpha}{2}}$, and $d_{BU}(\phi, d_{BU}^h) = \sqrt{(d_{BU}^h)^2 + (H_B - H_U)^2}$. We can also obtain that SNR γ exceeds threshold γ_{th} only if $d_{BU}^h \leq d_{th}(\phi)$.

Recall that the UE and the BS are in the same side of the RIS. Therefore, the cell coverage in some directions cannot achieve $d_{th}(\phi)$. Specifically, define ψ as the angle between the RIS orientation and direction from the BS to the RIS, where $\psi \in (0, \pi)$, and we have the following two cases,

- $\phi \in [0, \psi) \cup (\psi + \pi, 2\pi)$: To guarantee that the UE and the BS are in the same side of the RIS, the horizontal distance between the BS and the UE has to satisfy

$$d_{BU}^{h} \leq l(\phi), \tag{2.70}$$

  where $l(\phi) = D^h/(\cos(\phi) - \sin(\phi)\cot(\psi))$ denotes the distance between the BS and the RIS in direction ϕ, as shown in Fig. 2.14. Therefore, the cell coverage is given by

$$c(\phi) = \min\left(d_{th}(\phi), l(\phi)\right). \tag{2.71}$$

- $\phi \in [\psi, \psi + \pi]$: The UE and the BS are always in the same side of the RIS. Therefore, the cell coverage is

$$c(\phi) = d_{th}(\phi). \tag{2.72}$$

To have a better understanding of the cell coverage in directions $\phi \in [0, \psi) \cup (\psi + \pi, 2\pi)$, we compare $d_{th}(\phi)$ with $l(\phi)$ in the following remark.

Remark 2.7 When $\phi \in [0, \psi) \cup (\psi + \pi, 2\pi)$, there are two angles satisfying $d_{th}(\phi) = l(\phi)$, denoted by ϕ_u and ϕ_l, i.e.,

$$d_{th}(\phi_l) = l(\phi_l), \phi_l < \pi, \tag{2.73}$$

$$d_{th}(\phi_u) = l(\phi_u), \phi_u > \pi. \tag{2.74}$$

Besides, $d_{th}(\phi) > l(\phi)$ only when $\phi \in [0, \phi_l) \cup (\phi_u, 2\pi)$.

Proof First, we show that as ϕ approaches to π, $d_{th}(\phi)$ will become smaller while $l(\phi)$ will be larger. Based on cosine theorem, the horizontal distance between the UE and the RIS can be expressed as $d^h = \sqrt{(D^h)^2 + (d_{BU}^h)^2 - 2D^h d_{BU}^h \cos(\phi)}$.

Therefore, when ϕ approaches to π, d^h will become larger, which leads to a larger distance between the UE and the RIS, and thus, a lower SNR γ. Therefore, $d_{th}(\phi)$ becomes smaller.

When $\phi = \psi$ or $\phi = \psi + \pi$, $l(\phi) \rightarrow \infty$ while $d_{th}(\phi)$ is limited. On the contrary, when $\phi = 0$, we have $l(\phi) = D^h < d_{th}(\phi)$. Therefore, there are two angles satisfying $d_{th}(\phi) = l(\phi)$, and they are located within $[0, \psi)$ and $(\psi + \pi, 2\pi)$, respectively. Besides, $d_{th}(\phi)$ is larger than $l(\phi)$ only when $\phi \in [0, \phi_l) \cup (\phi_u, 2\pi)$. □

2. *Area of cell coverage:* Based on (2.71), (2.72), and Remark 2.7, we can derive the area of the cell coverage, given in the following theorem.

Theorem 2.2 *The area of the cell coverage is given by*

$$S=\int_{\phi_l}^{\phi_u}\frac{1}{2}d_{th}^2(\phi)d\phi+\frac{1}{2}\sin(\phi_l-\phi_u)l(\phi_l)l(\phi_u). \tag{2.75}$$

Proof The area of the cell coverage can be expressed as $S=\int_0^{2\pi}\frac{c^2(\phi)}{2}d\phi$. According to Remark 2.7, when $\phi\in[0,\phi_l)\cup(\phi_u,2\pi)$, $c(\phi)=l(\phi)$. Otherwise, $c(\phi)=d_{th}(\phi)$. Therefore, we have

$$S=\int_{\phi_l}^{\phi_u}\frac{d_{th}^2(\phi)}{2}d\phi+\int_0^{\phi_l}\frac{l^2(\phi)}{2}d\phi+\int_{\phi_u}^{2\pi}\frac{l^2(\phi)}{2}d\phi. \tag{2.76}$$

Define $F(\phi)=\frac{(D^h)^2}{2}\frac{\sin\phi}{\cos\phi-\sin\phi\cot\psi}$. Since $F'(\phi)=\frac{l^2(\phi)}{2}$, we have

$$\int_0^{\phi_l}\frac{l^2(\phi)}{2}d\phi+\int_{\phi_u}^{2\pi}\frac{l^2(\phi)}{2}d\phi=\frac{\sin(\phi_l-\phi_u)l(\phi_l)l(\phi_u)}{2}, \tag{2.77}$$

which ends the proof. □

2.3.4 *RIS Placement Optimization*

In the following, we first formulate a coverage maximization problem. A coverage maximization algorithm (CMA) is then proposed to solve the formulated problem.

2.3.4.1 Coverage Maximization Problem Formulation

Since the horizontal distance D^h between the RIS and the BS, and the orientation ψ of the RIS will influence the cell coverage, our aim is to maximize the area of the cell coverage S by jointly optimizing (D^h, ψ), as formulated below:

$$\max_{D^h,\psi} S \quad s.t.\ (2.67). \tag{2.78}$$

2.3.4.2 Coverage Maximization Algorithm Design

To solve the formulated problem, we first find the optimal RIS orientation for any horizontal distance D^h. The result is given by the following theorem.

Theorem 2.3 *Regardless of the horizontal distance, the optimal RIS orientation is always* $\psi=\frac{\pi}{2}$, *i.e., the RIS is deployed vertical to the direction from the BS to the RIS.*

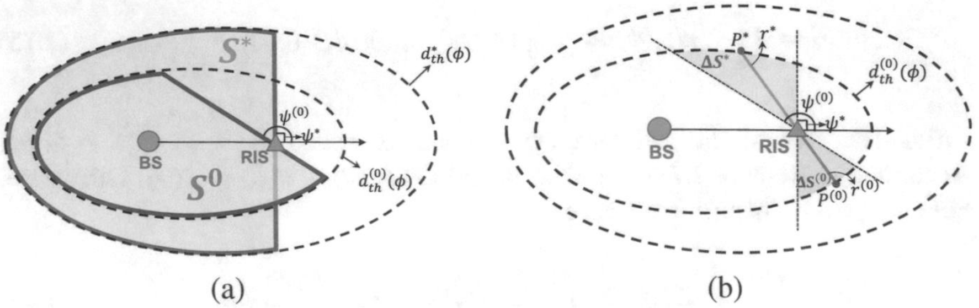

Fig. 2.15 Graphical interpretation for proof of Theorem 2.3

Proof To show that $\psi^* = \frac{\pi}{2}$ is optimal, we compare ψ^* with another RIS orientation $\psi^{(0)}$. As shown in Fig. 2.15a, the cell coverage under ψ^* and that under $\psi^{(0)}$ correspond to the yellow region and green region, respectively, where the area of the two regions are denoted by S^* and $S^{(0)}$, respectively. It is worthwhile noting that d^*_{th} is larger than $d^{(0)}_{th}$ since as ψ deviates from $\frac{\pi}{2}$, reflection amplitude Γ decreases, and thus, the SNR at the UE degrades. In the following, we show that $S^{(0)} \leq S^*$, from which we can obtain that $\frac{\pi}{2}$ is the optimal RIS orientation.

Note that $S^* - S^{(0)} > \Delta S^* - \Delta S^{(0)}$, where ΔS^* and $\Delta S^{(0)}$ are the areas of the two sectors shown in Fig. 2.15b. Therefore, we will focus on comparing ΔS^* with $\Delta S^{(0)}$ in the following. To compare ΔS^* with $\Delta S^{(0)}$, we arbitrarily select two opposite radius of the sectors, denoted by r^* and $r^{(0)}$, respectively, and we use P^* and $P^{(0)}$ to represent the intersection between r^* and $d^{(0)}_{th}$, and that between $r^{(0)}$ and $d^{(0)}_{th}$, respectively. Compared with the case where the UE is in $P^{(0)}$, the UE is closer to the BS when it is in P^*. However, SNR γ is equal to γ_{th} when the UE is in $P^{(0)}$ or P^*. Therefore, when the UE is in $P^{(0)}$, it is closer to the RIS compared with the case where the UE is in P^*, i.e., $r^{(0)} < r^*$. As a result, we have $\Delta S^{(0)} < \Delta S^*$, which indicates that $S^* \geq S^{(0)}$. □

Then, we optimize the horizontal distance D^h given the optimal RIS orientation ψ, i.e.,

$$\max_{D^h} S \quad s.t. \ (2.67). \tag{2.79}$$

It can be found that D^h has an influence on ϕ_l, ϕ_u, and $d_{th}(\phi)$ in the objective function. However, it is hard to depict the influence in a closed-form expression. Inspired by the technique for solving the max–min problem [35], by regarding ϕ_l, ϕ_u, and $d_{th}(\phi)$ as additional optimization variables, their closed-form expressions are no longer needed. However, due to the integral in the objective function, an infinite number of additional variables are needed. To cope with this challenge, we discretize the integral in the objective function as below,

$$\int_{\phi_l}^{\phi_u} \frac{1}{2} d_{th}^2(\phi) d\phi \approx \sum_{i=0}^{K-1} \frac{1}{2} d_{th}^2(\phi_l + i\Delta)\Delta, \tag{2.80}$$

where the integral interval is divided equally into K parts, each of which has a dimension of $\Delta = \frac{\phi_u - \phi_l}{K}$. For simplicity, we use y_i to replace $d_{th}(\phi_l + i\Delta)$. Therefore, horizontal distance optimization problem (2.79) can be transformed into its equivalent form,

$$\max_{\substack{\phi_l, \phi_u, D^h \\ \{y_0, \ldots, y_K\}}} \sum_{i=0}^{N-1} \frac{1}{2} y_i^2 \Delta + \frac{\sin(\phi_l - \phi_u) l(\phi_l) l(\phi_u)}{2} \tag{2.81a}$$

$$s.t.\ g(\phi_l + i\Delta, y_i) = \gamma_{th}, i = 0, \ldots, K,$$

$$(2.67), (2.73), \text{and } (2.74), \tag{2.81b}$$

where constraint (2.81b) is derived based on (2.68). Denote an optimal solution for problem (2.81) by $(\phi_l^*, \phi_u^*, (D^h)^*, \boldsymbol{y}^*)$, where $\boldsymbol{y}^* = \{y_0^*, \ldots, y_K^*\}$. We have $\left((D^h)^*, \frac{\pi}{2}\right)$ is optimal for RIS deployment problem (2.78). This is because problem (2.81) is equivalent to (2.79). Therefore, $(D^h)^*$ is optimal for horizontal problem (2.79). Besides, in (2.79), the horizontal distance is optimized given the RIS orientation as $\frac{\pi}{2}$, and thus, $((D^h)^*, \frac{\pi}{2})$ is an optimal solution for RIS deployment problem (2.78).

Problem (2.81) is a constrained optimization problem, which can be solved by the interior point method [35]. The basic idea can be summarized as follows: in each iteration, we approximate problem (2.81) by removing inequality constraints in (2.81), and adding an additional logarithmic barrier function consisting of the inequality constraint functions to the objective function of (2.81). As such, the approximated problem is then solved utilizing the Newton method.

2.3.5 Simulation Results

In this section, we verify our theoretic results on the cell coverage by simulations. The parameters are based on the 3GPP standard [21] and existing work [25], which are summarized in Table 2.1. To show the effectiveness of the proposed CMA, we compare it with another two algorithms, i.e., random algorithm and BS side scheme (BSS). In the random algorithm, the horizontal distance and RIS orientation are determined in a random manner. Based on the conclusion that the received SNR at the receiver will be maximized when the RIS is close to the receiver or transmitter [20], the RIS is placed close to the BS at the edge of the far field in scheme BSS. Figure 2.16 shows the cell coverage S versus RIS orientation ψ. The theoretic result is based on (2.75), and the simulated result is obtained by 10^5 Monte

Table 2.1 Simulation parameters

Parameters	Values
Height of the BS H_B	35 m
Height of the UE H_U	1.5 m
Height of the RIS H_R	2 m
Size of the RIS element $s_M = s_N$	0.04 m
Noise power σ^2	−96 dBm
Wavelength λ	0.1 m
Antenna gain G	1
Pathloss exponent α	2
UE sensitivity γ_s	8 dB
Margin for penetration loss L_{mar}	28 dB
SNR threshold γ_{th}	36 dB
Discretization level K	50

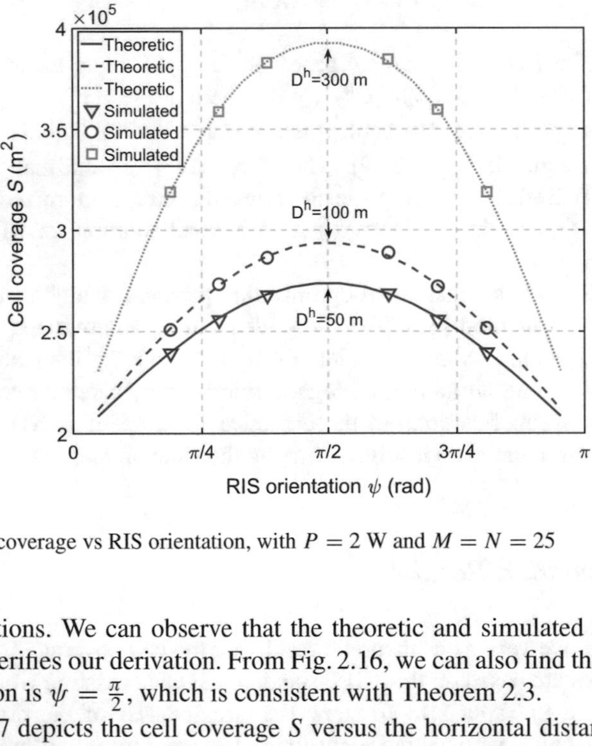

Fig. 2.16 Cell coverage vs RIS orientation, with $P = 2$ W and $M = N = 25$

Carlo simulations. We can observe that the theoretic and simulated results match well, which verifies our derivation. From Fig. 2.16, we can also find that the optimal RIS orientation is $\psi = \frac{\pi}{2}$, which is consistent with Theorem 2.3.

Figure 2.17 depicts the cell coverage S versus the horizontal distance D^h under the optimal RIS orientation. We can observe that the theoretic and simulated results are consistent, which further verifies our derivation. From Fig. 2.17, we can also find that when the RIS moves away from the BS, the coverage will become larger first since the reflection coefficient increases. However, when D^h further increases, the cell coverage degrades since the received SNR is negatively correlated with the distance between the BS and the RIS. Therefore, the RIS should be placed at a moderate distance from the BS. Besides, it can be found that the optimal RIS

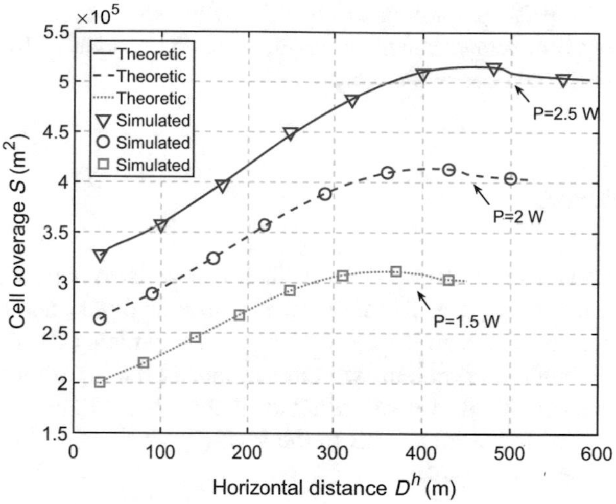

Fig. 2.17 Cell coverage vs horizontal distance, with $\psi = \frac{\pi}{2}$ and $M = N = 25$

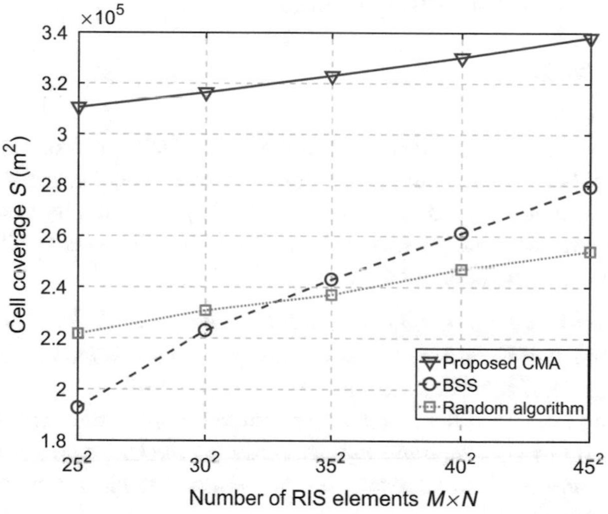

Fig. 2.18 Cell coverage vs number of RIS elements, with $P = 1.5$ W

deployment is close to the cell edge, and thus, the RIS can significantly benefit UEs at the cell edge. According to Fig. 2.17, it can also be observed that the cell coverage increases with the transmit power, due to improvement of the received SNR.

Figure 2.18 shows the cell coverage S versus the number of RIS elements MN. From Fig. 2.18, we can find that the proposed CMA achieves larger cell coverage than the random algorithm. Besides, it can be observed that the CMA outperforms the BSS, since the influence of incidence angle on the reflection coefficient is not considered in the BSS. According to Fig. 2.18, when the number of RIS elements increases, the cell coverage becomes larger, since the received SNR is positively correlated with the number of elements.

2.3.6 Summary

In this section, we have considered a downlink RIS-assisted network with one BS and one UE. The coverage of this network has been analyzed, and a problem has been formulated to maximize the cell coverage by optimizing the RIS placement. To solve the formulated problem, we have proposed the algorithm CMA. From the analysis and simulation, we can conclude that (1) the RIS should be deployed vertical to the direction from the BS to the RIS; (2) we should place the RIS at a moderate distance from the BS.

2.4 Hybrid Beamforming Design

2.4.1 Motivations

To exploit the potential of RIS techniques, many existing works have considered the RIS as a reflection-type surface deployed between sources and destinations in either point-to-point communications [8, 17, 22, 36] or multi-user systems [37, 38] for higher data rates or energy saving. However, two major issues still remain to be further discussed in the open literature.

- *For the multi-user case, how to determine the limited discrete phase shifts directly such that the inter-user interference can be eliminated? How does the quantization level influence the sum rate of the system?*
- *Considering the strengthened coupling between propagation and discrete phase shifts brought by the dominant reflection ray via the RIS, how do we design the size of RIS and perform beamforming in a multi-antenna system to achieve the maximum sum rate?*

To address the above issues, in this section, we consider a downlink multi-user multi-antenna system where the direct links between the multi-antenna BS and users suffer from deep shadowing. To provide high-quality data services, a RIS with limited discrete phase shifts is deployed between the BS and users such that the signals sent by the BS are reflected by the RIS towards the users. Since the incident waves are reflected rather than scattered at the RIS, it is the reflection-based one-hop ray [39] via the RIS that dominates the propagation between the BS and users. Therefore, the propagation model in the RIS-based system differs from those for traditional two-hop relays and one-hop direct links in MIMO systems, revealing an inner connection between the phase shifts and the propagation paths.

To achieve better directional reflection rays towards the desired users, it is vitally important to determine the phase shifts of all RIS elements, the process of which is also known as the *RIS configuration*. Such built-in programmable configuration [8] is actually equivalent to analog beamforming, realized by the RIS inherently. Since the RIS elements do not have any digital processing capability, we consider the hybrid beamforming (HBF) [40] consisting of the digital beamforming at the BS and RIS configuration based analog beamforming. A novel HBF scheme for RIS-based communications with discrete phase shifts is thus required for better shaping the propagation environment and sum rate maximization.

Designing an HBF scheme presents several major challenges. *First*, the reflection-dominated one-hop propagation and the RIS configuration based analog beamforming are coupled with each other, rendering the optimal scheme very hard to be obtained. The traditional beamforming schemes with separate channel matrix and analog beamformer do not work any more. *Second*, discrete phase shifts required by the RIS renders the sum rate maximization to be a mixed integer programming problem, which is non-trivial to be solved especially in the complex domain. *Third*, given the dense placing of the RIS elements, the correlation between elements may degrade the data rate performance. Thus, it is necessary to explore how the achievable rate is influenced by the size of RIS, which is challenging due to the complicated propagation environments especially in a multi-user case.

Through solving the above challenges, we aim to design an HBF scheme for the RIS-based multi-user system with limited discrete phase shifts to maximize the sum rate as listed below.

- We consider a downlink RIS-based multi-user system where a RIS with limited discrete phase shifts reflects signals from the BS towards various users. Given a reflection-based one-hop propagation model, we design an HBF scheme where the digital beamforming is performed at the BS, and the RIS-based analog beamforming is conducted at the RIS.
- A mixed-integer sum rate maximization problem for RIS-based HBF is formulated and decomposed into two subproblems. We propose an iterative algorithm in which the digital beamforming subproblem is solved by zero-forcing (ZF) beamforming with power allocation and the RIS-based analog beamforming is solved by the outer approximation.
- We prove that the proposed RIS-based HBF scheme can save as much as half of the radio frequency (RF) chains compared to traditional HBF schemes. Extending from our theoretical analysis on the pure LoS case, we reveal the influence of the size of RIS and the number of discrete phase shifts on the sum rate both theoretically and numerically.

The rest of this section is organized as follows. In Sect. 2.4.2, we introduce the system model of the downlink RIS-based MU multi-antenna system. The frequency-response model of the RIS and the channel model are derived. In Sect. 2.4.3, the HBF scheme for the RIS-based system is proposed. A sum rate maximization problem is formulated and decomposed into two subproblems: digital beamforming and RIS configuration based analog beamforming. An iterative algorithm is developed in Sect. 2.4.4 to solve the above two subproblems and a sub-optimal solution is obtained. In Sect. 2.4.5, we compare the RIS-based HBF scheme with the traditional one theoretically, and discuss how to achieve the maximum sum rate in the pure LoS case. The complexity and convergence of the proposed algorithm are also analyzed. Numerical results in Sect. 2.4.6 evaluate the performance of the proposed algorithm and validate our analysis. Finally, conclusions are drawn in Sect. 2.4.7.

2.4.2 System Model

In this section, we first introduce the RIS-based downlink multi-user multi-antenna system in which the BS with multiple antennas serves various single-antenna users via the RIS such that the propagation environment can be pre-designed and configured to optimize the system performance [25]. The discrete phase shift model of RIS and the channel model are then constructed, respectively.

2.4.2.1 Scenario Description

Consider a downlink multi-user communication system as shown in Fig. 2.19 where a BS equipped with N_t antennas transmits to K single-antenna users. Due to the complicated and dynamic wireless environment involving unexpected fading and potential obstacles, the BS—user link may not be stable enough or even be in outage. To alleviate this issue, we consider to deploy an RIS between the BS and users, which reflects the signals from the BS and directly projects to the users by actively shaping the propagation environment into a desirable form.

An RIS consists of $N_R \times N_R$ electrically controlled RIS elements as shown in Fig. 2.19, each of which is a sub-wavelength meta-material particle with very small features. An RIS controller, equipped with several PIN diodes, can control the ON/OFF state of the connection between adjacent metal plates where the RIS elements are laid, thereby manipulating the electromagnetic response of RIS elements towards incident waves. Due to these built-in programmable elements, the RIS requires no extra active power sources nor do not have any signal processing capability such as decoding [8]. In other words, it serves as a low-cost reconfigurable phased array that only relies on the combination of multiple programmable radiating elements to realize a desired transformation on the transmitted, received, or reflected waves [41].

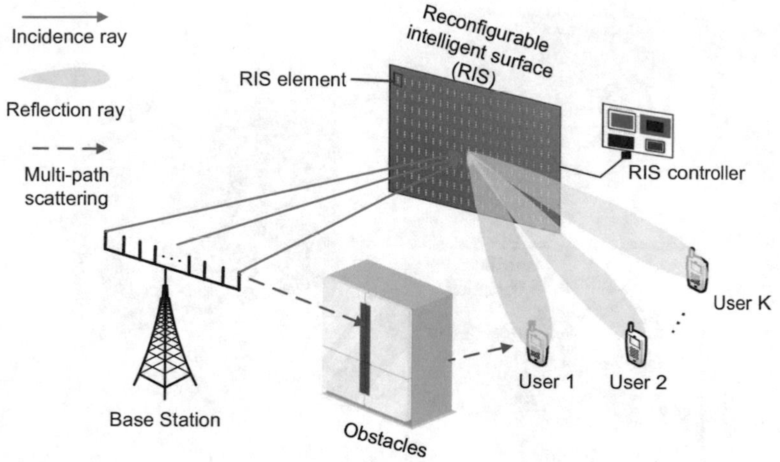

Fig. 2.19 System model of the RIS-based downlink multi-user communication system

2.4.2.2 Reconfigurable Intelligent Surface with Limited Discrete Phase Shifts

As shown in Fig. 2.20, the RIS is achieved by the b-bit re-programmable meta-material, which has been implemented as a set of radiative elements layered on a guiding structure following the wave guide techniques, forming a two-dimensional (2D) planar antenna array [14]. Being a miniature radiative element, the field radiated from the RIS element has a *phase* and *amplitude* determined by this element's polarizability, which can be tuned by the RIS controller via multiple PIN diodes (ON/OFF) [8]. However, the phase and amplitude introduced by an RIS element are not generated randomly; instead they are constrained by the Lorentzian resonance response [42], which greatly limits the range of phase values. Based on such constraints, one common manner of implementation is to constrain the amplitude and sample the phase values from the finite feasible set [42] such that the voltage-controlled diodes can easily manipulate a *discrete* set of phase values at a very low cost.

Specifically, we assume that each RIS element is encoded by the controller (e.g., via PIN diodes) to conduct 2^b possible phase shifts to reflect the radio wave. Due to the frequency-selective nature of the meta-materials, these elements only vibrate in resonance with the incoming waves over a narrow band centering at the resonance frequency. Without loss of generality, we denote the frequency response of each element (l_1, l_2) at the l_1-th row and l_2-th column of the 2D RIS within the considered frequency range as q_{l_1,l_2}, $0 \leq l_1, l_2 \leq N_R - 1$. Since the RIS is b-bit controllable, 2^b possible configuration modes (i.e., phases) of each q_{l_1,l_2} can be defined according to the Lorentzian resonance response [42].

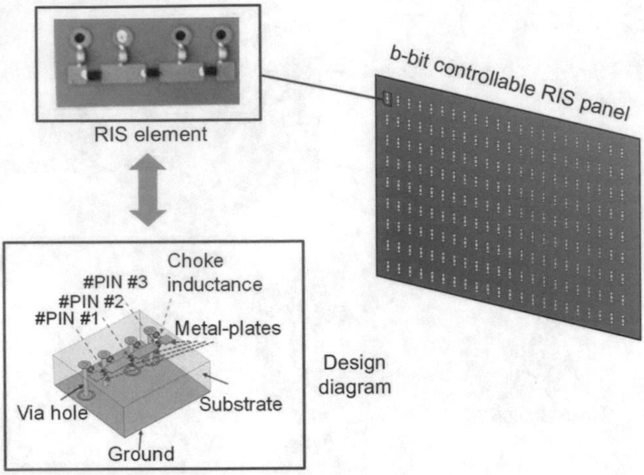

Fig. 2.20 Schematic structure of *b*-bit encoded RIS

$$q_{l_1,l_2}=\frac{j+e^{j\theta_{l_1,l_2}}}{2},\ \theta_{l_1,l_2}=\frac{m_{l_1,l_2}\pi}{2^{b-1}},\ m_{l_1,l_2}\in\left\{0,1,\ldots,2^b-1\right\},\ 0\leq l_1,l_2\leq N_R-1, \tag{2.82}$$

where θ_{l_1,l_2} denotes the phase shift of RIS element (l_1, l_2). For convenience, we refer to b as the number of *quantization bits*.

Since the 2D RIS is constructed based on the ultra-thin wave guides, the propagation inside the 2D RIS is influenced by both the location and the wave-number of each meta-surface element. The latter one is also a function of the frequency, which reflects the frequency-selective nature of RIS. Denote the position of the element (l_1, l_2) as p_{l_1,l_2}. The propagation introduced by this element can then be given by $e^{-j\beta_{l_1,l_2}(\lambda)\frac{2\pi}{\lambda}p_{l_1,l_2}}$, where $\beta_{l_1,l_2}(\lambda)$ is the wave-number of RIS element (l_1, l_2), and λ is the corresponding wave length. For simplicity, we assume that the wave-number $\beta_{l_1,l_2}(\lambda)$ remains the same for all RIS element within the considered narrow band.[8]

2.4.2.3 Reflection-Dominated Channel Model

In this subsection, we model the channel between antenna $0 \leq n \leq N_t - 1$ of the BS and user k. Specifically, instead of the traditional two-hop channel for relays [36], we use the one-hop reflection ray to model the dominant channel between the BS and users via the RIS which only passively reflects the received signals. The key

[8]This model can be easily extended to a frequency-selective case where $\beta_{l_1,l_2}(\lambda)$ varies with the working frequency. The propagation can then be modelled by a filter with finite impulse response [41].

reason can be detailed below. Due to small spacing between adjacent RIS elements (usually much less than the wavelength), the signals projected onto the surface are no longer just scattered randomly into the open space like those signals spread by the traditional antennas. Instead, the superposition of spherical waves facilitated by a number of miniature scatters enables refracted and reflected waves [43] without any extra decoding or signal forwarding procedures. Therefore, unlike the traditional scattering-based propagation model where signals travel independently along the BS–RIS and RIS—user paths, in the considered scenario signals are only passively reflected by the RIS along the reflection-based path due to the coupling effect of RIS elements.

Moreover, benefited from the directional reflections of the RIS, the BS–RIS—user link is usually stronger than other multipaths as well as the degraded direct link between the BS and the user [39]. Therefore, we model the channel between the BS and each user k as a Rician model such that the BS–RIS—user link acts as the dominant "LoS" component and all the other paths together form the "NLOS" component.

Specifically, let $D_{l_1,l_2}^{(n)}$ and $d_{l_1,l_2}^{(k)}$ denote the distance between antenna n and RIS element (l_1, l_2), and that between user k and RIS element (l_1, l_2), respectively. The "LoS" channel between the signal transmitted by the BS at antenna $1 \leq n \leq N_t$ to user k via RIS element (l_1, l_2) can be given by

$$h_{l_1,l_2}^{(k,n)} = \left[D_{l_1,l_2}^{(n)} + d_{l_1,l_2}^{(k)}\right]^{-\alpha} \cdot e^{-j\beta_{l_1,l_2}(\lambda)\frac{2\pi}{\lambda}\left[D_{l_1,l_2}^{(n)}+d_{l_1,l_2}^{(k)}\right]}, \tag{2.83}$$

where α is the path loss parameter. Therefore, the channel model between each antenna n of the BS and user k via RIS element (l_1, l_2) can be written by

$$\tilde{h}_{l_1,l_2}^{(k,n)} = \sqrt{\frac{\kappa}{1+\kappa}} h_{l_1,l_2}^{(k,n)} + \sqrt{\frac{1}{1+\kappa}} PL\left(D_{l_1,l_2}^{(n)} + d_{l_1,l_2}^{(k)}\right) h_{NLOS,(l_1,l_2)}^{(k,n)}, \tag{2.84}$$

where κ is the Rician factor, $PL(\cdot)$ is the path loss model for NLOS transmissions, and $h_{NLOS,(l_1,l_2)}^{(k,n)} \sim \mathscr{CN}(0, 1)$ is the small-scale NLOS component. Here we assume that the perfect channel state information is known to the BS via communicating with the RIS controller over a dedicated wireless link. For the case where channel information is partially known to the BS, we will consider the pure LoS transmission and discuss it in detail in Sect. 2.4.5.

Since the distance between any two antennas or RIS elements is much smaller than the distance between the BS and the user, i.e., $D_{l_1,l_2}^{(n)} + d_{l_1,l_2}^{(k)} \gg d_R^{(1)}, d_R^{(2)}, d_B$, we assume that the path loss of each BS-user link is the same, ignoring the influence brought by different antennas or RIS elements. Therefore, the channel propagation $h_{l_1,l_2}^{(k,n)}$ can be rewritten as

$$h_{l_1,l_2}^{(k,n)} = \left[D_{0,0}^{(0)} + d_{0,0}^{(k)}\right]^{-\alpha} e^{-j\beta_{l_1,l_2}(\lambda)\frac{2\pi}{\lambda}\left[D_{l_1,l_2}^{(n)}+d_{l_1,l_2}^{(k)}\right]}. \tag{2.85}$$

$D_{0,0}^{(0)}$ and $d_{0,0}^{(k)}$ have been derived in Sect. 2.1.2 and are omitted here.

Based on the propagation characteristics introduced above, we will investigate how RIS can be utilized to assist multi-user transmissions in the following section.

2.4.3 *RIS-Based Hybrid Beamforming and Problem Formulation for Multi-User Communications*

Note that the RIS usually consists of a large number of RIS elements, which can be viewed as antenna elements fay away from the BS, inherently capable of realizing analog beamforming via RIS configuration. However, these RIS elements do not have any digital processing capacity, requiring signal processing to be carried out at the BS.

In this part, to realize reflected waves towards preferable directions, we present an HBF scheme for RIS-based multi-user communications. As shown in Fig. 2.21, the digital beamforming is performed at the BS while the analog beamforming is achieved by the RIS with discrete phase shifts. Based on the considered HBF scheme, we formulate a sum rate maximization problem, and then decompose it into the digital beamforming subproblem and the RIS configuration based analog beamforming subproblem.

2.4.3.1 Hybrid Beamforming Scheme

1. Digital Beamforming at the BS: The BS first encodes K different data steams via a digital beamformer, $\mathbf{V}_D$, of size $N_t \times K$, satisfying $N_t \geq K$. After up-converting the encoded signal over the carrier frequency and allocating the transmit powers, the BS sends users' signals directly through N_t antennas. Denote the intended signal vector for K users as $\boldsymbol{s} \in \mathbb{C}^{K\times 1}$. The transmitted signals of the BS can be given by

$$\boldsymbol{x} = \mathbf{V}_D \boldsymbol{s}. \tag{2.86}$$

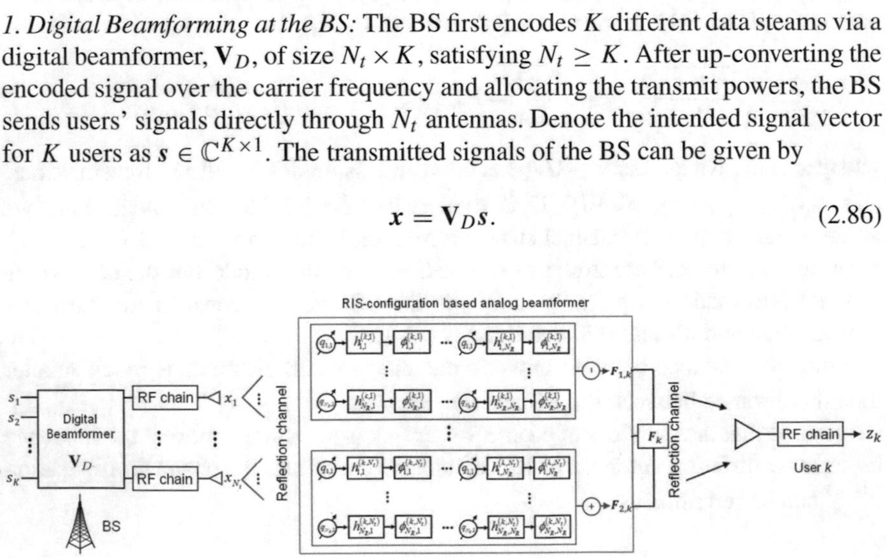

Fig. 2.21 Block diagram of the RIS-based transmission between the BS and user k

2. RIS Configuration based Analog Beamforming: After travelling through the reflection-dominated channel introduced in Sect. 2.4.2.3, the received signal at the antenna of user k can then be expressed as

$$z_k = \sum_n \sum_{l_1,l_2} \phi_{l_1,l_2}^{(k,n)} h_{l_1,l_2}^{(k,n)} q_{l_1,l_2} V_{D_{k,n}} s_k + \sum_{k \neq k'} \sum_n \sum_{l_1,l_2} \phi_{l_1,l_2}^{(k,n)} h_{l_1,l_2}^{(k,n)} q_{l_1,l_2} V_{D_{k',n}} s_{k'} + w_k, \tag{2.87}$$

where $w_k \sim \mathscr{CN}\left(0, \sigma^2\right)$ is the additive white Gaussian noise and $V_{D_{k,n}}$ denotes the k-th element in row n of matrix $\mathbf{V}_D$. In (2.87), $\phi_{l_1,l_2}^{(k,n)}$ denotes the reflection coefficient of the RIS element (l_1, l_2) with respect to the transmitting antenna n and user k. In practice, it is a function of the incidence and reflection angels, but here without loss of generality we assume that $\phi_{l_1,l_2}^{(k,n)} = \phi^{(k)}, \forall n, l_1, l_2$. We ignore the coupling between any two RIS elements here for simplicity, and thus the received signal of each user k comes from the accumulated radiations of all RIS elements, as shown in (2.87). This is a common assumption widely used in the literature on both meta-surfaces [42] and traditional antenna arrays [44].

3. Received Signal at the User: For each user k, after it receives the signal z_k, it down converts the signal to the baseband and then recovers the final signal. The whole transmission model of K users can be formulated by

$$\tilde{z}_k = \mathbf{F}\mathbf{V}_D \boldsymbol{s} + \boldsymbol{w}, \tag{2.88}$$

where $\boldsymbol{w} = [w_1, \cdots, w_K]^T$ is the noise vector. The transmission matrix $\mathbf{F}$ in (2.88) is defined as

$$\boldsymbol{F} = \sum_{l_1,l_2} q_{l_1,l_2} \left(\boldsymbol{H}_{l_1,l_2} \circ \boldsymbol{\Phi}\right), \tag{2.89}$$

where $\boldsymbol{H}_{l_1,l_2}$ and $\boldsymbol{\Phi}$ are both $K \times N_t$ matrices consisting of elements $\left\{h_{l_1,l_2}^{(k,n)}\right\}$ and $\{\phi^{(k)}\}$, respectively. The notion $\circ$ implies the element-by-element multiplication of two matrices.

2.4.3.2 Sum Rate Maximization Problem Formulation

To explore how the HBF design influences the sum-rate performance, we evaluate the achievable data rates of all users in the RIS-based system. Based on (2.87) and (2.88), we first rewrite the received signal of user k in matrix form as

$$z_k = \boldsymbol{F}_k^H \boldsymbol{V}_{D,k} s_k + \underbrace{\sum_{k' \neq k} \mathbf{F}_k^H \boldsymbol{V}_{D,k'} s_{k'}}_{\text{inter-user interference}} + w_k, \tag{2.90}$$

where $\mathbf{F}_k$ and $\mathbf{V}_{D,k}$ denote the k-th columns of matrices $\mathbf{F}$ and $\mathbf{V}_D$, respectively. The achievable rate of user k can then be given by

$$R_k = \log_2\left(1 + \frac{\left|\mathbf{F}_k^H \mathbf{V}_{D,k}\right|^2}{\sum\limits_{k' \neq k} \left|\mathbf{F}_k^H \mathbf{V}_{D,k'}\right|^2 + \sigma^2}\right). \tag{2.91}$$

We aim to maximize the achievable rates of all users by optimizing the digital beamformer $\mathbf{V}_D$ and the RIS configuration $\left\{q_{l_1,l_2}\right\}$, as formulated below:

$$\underset{\mathbf{V}_D, \left\{q_{l_1,l_2}\right\}}{\text{maximize}} \quad \sum_{1 \leq k \leq K} R_k \tag{2.92a}$$

$$\text{subject to} \quad \text{Tr}\left(\mathbf{V}_D^H \mathbf{V}_D\right) \leq P_T, \tag{2.92b}$$

$$q_{l_1,l_2} = \frac{j + e^{j\theta_{l_1,l_2}}}{2}, 0 \leq l_1, l_2 \leq N_R - 1, \tag{2.92c}$$

$$\theta_{l_1,l_2} = \frac{m_{l_1,l_2}\pi}{2^{b-1}}, m_{l_1,l_2} \in \left\{0, 1, \ldots, 2^b - 1\right\}, \tag{2.92d}$$

where P_T is the total transmit power of the BS.

2.4.3.3 Problem Decomposition

Note that problem (2.92) is a mixed integer non-convex optimization problem which is very challenging due to the large number of discrete variables $\left\{q_{l_1,l_2}\right\}$ as well as the coupling between propagation and RIS configuration based analog beamforming. Traditional analog beamforming design methods with finite resolution phase shifters [40] may not fit well since it is non-trivial to decouple the transmission matrix $\mathbf{F}$ into the product of a channel matrix and a beamformer matrix in our case. To solve this problem efficiently, we decouple it into two subproblems as shown below.

1. Digital Beamforming: Given RIS configuration $\{q_{l_1,l_2}\}$, the digital beamforming subproblem can be written by

$$\underset{V_D}{\text{maximize}} \quad \sum_{1 \leq k \leq K} R_k, \tag{2.93a}$$

$$\text{subject to } \text{Tr}\left(\mathbf{V}_D^H \mathbf{V}_D\right) \leq P_T, \tag{2.93b}$$

where $\boldsymbol{F}$ is fixed.

2. RIS Configuration based Analog Beamforming: Based on constraint (2.92c), the RIS configuration subproblem with fixed beamformer $\boldsymbol{V}_D$ is equivalent to

$$\underset{\{\theta_{l_1,l_2}\}}{\text{maximize}} \sum_{1\le k\le K} R_k, \tag{2.94a}$$

$$\text{subject to } \theta_{l_1,l_2} = \frac{m_{l_1,l_2}\pi}{2^{b-1}}, m_{l_1,l_2} \in \left\{0, 1, \ldots, 2^b - 1\right\}. \tag{2.94b}$$

In the next section, we will design two algorithms to solve these subproblems, respectively.

2.4.4 Sum Rate Maximization Algorithm Design

In this part, we will develop a sum rate maximization (SRM) algorithm to obtain a suboptimal solution of problem (2.92) in Sect. 2.4.3. Specially, we iteratively solve subproblem (2.93) given RIS configuration $\{q_{l_1,l_2}\}$, and solve subproblem (2.94) given beamformer $\boldsymbol{V}_D$. Finally, we will summarize the overall algorithm and provide convergence and complexity analysis.

2.4.4.1 Digital Beamforming Algorithm

Subproblem (2.93) is a well-known digital beamforming problem. According to the results in [45], the ZF digital beamformer can obtain a near optimal solution. Therefore, we consider ZF beamforming together with power allocation as the beamformer at the BS to alleviate the interference among users. Based on the results in [46], the beamformer can be given by

$$\boldsymbol{V}_D = \boldsymbol{F}^H(\boldsymbol{F}\boldsymbol{F}^H)^{-1}\boldsymbol{P}^{\frac{1}{2}} = \tilde{\boldsymbol{V}}_D\boldsymbol{P}^{\frac{1}{2}}, \tag{2.95}$$

where $\tilde{\boldsymbol{V}}_D = \boldsymbol{F}^H(\boldsymbol{F}\boldsymbol{F}^H)^{-1}$ and $\boldsymbol{P}$ is a diagonal matrix whose k-th diagonal element is the received power at the k-th user, i.e., p_k.

In the ZF beamforming, we have the following constraints:

$$\begin{aligned} |\boldsymbol{F}_k^H(\boldsymbol{V}_D)_k| &= \sqrt{p_k}, \\ |\boldsymbol{F}_k^H(\boldsymbol{V}_D)_{k'}| &= 0, \forall k' \neq k. \end{aligned} \tag{2.96}$$

With these constraints, subproblem (2.93) can be reduced to the following power allocation problem:

$$\max_{\{p_k\ge 0\}} \sum_{1\le k\le K} \log_2\left(1+\frac{p_k}{\sigma^2}\right), \tag{2.97a}$$

$$\text{subject to } \operatorname{Tr}\left(\boldsymbol{P}^{\frac{1}{2}}\tilde{\boldsymbol{V}}_D^H\tilde{\boldsymbol{V}}_D\boldsymbol{P}^{\frac{1}{2}}\right) \le P_T. \tag{2.97b}$$

Algorithm 1: Digital beamforming algorithm

1 **begin**
2 Solve power allocation problem (2.97);
3 Obtain the optimal power allocation result (2.98);
4 Derive the beamformer matrix from the optimal power allocation based on (2.95);
5 **end**

The optimal solution of this problem can be obtained by water-filling [26] as

$$p_k = \frac{1}{\nu_k} \max\left\{\frac{1}{\mu} - \nu_k \sigma^2, 0\right\}, \tag{2.98}$$

where ν_k is the k-th diagonal element of $\tilde{\boldsymbol{V}}_D^H \tilde{\boldsymbol{V}}_D$ and μ is a normalized factor which is selected such that $\sum\limits_{1 \le k \le K} \max\{\frac{1}{\mu} - \nu_k \sigma^2, 0\} = P_T$. The algorithm can be summarized in Algorithm 1.

2.4.4.2 RIS Configuration Based Analog Beamforming Algorithm

Since we iterate between the digital beamforming and RIS configuration based analog beamforming, the latter can be optimized assuming ZF precoding as shown in (2.95). Since the data rate with ZF precoding in (2.97) only depends on the RIS configuration through the power constraint (2.97b), the RIS configuration based analog beamforming problem can be reformulated as a power minimization problem:

$$\min_{\theta_{l_1,l_2}} f(\boldsymbol{F}), \tag{2.99a}$$

$$\text{subject to } \theta_{l_1,l_2} = \frac{m_{l_1,l_2}\pi}{2^{b-1}}, m_{l_1,l_2} \in \{0, 1, \ldots, 2^b - 1\}, \tag{2.99b}$$

where

$$\begin{aligned} f(\boldsymbol{F}) &= \mathrm{Tr}(\tilde{\boldsymbol{V}}_D \boldsymbol{P} \tilde{\boldsymbol{V}}_D^H) = \mathrm{Tr}(\boldsymbol{P}^{\frac{1}{2}} \tilde{\boldsymbol{V}}_D^H \tilde{\boldsymbol{V}}_D \boldsymbol{P}^{\frac{1}{2}}) \\ &= \mathrm{Tr}((\tilde{\boldsymbol{F}} \tilde{\boldsymbol{F}}^H)^{-1}). \end{aligned} \tag{2.100}$$

Here, $\tilde{\boldsymbol{F}} = \boldsymbol{P}^{-\frac{1}{2}} \boldsymbol{F}$.

Since $\tilde{\boldsymbol{F}} \tilde{\boldsymbol{F}}^H$ is a symmetric, positive semi-definite matrix, we can transform this problem into a semi-definite program (SDP). Let $\mathrm{Tr}((\tilde{\boldsymbol{F}} \tilde{\boldsymbol{F}}^H)^{-1}) = \mathrm{Tr}(\frac{w}{K} \boldsymbol{I}_K)$. According to Schur complement [47], the problem can be rewritten by

$$\min_{\theta_{l_1,l_2},w} w, \tag{2.101a}$$

$$\text{subject to } \mathbf{Z} = \begin{bmatrix} \frac{w}{K}\mathbf{I}_K & \mathbf{I}_K \\ \mathbf{I}_K & \tilde{\mathbf{F}}\tilde{\mathbf{F}}^H \end{bmatrix} \succeq 0, \tag{2.101b}$$

$$\theta_{l_1,l_2} = \frac{m_{l_1,l_2}\pi}{2^{b-1}}, m_{l_1,l_2} \in \{0, 1, \ldots, 2^b - 1\}, \tag{2.101c}$$

where $\mathbf{X} \succeq 0$ means that matrix $\mathbf{X}$ is a symmetric and positive semi-definite matrix.

Remark on the tractability of the formulated problem: This problem is a mix-integer SDP, which is generally NP-hard. Moreover, any two discrete variables $\{\theta_{l_1,l_2}\}$ are coupled with each other via constraint (2.101b), which makes the problem even more complicated. One commonly used solution is to first relax the discrete variables into continuous ones and then round the obtained solution to satisfy the discrete constraints. However, for the RIS-based systems, the typical value of the number of quantization bits is usually very small (e.g., 2 or 3) such that the round-off methods will lead to inevitable performance degrade.

To avoid the above issue, we consider to solve the SDP discretely. In the following, we first present the following Proposition 2.10 to transform the nonlinear functions in (2.101) with respect to θ_{l_1,l_2} into linear ones. We then use the outer approximation method [48] to solve this problem.

Proposition 2.10 *Let*

$$\mathbf{a} = \left[-\frac{(2^b-1)\pi}{2^{b-1}}, \ldots, -\frac{m\pi}{2^{b-1}}, \ldots, \frac{m\pi}{2^{B-1}}, \ldots, \frac{(2^b-1)\pi}{2^{b-1}} \right],$$

$$\mathbf{c} = \left[\cos\left(\frac{(2^b-1)\pi}{2^{b-1}} \right), \ldots, \cos\left(-\frac{m\pi}{2^{b-1}} \right), \ldots, \cos\left(\frac{m\pi}{2^{b-1}} \right), \ldots, \cos\left(\frac{(2^b-1)\pi}{2^{b-1}} \right) \right],$$

$$\mathbf{s} = \left[\sin\left(-\frac{(2^b-1)\pi}{2^{b-1}} \right), \ldots, \sin\left(-\frac{m\pi}{2^{b-1}} \right), \ldots, \sin\left(\frac{m\pi}{2^{b-1}} \right), \ldots, \sin\left(\frac{(2^b-1)\pi}{2^{b-1}} \right) \right].$$

We introduce a binary vector $\mathbf{x}^{l_1,l_2}$*, where* $x_i^{l_1,l_2}$ *indicates whether* $\theta_{l_1,l_2} = a_i$*, and a binary vector* $\mathbf{y}^{l_1,l_2,l_1',l_2'}$ *for phase difference* $\Delta\theta_{l_1,l_2,l_1',l_2'} = \theta_{l_1,l_2} - \theta_{l_1',l_2'}$*. Therefore, problem (2.101) can be rewritten by*

$$\min_{\mathbf{x}^{l_1,l_2}, \mathbf{y}^{l_1,l_2,l_1',l_2'}, w} \quad w, \tag{2.102a}$$

$$\text{subject to } \mathbf{Z} = \begin{bmatrix} \frac{w}{K}\mathbf{I}_K & \mathbf{I}_K \\ \mathbf{I}_K & \tilde{\mathbf{F}}\tilde{\mathbf{F}}^H \end{bmatrix} \succeq 0, \tag{2.102b}$$

$$\|\mathbf{x}^{l_1,l_2}\|_1 = 1, \mathbf{e}^T\mathbf{x}^{l_1,l_2} = 0, \tag{2.102c}$$

$$\mathbf{a}^T(\mathbf{x}^{l_1,l_2} - \mathbf{x}^{l_1',l_2'}) = \mathbf{a}^T\mathbf{y}^{l_1,l_2,l_1',l_2'}. \tag{2.102d}$$

*Here, **e** is a constant vector whose first $2^B - 1$ elements are 1 and others are 0.*

Proof Note that

$$\begin{aligned}
\tilde{\boldsymbol{F}}\tilde{\boldsymbol{F}}^H &= \boldsymbol{P}^{-\frac{1}{2}} \sum_{l_1,l_2} q_{l_1,l_2}(\boldsymbol{H}_{l_1,l_2} \circ \boldsymbol{\Phi}) \sum_{l_1',l_2'} q^H_{l_1',l_2'}(\boldsymbol{H}_{l_1',l_2'} \circ \boldsymbol{\Phi})^H \boldsymbol{P}^{-\frac{1}{2}} \\
&= \sum_{l_1,l_2} q_{l_1,l_2} \sum_{l_1',l_2'} q^H_{l_1',l_2'} \boldsymbol{P}^{-\frac{1}{2}}(\boldsymbol{H}_{l_1,l_2} \circ \boldsymbol{\Phi})(\boldsymbol{H}_{l_1',l_2'} \circ \boldsymbol{\Phi})^H \boldsymbol{P}^{-\frac{1}{2}} \\
&= \sum_{l_1,l_2} \sum_{l_1',l_2'} \frac{(j + e^{j\theta_{l_1,l_2}})(-j + e^{-j\theta_{l'_1,l'_2}})}{4} \boldsymbol{A}_{l_1,l_2,l_1',l_2'} \\
&= \sum_{l_1,l_2} \sum_{l_1',l_2'} \frac{\boldsymbol{A}_{l_1,l_2,l_1',l_2'}}{4} \left(\cos(\theta_{l_1,l_2} - \theta_{l_1',l_2'}) + j\sin(\theta_{l_1,l_2} - \theta_{l_1',l_2'})\right) \\
&\quad + \sum_{l_1,l_2} \sum_{l_1',l_2'} \frac{\boldsymbol{A}_{l_1,l_2,l_1',l_2'}}{4} \left(\cos(\theta_{l_1',l_2'}) + \cos(\theta_{l_1,l_2}) + 1\right) \\
&= \sum_{l_1,l_2} \sum_{l_1',l_2'} \frac{\boldsymbol{A}_{l_1,l_2,l_1',l_2'}}{4} \left(\cos(\Delta\theta_{l_1,l_2,l_1',l_2'}) + j\sin(\Delta\theta_{l_1,l_2,l_1',l_2'})\right) \\
&\quad + \sum_{l_1,l_2} \sum_{l_1',l_2'} \frac{\boldsymbol{A}_{l_1,l_2,l_1',l_2'}}{4} \left(\cos(\theta_{l_1',l_2'}) + \cos(\theta_{l_1,l_2}) + 1\right),
\end{aligned} \tag{2.103}$$

is not linear with respect to θ_{l_1,l_2}. Taking the advantage of the discrete property of θ_{l_1,l_2}, we can further transform the non-linear functions into linear ones.

With the definitions of $\boldsymbol{x}^{l_1,l_2}$, we have

$$\cos(\theta_{l_1,l_2}) = \boldsymbol{x}^{l_1,l_2}\boldsymbol{c}^T, \quad \sin(\theta_{l_1,l_2}) = \boldsymbol{x}^{l_1,l_2}\boldsymbol{s}^T, \tag{2.104}$$

with $\|\boldsymbol{x}^{l_1,l_2}\|_1 = 1$. It is also worthwhile to point out that the value of θ_{l_1,l_2} only falls in the range $[0, 2\pi)$, and thus, we have

$$\boldsymbol{e}^T\boldsymbol{x}^{l_1,l_2} = 0. \tag{2.105}$$

Similarly, according to the definitions of $\boldsymbol{y}^{l_1,l_2,l_1',l_2'}$, we have

$$\cos(\Delta\theta_{l_1,l_2,l_1',l_2'}) = \boldsymbol{y}^{l_1,l_2,l_1',l_2'}\boldsymbol{c}^T, \quad \sin(\Delta\theta_{l_1,l_2,l_1',l_2'}) = \boldsymbol{y}^{l_1,l_2,l_1',l_2'}\boldsymbol{s}^T, \tag{2.106}$$

with

$$\boldsymbol{a}^T(\boldsymbol{x}^{l_1,l_2} - \boldsymbol{x}^{l_1',l_2'}) = \boldsymbol{a}^T\boldsymbol{y}^{l_1,l_2,l_1',l_2'}. \tag{2.107}$$

Algorithm 2: RIS configuration based analog beamforming Algorithm

begin
- Remove the semi-definite constraint of problem (2.102) and solve an initial solution;
- **repeat**
  - Use the branch-and-bound method to solve problem (2.102) and obtain the optimal solution x^{l_1,l_2} for each RIS element;
  - Check the feasibility;
  - If the obtained solution is not feasible, add a cut according to (2.108);
- **until** *The obtained solution is feasible*;
- Derive the phase shits according to the obtained solution;

end

With these transformations, $\tilde{F}\tilde{F}^H$ is linear with respective to x^{l_1,l_2} and $y^{l_1,l_2,l_1',l_2'}$. □

Problem (2.102) is a mix-integer SDP with linear constraints, which can be solved by the outer approximation method. The basic idea of the outer approximation method is to enforce the SDP constraint via linear cuts and transform the original problem into a mix-integer linear programming one, which can be solved by the branch-and-bound algorithm [49]. In the following, we will elaborate on how to enforce the SDP constraints via linear cuts.

Assume that a solution is $\bar{w}$, $\bar{x}^{l_1,l_2}$, $\bar{y}^{l_1,l_2,l_1',l_2'}$. In most mix-integer programming problems, it is very common to use the gradient cuts to approach the feasible set. However, the function of smallest eigenvalues is not always differentiable. Therefore, we use the characterization instead [48]. Note that $Z \succeq 0$ is equivalent to $u^T Z u \geq 0$ for arbitrary u. If Z with $\bar{w}$, $\bar{x}^{l_1,l_2}$, $\bar{y}^{l_1,l_2,l_1',l_2'}$ is not positive semi-definite, we compute eigenvector u associated with the smallest eigenvalue. Then

$$u^T Z u \geq 0 \tag{2.108}$$

is a valid cut that cuts off $\bar{w}$, $\bar{x}^{l_1,l_2}$, $\bar{y}^{l_1,l_2,l_1',l_2'}$. The RIS configuration based analog beamforming algorithm can be summarized in Algorithm 2.

2.4.4.3 Overall Algorithm Description

Based on the results presented in the previous two subsections, we propose an overall iterative algorithm, i.e., the SRM algorithm, for solving the original problem in an iterative manner. Specially, the beamformer V_D is solved by Algorithm 1 while keeping the RIS configuration fixed. After obtaining the results, we will optimize the RIS configuration θ^{l_1,l_2} by Algorithm 2. Those obtained results are set as the initial solution for subsequent iterations. Define R as the value of the objective function. The two subproblems will be solved alternatively until in iteration t the value difference of the objective functions between two adjacent iterations is less than a predefined threshold π, i.e., $R^{(t+1)} - R^{(t)} \leq \pi$.

2.4.4.4 Convergence and Complexity Analysis

We now analyze the convergence and complexity of our proposed SRM algorithm.

1. Convergence: First, according to Algorithm 1, in the digital beamforming subproblem, we can obtain a better result given RIS configuration $\boldsymbol{\theta}^{(t)}$ in the $(t+1)$-th iteration. Therefore, we have

$$R(\boldsymbol{V}_D^{(t+1)}, \boldsymbol{\theta}^{(t)}) \geq R(\boldsymbol{V}_D^{(t)}, \boldsymbol{\theta}^{(t)}). \tag{2.109}$$

Second, given the beamforming result $\boldsymbol{V}_D^{(t+1)}$, we maximize sum rate of all users, and thus, the following inequality holds:

$$R(\boldsymbol{V}_D^{(t+1)}, \boldsymbol{Q}^{(t+1)}) \geq R(\boldsymbol{V}_D^{(t+1)}, \boldsymbol{\theta}^{(t)}). \tag{2.110}$$

Based the above inequalities, we can obtain

$$R(\boldsymbol{V}_D^{(t+1)}, \boldsymbol{\theta}^{(t+1)}) \geq R(\boldsymbol{V}_D^{(t)}, \boldsymbol{Q}^{(t)}). \tag{2.111}$$

which implies that the objective value of the original problem is non-decreasing after each iteration of the SRM algorithm. Since the objective value is upper bounded, the proposed SRM algorithm is guaranteed to converge.

2. Complexity: We consider the complexity of the proposed algorithms for two subproblems separately.

- In the digital beamforming subproblem, we need to optimize the received power for each user according to (2.98). Therefore, its computational complexity is $O(K)$.
- In the RIS configuration based analog beamforming subproblem, we solve a series of linear programs by the branch-and-bound method. Since only one element in $\boldsymbol{x}^{l_1,l_2}$ can be 1, $\boldsymbol{x}^{l_1,l_2}$ can have at most 2^b possible solutions. Thus, the scale of the computational complexity of each linear program is $O(2^{bN_T^2})$.

2.4.5 *Performance Analysis of RIS-Based Multi-User Communications*

In this section, we compare the RIS-based HBF scheme with the traditional ones in terms of the minimum number of required RF chains. A special case, i.e., the pure

LoS transmission, is also considered to explore theoretically how the size of RIS and its placement influence the achievable rates.

2.4.5.1 Comparison with Traditional Hybrid Beamforming

We adopt the fully digital beamforming scheme as a benchmark to compare the traditional and RIS-based HBF schemes. In the traditional HBF, it has already been proved that when the number of RF chains is not smaller than twice the number of target data streams, any fully digital beamforming matrix can be realized [40]. However, in the RIS-based HBF, the inherent analog beamforming (i.e., the RIS configuration) is closely coupled with propagation, which offers more freedom for shaping the propagation environment than the traditional scheme. To capture this characteristic, we explore a new condition for the RIS-based system to achieve fully digital beamforming.

We start by describing the fully digital beamforming scheme in an RIS-based system. Consider an ideal case where each RIS element directly connects with an RF chain and analog-digital converter (ADC) as if it is part of the BS.[9] The fully digital beamformer can then be denoted by $V_{FD} \in {}^{N_R^2 \times K}$, based on which we present the following proposition.

Proposition 2.11 *For the RIS-based HBF scheme with $N_R^2 \geq KN_t$, to achieve any fully digital beamforming scheme, the number of transmit antennas at the BS should not be smaller than the number of single-antenna users, i.e., $N_t \geq K$.*

Proof We first prove that when $N_t = K$, any fully digital beamforming scheme can be achieved by the RIS-based HBF scheme if $N_R^2 \geq KN_t$ holds. We then state that it is not possible to achieve digital beamforming when $N_t < K$. Therefore, $N_t \geq 2K$ is a sufficient condition.

(i) Denote the channel matrix between the RIS and the users as $H_{FD} \in {}^{K \times N_R^2}$. Since $N_t = K$, the kth column of the digital beamformer can be expressed by $\mathbf{V}_{D,k} = \left[0^T, v_{k,k}, 0^T\right]^T$, where $v_{k,k}$ is the kth element of the the kth column in $\mathbf{V}_D \in \mathbb{R}^{N_t \times K}$. To satisfy $\mathbf{H}_{FD}\mathbf{V}_{FD} = \mathbf{F}\mathbf{V}_D$, we have

$$\left[\ldots, f_{m,k}, \ldots\right] \begin{bmatrix} 0 \\ \vdots \\ v_{k,k} \\ \vdots \\ 0 \end{bmatrix} = (\mathbf{H}_{FD}\mathbf{V}_{FD})_{m,k}, \tag{2.112}$$

[9] In practice, the RIS does not connect to the RF chain directly unless it is installed at the BS, which is not the truth in our case where RIS actually only reflects signals. Therefore, we only consider such an ideal scheme as a benchmark to evaluate the effectiveness of RIS-based HBF.

i.e.,

$$\sum_{l_1,l_2} \frac{1}{2}\phi^{(k)} \left[\left(j + e^{j\theta_{l_1,l_2}} \right) g_{l_1,l_2}^{m,k} \right] v_{k,k} = (H_{FD} V_{FD})_{m,k}, \tag{2.113}$$

for all $0 \leq m, k \leq K$, where $f_{m,k}$ is the element of matrix $\mathbf{F}$. Note that the term multiplied by $v_{k,k}$ in (2.113) can achieve different magnitudes owning to the linear combination of channel coefficients, which is different from the traditional HBF [44]. At least one solution can be found for this set of equations if the number of equations is no smaller than the number of variables, i.e., $N_R^2 \geq K N_t$. This also holds for $N_t > K$ since we can always use the solution for $N_t = K$ and set those extra variables to be zero.

(ii) We observe that $rank\,(\mathbf{H}_{FD}\mathbf{V}_{FD}) = K$ and $rank\,(\mathbf{F}\mathbf{V}_D) = \min\{K, N_t\}$. If $N_t < K$, then $rank\,(\mathbf{H}_{FD}\mathbf{V}_{FD}) > rank\,(\mathbf{F}\mathbf{V}_D)$, implying that the RIS-based HBF cannot implement the fully digital beamforming scheme. This completes the proof. □

This implies two conditions for the proposed RIS-based scheme to achieve the fully digital beamforming scheme. First, the size of RIS should be no smaller than the product of the number of users and the size of the antenna array at the BS. Second, the number of transmit antennas at the BS should be no smaller than the number of single-antenna users.

Remark on dedicated hardware reduction for analog beamforming: In the traditional HBF, to offload part of the digital baseband processing to the analog domain, a number of RF chains are required at the BS to feed the analog beamformer, equipped with necessary hardwares such as mixers, filters, and phase shifters [50]. In contrast, as shown in Proposition 2.11, the number of minimum RF chains required by the RIS-based HBF to achieve the fully digital beamforming has already been reduced by half compared to the traditional scheme. Moreover, the phase shifters can be saved since the RIS inherently realizes analog beamforming owning to its flexible physical structure [51].

2.4.5.2 Special Case: Pure Line-of-Sight Transmissions

We consider the data rate obtained by the pure LoS case as a lower bound of the achievable rate and analyze optimal RIS placement to provide orthogonal communication links. The LoS case also reveals insights on how the achievable data rate is influenced by RIS design and placement.

Since a multitude of RIS elements are placed in a sub-wavelength order, spacial correlation between these elements are inevitable. In this case, the channel matrix $\mathbf{F}$ approaches a low-rank matrix, leading to a degraded performance in terms of the achievable data rate. Traditionally, we can utilize the multi-path effect to decorrelate different channel links between the transceiver antennas. However, it is also important to understand how the system works in the pure LoS case, especially when it comes to the RIS-based systems where the reflection-based one-hop link between the BS and users acts as the dominated "LoS" component and is usually much stronger than other multi-paths as well as the degraded direct links.

In the pure LoS case, we aim to achieve a high-rank LoS channel matrix by designing the size of RIS and the antenna array at the BS. Different from those existing works on antenna array design for LoS MIMO systems, the discrete phase shifts depicted by $\{q_{l_1,l_2}\}$ needs to be considered in the RIS-based propagation. We thus present the results in Proposition 2.12 below.

Proposition 2.12 *In the RIS-based system, to make different links between the BS and user k via the RIS orthogonal to each other, the RIS design should satisfy the following conditions:*

$$d_R^{(1)} d_B = \frac{\lambda D_{0,0}^{(0)}}{N_R \cos\theta_R \cos\theta_B}, \tag{2.114a}$$

$$\sum_{l_2=0}^{N_R-1} \left(1+\sin\theta_{l_1,l_2}\right) = \sum_{l_2=0}^{N_R-1} \left(1+\sin\theta_{l_1',l_2}\right), \forall 0 \le l_1, l_1' \le N_R - 1. \tag{2.114b}$$

Proof Note that each user k goes through different path losses due to various positions, which naturally varies the corresponding channel coefficients. Therefore, we only focus on orthogonalizing different links from the BS to the same user. Since the path loss is the same for different links with respect to user k, we consider the channel response between one BS antenna n and all RIS elements as

$$\begin{aligned} \mathbf{f}^{(n,k)} = \Big[& q_{0,0} e^{-j\frac{2\pi}{\lambda}\left(D_{0,0}^{(n)}+d_{0,0}^{(k)}\right)}, q_{0,1} e^{-j\frac{2\pi}{\lambda}\left(D_{0,1}^{(n)}+d_{0,1}^{(k)}\right)}, \cdots, \\ & q_{N_R-1,N_R-1} e^{-j\frac{2\pi}{\lambda}\left(D_{N_R-1,N_R-1}^{(n)}+d_{N_R-1,N_R-1}^{(k)}\right)} \Big]. \end{aligned} \tag{2.115}$$

To keep any two different links orthogonal to each other, the following condition should be satisfied,

$$\left[\mathbf{f}^{(n_a,k)}\right]^H \cdot \mathbf{f}^{(n_b,k)} = 0, \forall n_a \ne n_b, \tag{2.116}$$

where n_a and n_b denote two different transmit antennas.

By substituting (2.115), (2.116) can be rewritten by

$$\sum_{l_2=0}^{N_R-1} \sum_{l_1=0}^{N_R-1} q_{l_1,l_2} \cdot \left(q_{l_1,l_2}\right)^* e^{j\frac{2\pi}{\lambda}\left(D_{l_1,l_2}^{(n_a)} - D_{l_1,l_2}^{(n_b)}\right)} = 0. \tag{2.117}$$

We substitute (2.82) and the expression of $D_{l_1,l_2}^{(n_a)}$ given in Sect. 2.1.2 into (2.117) and obtain the following

$$\sum_{l_1=0}^{N_R-1}\left[e^{j\frac{2\pi}{\lambda D_{0,0}^{(0)}}l_1 d_R^{(1)} d_B(n_b-n_a)\cos\theta_R\cos\theta_B}\cdot\sum_{l_2=1}^{N_R-1}\left(1+\sin\theta_{l_1,l_2}\right)\right]=0, \tag{2.118}$$

where $n_b - n_a \in \mathbb{Z}$. Since this condition should hold for any two transmit antennas n_a and n_b, we then have (2.114) based on the principle of geometric sums. □

This proposition shows that the achievable data rate is highly related to the size of the RIS and its placement. For convenience, when all other parameters are fixed, we refer to the *threshold* of the RIS size as

$$N_R^{th}=\frac{\lambda D_{0,0}^{(0)}}{d_R^{(1)} d_B\cos\theta_R\cos\theta_B}. \tag{2.119}$$

For the pure LoS case, the sum rate maximization problem can still be formulated as (2.92) with one extra constraint (2.114b). Our proposed SRM algorithm can be utilized to solve this problem after we convert the extra constraint (2.114b) into a linear one shown below by following the transformations in Proposition 1,

$$\sum_{l_2=0}^{N_r-1}(\boldsymbol{x}^{l_1,l_2}-\boldsymbol{x}^{l_1',l_2})\boldsymbol{s}^T=0. \tag{2.120}$$

2.4.6 Simulation Results

In this section, we evaluate the performance of our proposed algorithm for RIS-based HBF in terms of the sum rate. We show how the system performance is influenced by the SNR, number of users, the size of RIS, and the number of quantization bits for discrete phase shifts. For comparison, the following algorithms are performed as well.

- *Simulated annealing*: We utilize the simulated annealing method [52] to approach the global optimal solution of the sum rate maximization problem with discrete phase shifts. The maximum number of iterations is set as 10^7.
- *Pure LoS case*: We consider the pure LoS case as a lower bound to evaluate the performance of the HBF scheme. The proposed iterative SRM algorithm can still be utilized to solve the corresponding optimization problem.
- *Random phase shift*: Algorithm 1 is first performed, followed by a random algorithm to solve the RIS-based analog beamforming subproblem (2.94).

- *RIS-based HBF with continuous phase shifts*: This scheme only serves as a benchmark when we investigate how the discreteness level (i.e., the number of quantization bits, b) influences the sum rate. The discrete constraint in the original subproblem (2.94) is relaxed to $\theta_{l_1,l_2} \in [0, 2\pi]$. The HBF solution is obtained by iteratively performing Algorithm 1 and the gradient descent method.

In our simulation, we set the distance between the BS and the RIS, $D_{0,0}^{(0)}$, as 20 m, and users are randomly deployed within a half circle of radius 60 m centering at the RIS. The antenna array at the BS and the RIS are placed at angles of 15° and 30° to the x axis, respectively. The transmit power of the BS P_T is 20 W, the carrier frequency is 5.9 GHz, the antenna separation at the BS d_B is 1 m, the RIS element separation d_R is 0.03 m, and the Rician fading parameter κ is 4 [53]. We set the size of the RIS N_R^2 ranging between $5^2 \sim 65^2$, the number of antennas at the BS N_t and the number of users K between $5 \sim 15$, the discreteness level of RIS b between $1 \sim 5$, and the SNR (defined as P_T/σ^2) between -2 dB $\sim$ 10 dB. Specifically, for the LoS case, given the above parameters the designing rule in (2.114a) is only satisfied when the size of RIS is set as 40 or the antenna separation at the BS is set as 6.75 m.

We present Fig. 2.22 to verify Proposition 2.12 in the pure LoS case. The figure shows the sum rate of all users versus the antenna separation at the BS, d_B, with different numbers of quantization bits in the pure LoS case. Given the parameters set above, according to Proposition 2.12, the optimal value of d_B should be 6.75 m so as to orthogonalize the channel links between each antenna of the BS and a user k via any RIS element. We observe that the optimal sum rate can be achieved when

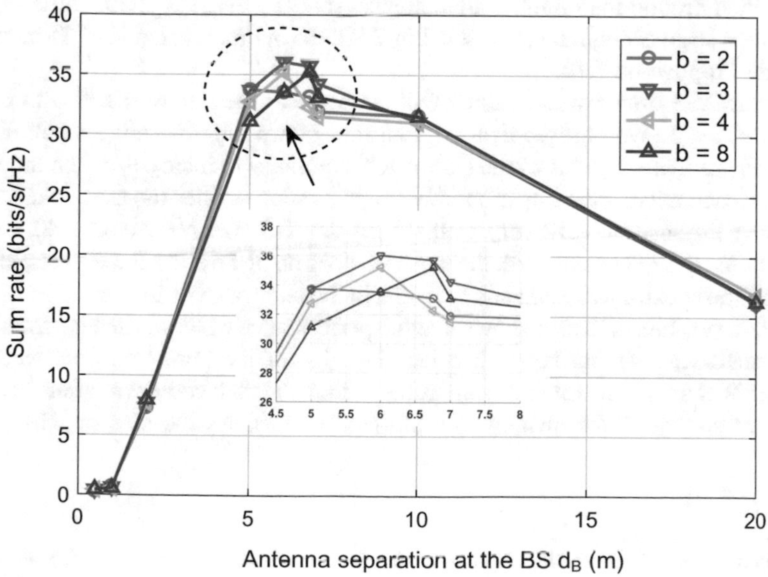

Fig. 2.22 Sum rate v.s. antenna separation at the BS (SNR = 2 dB, $N_t = K = 5$, $N_R = 6$)

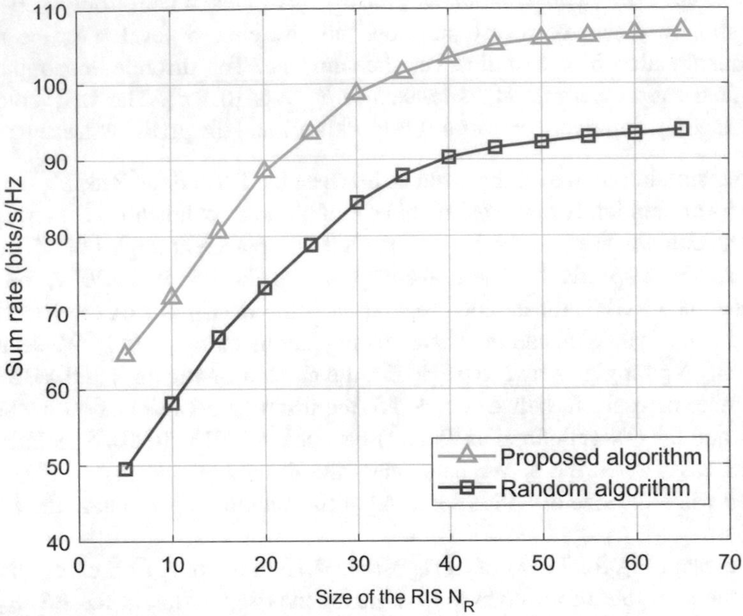

Fig. 2.23 Sum rate v.s. size of RIS N_R (SNR = 2dB, $K = N_t = 5, b = 2$)

d_B is around $5 \sim 6.75$ m. The numerically optimal value of d_B approaches the theoretical result as the number of quantization bits, b, grows, implying that such fluctuation around the optimal value comes from the discrete phase shifts of RIS. When b is large enough, i.e., $b = 8$ in Fig. 2.22, the optimal d_B equals 6.75 m, which justifies Proposition 2.12.

Figure 2.23 shows the sum rate of all users versus the size of RIS[10] with $b = 2$, $N_t = K = 5$. We observe that the sum rate grows rapidly with a small size of RIS and gradually flattens as the size of RIS continues to increase.[11] The inflection point of each curve shows up around $N_R = 40$, which verifies the threshold (2.119) given by Proposition 2.12 very well. When the RIS size N_R exceeds 40, though Proposition 2.12 does not hold, the sum rate does not drop since the RIS can always turn off those extra RIS elements to maintain the sum-rate performance.

Moreover, Fig. 2.23 also shows that the performance of RIS-based beamforming with small-scale fading is much better than that of the pure LoS case when the size of RIS is small. Such a gain comes from the reduced correlation between different channel links owning to multi-path effects. As the size of RIS grows,

[10]For convenience, here we adopt N_R to represent the size of RIS to better display the curves.

[11]We do not show the simulated annealing algorithm in this figure due to its high complexity with a large size of RIS.

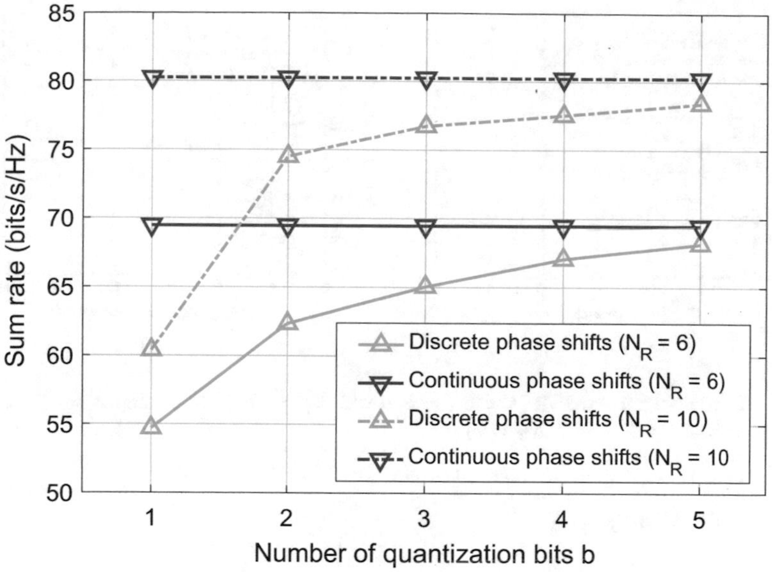

Fig. 2.24 Sum rate v.s. number of quantization bits b (SNR = 2dB, $K = N_t = 5$, $N_R = 6$)

Proposition 2.12 is satisfied such that the channel links in the pure LoS case are orthogonalized, making the gap between these two cases smaller.

Figure 2.24 depicts the sum rate of all users versus the number of quantization bits b for discrete phase shifts in RIS configuration with SNR = 2 dB, $N_t = K = 5$, and $N_R = 6$. As the number of quantization bits increases, the sum rate obtained by our proposed algorithm with discrete phase shifts approaches that in the continuous case. When the size of RIS grows, the gap between the discrete and continuous cases shrinks since a larger RIS usually provides more freedom of generating directional beams. Note that the implementation difficulty increases dramatically in practice with the number of quantization bits. A trade-off can then be achieved between the sum rate and number of quantization bits.

Figure 2.25 show the sum rate of all users versus SNR and the number of users, respectively, obtained by different algorithms with a RIS of size 6×6 (i.e., $N_R = 6$), $b = 2$ quantization bits for phase shifts, equal number of transmit antennas at the BS and the downlink users. In Fig. 2.25a, the sum rate increases with SNR since more power resources are allocated by the BS. In Fig. 2.25b, the sum rate grows with the number of users since a higher diversity gain is achieved. From both figures, we observe that the performance of our proposed algorithm is close to that of the simulated annealing method and much better than the random algorithm. This indicates the efficiency of our proposed algorithm to solve the RIS-based HBF problem.

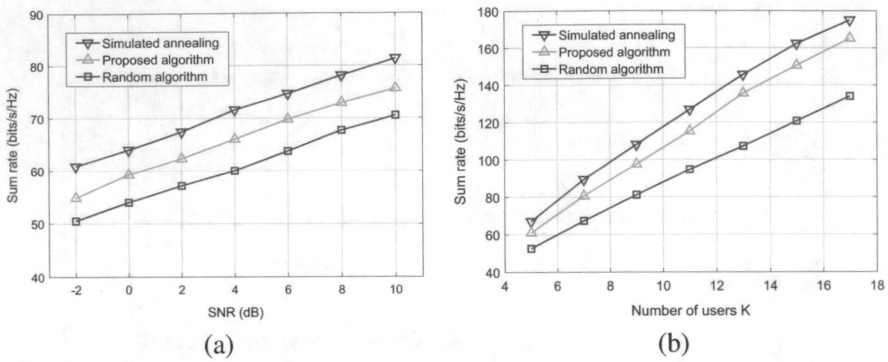

Fig. 2.25 (**a**) Sum rate v.s. SNR ($K = N_t = 5, b = 2, N_R = 6$); (**b**) Sum rate v.s. number of users (SNR = 2 dB, $N_t = K, b = 2, N_R = 6$)

2.4.7 Summary

In this section, we have studied a RIS-based downlink multi-user multi-antenna system in the absence of direct links between the BS and users. The BS transmits signals to users via the reflection-based RIS with limited discrete phase shifts. To better depict the close coupling between channel propagation and the RIS configuration pattern selection, we haved considered a reflection-dominated one-hop propagation model between the BS and users. Based on this model, we have carried out an HBF scheme for sum rate maximization where the continuous digital beamforming has been performed at the BS and the discrete analog beamforming has been achieved inherently at the RIS via configuration pattern selection. The sum rate maximization problem has been decomposed into two subproblems and solved iteratively by our proposed SRM algorithm.

Three remarks can be drawn from the theoretical analysis and numerical results, providing insights for RIS-based system design.

- *The sum rate of the RIS-based system with discrete phase shifts increases rapidly when the number of quantization bits b is small, and gradually approaches the sum rate achieved in the continuous case if b is large enough.*
- *The sum rate increases with the size of RIS and converges to a stable value as the size of RIS grows to reach the threshold determined by Proposition 2.12.*
- *The minimum number of transmit antennas at the BS required to achieve any fully digital beamforming scheme is only half of that in traditional HBF schemes, implying that the RIS-based HBF scheme can greatly reduce the cost of dedicated hardware.*

The above remarks have indicated that when designing the RIS-based systems, a moderate size of RIS and a very small number of quantization bits are enough to achieve the satisfying sum rate at low cost.

2.5 Full-Dimensional Coverage Extension

2.5.1 Motivations

In the literature, a widely studied example of RIS is referred to as intelligent reflecting surface (IRS), in which the metasurface is designed for reflecting the signals impinging upon one side of the surface towards users located on the same side of the surface [43]. An IRS is, therefore, a reflective surface. Examples of research works on IRS-aided transmission include the following. In [18], a joint power allocation and continuous phase shift design is studied for application to a reflective IRS-assisted system in order to maximize the energy efficiency. In [20], the achievable data rate of a reflective IRS-assisted communication system is evaluated and the effect of a limited number of phase shifts on the data rate is investigated. In [17], the IRS beamforming and phase shifts design are jointly optimized to maximize the sum rate in a reflective IRS-assisted point-to-point communication system. In [54], a reflective IRS is deployed at the cell boundary of multiple cells to assist the downlink transmission of cell-edge users, whilst mitigating the inter-cell interference.

In these research works, as mentioned, signals that impinge upon one of the two sides of a surface are completely reflected towards the same side. This implies that users located in the opposite side of the surface cannot be served by an IRS: They are out of coverage. To tackle this issue, we introduce an intelligent omni-surface (IOS)-assisted communication system. The proposed IOS is deployed in a general multi-user downlink communication system and, in contrast with an IRS, has the dual functionality of signal reflection and transmission [55]. More precisely, signals impinging upon one of both sides of the IOS can be simultaneously reflected and transmitted towards the mobile users (MUs) that are located on the same side and in the opposite side of the IOS, respectively [56]. Similar to an IRS, an IOS is made of multiple passive scattering elements and programmable PIN diodes, which are appropriately designed and configured, respectively, to customize the propagation environment [57]. Unlike an IRS that completely reflects all the received signals, an IOS is capable of simultaneously reflecting and diffracting (i.e., transmitting) the received signals [58]. The power ratio of the transmitted and reflected signals is determined by the hardware structure of the IOS. By enabling joint reflection and transmission, an IOS provides ubiquitous wireless coverage to the MUs on both sides of it, and the propagation environment of all the users can be jointly customized by adjusting the phase shifts of the IOS scattering elements [59]. As a result, the power of the received signals can be enhanced, and the QoS of the communication links can be improved.

To serve the MUs with good performance, it is of vital importance to design the amplitude and phase response of the passive scatterers of the IOS. As mentioned in [25], the optimization of an IRS-aided system can be viewed as a joint analog beamforming design at the RIS and a digital beamforming design at the BS, so as to shape the propagation environment and improve the sum-rate of the network. In an IOS-assisted communication system, the analog beamforming performed at the IOS and the digital beamforming performed at the BS provide directional reflective/transmissive radio waves to the MUs on both sides of the IOS concurrently.

However, when compared to an IRS-assisted communication system [25], the analog and digital beamforming design of an IOS-assisted communication system faces several new challenges. *First*, the power of the reflected and transmitted signals of the IOS may not be symmetric, i.e., the channel model of the reflected and transmitted signals can be different [56]. Therefore, the methods developed and the results obtained for IRSs may not be applicable to an IOS-assisted communication system. *Second*, the power ratio of the reflected and transmitted signals of an IOS can be appropriately optimized, which provides an extra degree of freedom for enhancing the communication performance [55]. In particular, the interplay between the spatial distribution of the MUs and the optimal power ratio of the reflected and transmitted signals plays an important role. *Third*, the LoS transmission link from the BS to the MUs may exist concurrently with the signals that are reflected and transmitted from the IOS, which implies that the IOS needs to be optimized in order to account for the LoS links as well.

In this section, motivated by these considerations, we aim to jointly optimize the digital beamforming at the BS and the analog reflective + transmissive beamforming at the IOS, in order to maximize the sum-rate of an IOS-assisted communication system. We first formulate a joint BS digital beamforming and IOS analog reflective + transmittive beamforming design problem in order to maximize the sum-rate of the considered IOS-assisted downlink communication system, and then propose an iterative algorithm to solve the obtained NP-hard optimization program. The convergence and complexity of the proposed algorithm are analyzed. The performance of the proposed multi-user IOS-assisted downlink communication system is analyzed theoretically and evaluated with the aid of simulations. The obtained numerical results unveil the impact of the optimal power ratio between the reflected and transmitted signals of the IOS as a function of the spatial distribution of the MUs.

The rest of this section is organized as follows. In Sect. 2.5.2, we introduce the structure and properties of an IOS. In Sect. 2.5.3, we illustrate the considered multi-user IOS-assisted downlink communication system, including the channel model and the beamforming design. The optimization problem for the joint design of the digital beamforming at the BS and the analog reflective + transmissive beamforming at the IOS is formulated in Sect. 2.5.4, and an iterative algorithm for solving the resulting non-convex problem is introduced in Sect. 2.5.5. The theoretical analysis of the optimal phase shifts and the optimal power ratio that maximizes the sum-rate of the network are elaborated in Sect. 2.5.6. Numerical results are illustrated in Sect. 2.5.7 in order to quantitatively evaluate the performance of the proposed algorithm. Finally, conclusions are drawn in Sect. 2.5.8.

2.5.2 Intelligent Omni-Surface

An IOS is a two-dimensional array of electrically controllable scattering elements, as illustrated in Fig. 2.26a. In particular, the considered IOS is made of M reconfigurable elements of equal size. The size of each element is δ_x and δ_y along the x and y axis, respectively. Each reconfigurable element consists of multiple metallic patches and N_D PIN diodes that are evenly distributed on a dielectric substrate. The metallic patches are connected to the ground via the PIN diodes that can be switched between their ON and OFF states according to predetermined bias voltages. The ON/OFF configuration of the PIN diodes determines the phase response applied by the IOS to the incident signals. In total, each metallic patch can introduce 2^{N_D} different phase shifts to the incident signals. For generality, we assume that a subset of the possible phase shifts is available, which is referred to as the *available phase shift set* and is denoted by $\mathscr{S}_a = \{1, \ldots, S_a\}$. The phase shift of the mth reconfigurable element is denoted by $s_m \in \mathscr{S}_a$. The S_a available phase shifts are uniformly distributed with a discrete phase shift step equal to $\Delta\psi_m = \frac{2\pi}{S_a}$ [20]. Therefore, the possible values of the phase shifts are $l_m \Delta\psi_m$, where l_m is an integer satisfying $0 \leq l_m \Delta\psi_m \leq S_a - 1$. The vector of phase shifts of the M elements of the IOS is denoted by $\boldsymbol{s} = (s_1, \ldots, s_M)$. When a signal impinges, from either sides of the surface, upon one of the M reconfigurable elements of the IOS, a fraction of the incident power is reflected and transmitted towards the same side and the opposite side of the impinging signal. This makes an IOS different from an IRS [60].

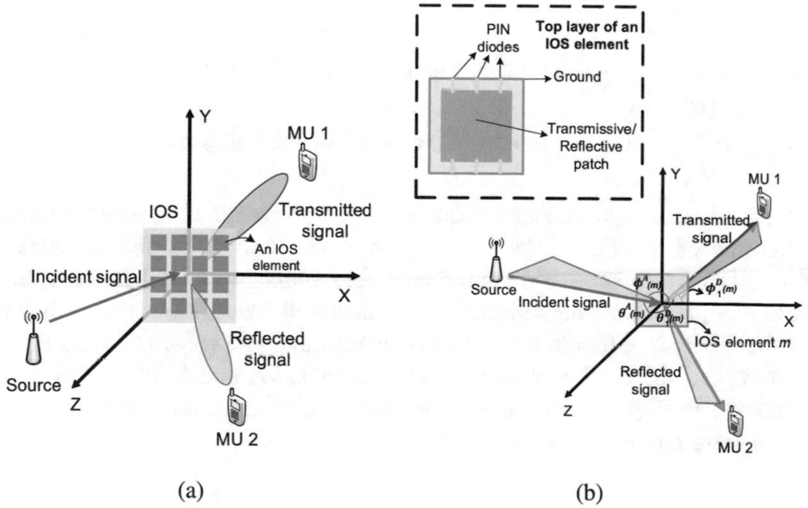

Fig. 2.26 Illustration of the reflected and transmitted signals in an IOS-assisted communication system. (**a**) Reflected and transmitted signals from an IOS. (**b**) Reflected and transmitted signals for a single reconfigurable element of the IOS

The direction of the signal that is emitted by the transmitter and that impinges upon the mth reconfigurable element of the IOS is denoted by $\xi^A(m) = (\theta^A(m), \phi^A(m))$, and the direction of the signal that is re-emitted by the mth reconfigurable element of the IOS towards the ith MU is denoted by $\xi_i^D(m) = (\theta_i^D(m), \phi_i^D(m))$, respectively, as illustrated in Fig. 2.26b. The response of the mth reconfigurable element of the IOS to the incident signal is denoted by the complex coefficient g_m, which is referred to as the *amplitude gain* of the signal. In particular, g_m depends on the direction of incidence i.e., $\xi^A(m)$, the direction of departure (either in reflection or in transmission), i.e., $\xi_i^D(m)$, and the phase shift s_m. In mathematical terms, we have

$$g_m(\xi^A(m), \xi_i^D(m), s_m) = \sqrt{G_m K^A(m) K_i^D(m) \delta_x \delta_y |\gamma_m|^2} \exp(-j\psi_m), \tag{2.121}$$

where G_m is the antenna power gain of the mth reconfigurable element, and ψ_m is the corresponding phase shift. The coefficient $|\gamma_m|^2$ is the power ratio between the power of the signal re-emitted by the IOS and the power of the incident signal. Depending on the implementation of the IOS, $|\gamma_m|^2$ can be either a function of s_m or a constant. In this section, for simplicity, we assume that $|\gamma_m|^2$ is independent of the phase shift s_m. $K^A(m)$ and $K_i^D(m)$ are the normalized power radiation patterns of the incident and the re-emitted (either reflected or transmitted) signal, respectively. An example for the normalized power radiation patterns is the following [32]:

$$K^A(m) = |\cos^3 \theta^A(m)|, \ \theta^A(m) \in (0, \pi), \tag{2.122}$$

$$K_i^D(m) = \begin{cases} \dfrac{1}{1+\epsilon} |\cos^3 \theta_i^D(m)|, \ \theta_i^D(m) \in (0, \pi/2), \\ \dfrac{\epsilon}{1+\epsilon} |\cos^3 (\theta_i^D(m))|, \ \theta_i^D(m) \in (\pi/2, \pi), \end{cases} \tag{2.123}$$

where ϵ is a constant parameter that quantifies the power ratio between the reflected and transmitted signals of the IOS, which is determined by the structure and hardware implementation of the reconfigurable elements [61]. It is worth noting that the strength of both the reflected and transmitted signals satisfy (2.123), where $\theta_i^D(m) \in (0, \pi/2)$ refers to the reflected signals, and $\theta_i^D(m) \in (\pi/2, \pi)$ refers to the transmitted signals. The normalized power radiation pattern of an IOS element is illustrated in Fig. 2.27, and it is compared against the same normalized power radiation pattern of a conventional IRS.

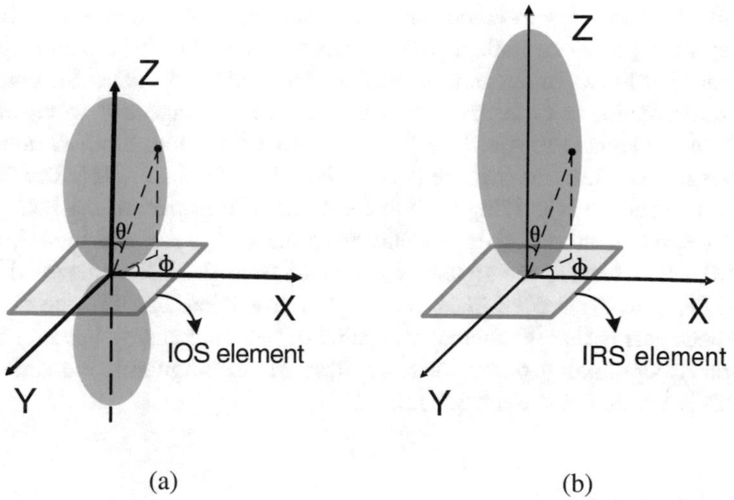

Fig. 2.27 Normalized power radiation pattern of (**a**) an IOS element, (**b**) an IRS element

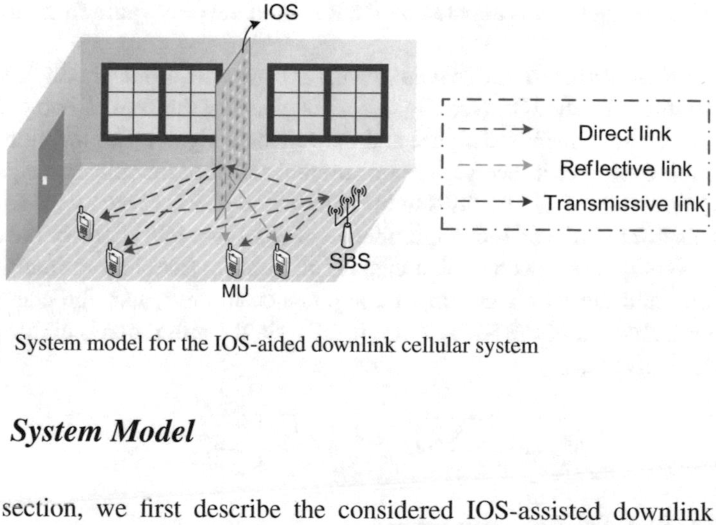

Fig. 2.28 System model for the IOS-aided downlink cellular system

2.5.3 System Model

In this section, we first describe the considered IOS-assisted downlink system model where a multi-antenna BS serves multiple MUs, and we then introduce the transmission channel model of the IOS-assisted system. Finally, the joint BS and IOS beamforming design for the considered transmission system is described.

2.5.3.1 Scenario Description

As shown in Fig. 2.28, we consider a downlink transmission scenario in an indoor environment, which consists of one small BS (SBS) with K antennas and N single-antenna MUs that are denoted by $\mathcal{N} = \{1, 2, \ldots, N\}$. Due to the complex scattering characteristics in indoor environments, some MUs that are far from the SBS may undergo severe fading, which leads to a low QoS for the corresponding communication links. To tackle this problem, we deploy an IOS in the considered indoor environment in order to extend the service coverage and to enhance the strength of the signals received by the MUs. As mentioned, the IOS consists of M reconfigurable elements that are denoted by $\mathscr{M} = \{1, 2, \ldots, M\}$. The MUs are split into two subsets according to their locations with respect to the IOS. The set of MUs that receive the signals *reflected* from the IOS is denoted by $\mathscr{N}_r$, and the set of MUs that receive the signals *transmitted* from the IOS is denoted by $\mathscr{N}_t$, with $\mathscr{N}_r \cap \mathscr{N}_t = \varnothing$, and $\mathscr{N}_r \cup \mathscr{N}_t = \mathscr{N}$. In this section, the IOS can be viewed as an antenna array that is inherently capable of realizing analog beamforming by appropriately optimizing phase shifts of these M reconfigurable elements of the IOS. This is elaborated in Sect. 2.5.3.3.

2.5.3.2 Channel Model

The channel from the SBS to each MU consists of two parts: the reflective-transmissive channel that is assisted by the IOS, and the direct path from the SBS to the MU.

1. Reflective-Transmissive Channel via the IOS: As mentioned in Sect. 2.5.2, the signal re-emitted by the IOS (see Fig. 2.26a) is given by the sum of two concurrent contributions: the transmitted signal and the reflected signal. The location of each MU determines wether it receives the transmitted signal or the reflected signal from the IOS. The channel from the SBS to the MU via the IOS is given by the sum of the M channels through the M reconfigurable elements of the IOS. Each of the M SBS-IOS-MU links is model as a Rician channel in order to take into account the LoS contribution and the NLoS multipath components. In particular, the channel gain from the kth antenna of the SBS to the ith MU via the mth reconfigurable element of the IOS is given as

$$h_{i,k}^{m} = \sqrt{\frac{\kappa}{1+\kappa}} h_{i,k}^{m,LoS} + \sqrt{\frac{1}{1+\kappa}} h_{i,k}^{m,NLoS}. \tag{2.124}$$

The LoS component of $h_{i,k}^{m}$ is expressed as

$$h_{i,k}^{m,LoS} = \frac{\lambda \sqrt{G_k^{tx} K^A(m) G_i^{rx} K_i^D(m)} \exp\left(\frac{-j2\pi(d_{k,m}+d_{m,i})}{\lambda}\right)}{(4\pi)^{\frac{3}{2}} d_{k,m}^{\alpha} d_{m,i}^{\alpha}} g_m(\xi_k^A(m), \xi_i^D(m), s_m), \tag{2.125}$$

where λ is the transmission wavelength, G_k^{tx} and G_i^{rx} are the power gains of the kth antenna of the SBS and the antenna of the ith MU, respectively. $K^A(m)$ is the normalized power gain of the kth antenna of the SBS in the direction of the mth reconfigurable element of the IOS, and $K_i^D(m)$ is the normalized power gain of the ith MU in the direction of the mth reconfigurable element of the IOS, which are given in (2.122) and (2.123), respectively. $d_{k,m}$ and $d_{m,i}$ are the transmission distances between the mth reconfigurable element of the IOS and the kth antenna of the SBS and the ith MU, respectively, and α is the corresponding path-loss exponent. In addition, $g_m(\xi_k^A(m), \xi_i^D(m), s_m)$ is given and defined in (2.121). In this section, for simplicity, the normalized radiation patterns of the SBS, the M reconfigurable elements of the IOS, and the MUs are assumed to be the same. The analysis can, however, be easily generalized to the case with different normalized radiation patterns.

Since the IOS is intended to be a passive device with no active power sources, the sum of the transmitted and reflected power cannot exceed the power of the incident signals. Therefore, the following constraint holds

$$\int_0^{2\pi}\int_0^{2\pi} |g_m(\xi^A(m), \xi^D(m), s_m)|^2 d\theta d\phi \le 1. \tag{2.126}$$

The NLoS component of $h_{i,k}^m$ is expressed as

$$h_{i,k}^{m,NLoS} = PL(k, m, i)h^{SS}, \tag{2.127}$$

where $PL(k, m, i)$ is the path-loss of the SBS-IOS-MU link given in (2.125), and $h^{SS} \sim \mathcal{CN}(0, 1)$ accounts for the cumulative effect of the large number of scattered paths that originate from the random scatterers available in the propagation environment.

2. *BS-MU Direct Path:* As far as the BS-MU channel is concerned, we assume a Rayleigh fading model. Therefore, the channel gain from the kth antenna of the SBS to the ith MU is

$$h_{i,k}^D = \sqrt{G_k^{tx} F_{k,i} G_i^{rx} d_{i,k}^{-\alpha}} h^{SS}, \forall i \in \mathcal{N}, \tag{2.128}$$

where $F_{k,i} = |\cos^3 \theta_{k,i}^{tx}||\cos^3 \theta_{k,i}^{rx}|$ is the normalized end-to-end power gain of the kth antenna of the SBS and the ith MU, where $\theta_{k,i}^{tx}$ is the angle between the kth antenna of the SBS and the direction to the ith MU, and $\theta_{k,i}^{rx}$ is the angle between the antenna of the ith MU and the direction to the kth antenna of the SBS. $d_{i,k}$ is the distance between the kth antenna of the SBS and the ith MU.

In summary, the channel gain from the kth antenna of the SBS to the ith MU can be written as

$$h_{i,k} = \sum_{m=1}^{M} h_{i,k}^m + h_{i,k}^D, \forall i \in \mathcal{N}, \tag{2.129}$$

where the first term represents the superposition of the transmissive-reflective channels of the M reconfigurable elements of the IOS, and the second term is the direct path.

2.5.3.3 IOS-Based Beamforming

In this section, we introduce the IOS-based beamforming that allows the IOS to reflect and transmit the incident signals towards specified locations of the MUs. Since the reconfigurable elements of the IOS have no digital processing capabilities, we consider a hybrid beamforming scheme, where the digital beamforming is performed at the SBS and the analog beamforming is performed at the IOS. Furthermore, due to practical implementation constraints, discrete phase shifts are assumed at the IOS. An example of the hybrid beamforming for $|\mathcal{N}_r| = 1$ and $|\mathcal{N}_t| = 1$ is shown in Fig. 2.29.

1. Digital Beamforming at the SBS: The SBS first encodes the N different data steams that are intended to the MUs via a digital beamformer, $\mathbf{V}_D$, of size $K \times N$, with $K \geq N$, and then emits the resulting signals through the K transmit antennas. We denote the intended signal vector for the N MUs by $\mathbf{x} = [x_1, x_2, \ldots, x_N]^T$. The transmitted vector of the SBS is given by

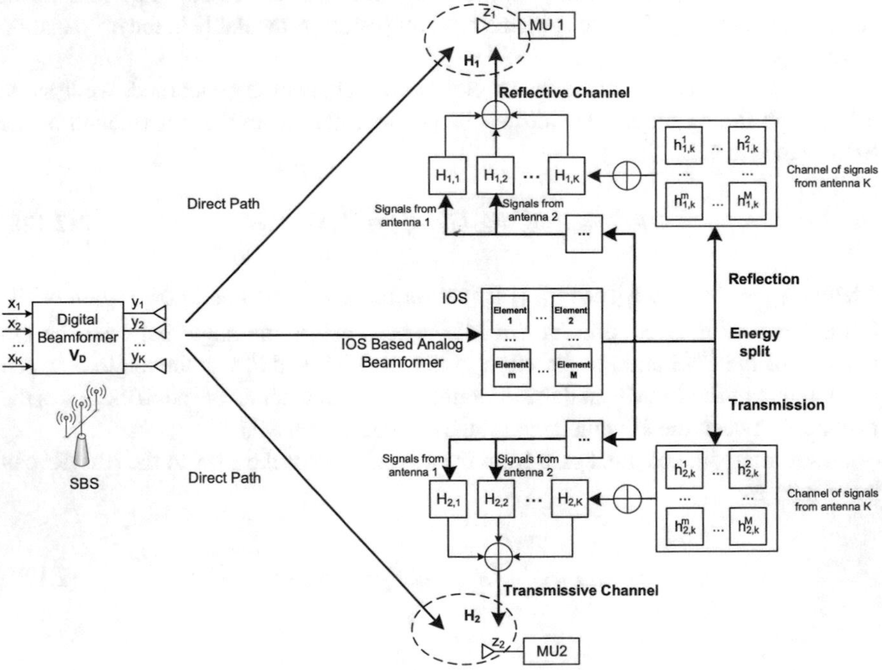

Fig. 2.29 Block diagram of the IOS-based beamforming for 2 MUs

$$\mathbf{y} = \mathbf{V}_D\mathbf{x}. \tag{2.130}$$

We denote the maximum transmission power of the SBS by P_B, hence the power constraint for the digital beamformer can be expressed as

$$\text{Tr}\left(\mathbf{V}_D\mathbf{V}_D^H\right) \leq P_B. \tag{2.131}$$

2. *IOS-Based Analog Beamforming:* The received signal at the ith MU can be expressed as

$$z_i = \sum_{k=1}^{K} h_{i,k}\mathbf{V}_D^{k,i} x_i + \sum_{i \neq i'}\sum_{k=1}^{K} h_{i,k}\mathbf{V}_D^{k,i'} x_{i'} + w_i, \tag{2.132}$$

where w_i is the additive white Gaussian noise (AWGN) at the ith MU whose mean is zero and whose variance is σ^2. Therefore, the N signals of the MUs in (2.132) can be cast in a vector $\mathbf{z} = [z_1, z_2, \ldots, z_N]^T$ as follows:

$$\mathbf{z} = \mathbf{H}\mathbf{V}_D\mathbf{x} + \mathbf{w}, \tag{2.133}$$

where $\mathbf{w} = [w_1, w_2, \ldots, w_N]^T$ is the noise vector, and $\mathbf{H} = \begin{bmatrix} h_{1,1} & \ldots & h_{1,K} \\ \ldots & \ldots & \ldots \\ h_{N,1} & \ldots & h_{1,K} \end{bmatrix}$ is the $N \times K$ compound channel matrix, which accounts of the propagation channel and for the phase shifts (analog beamforming) applied by the IOS (see (2.121) and (2.129)).

From (2.133), the downlink data rate of the ith MU can be formulated as

$$R_i = W_B \log_2\left(1 + \frac{|(\mathbf{H}^i)^H\mathbf{V}_D^i|^2}{|\sum_{i \neq i'}(\mathbf{H}^i)^H\mathbf{V}_D^{i'}|^2 + \sigma^2}\right), \tag{2.134}$$

where W_B is the bandwidth. It is worth noting that (2.134) is different from the rate in an IRS-assisted communication system, since the matrix $\mathbf{H}$ accounts for both the reflective and transmissive channels.

2.5.4 Problem Formulation and Decomposition

In this section, we first formulate a joint IOS phase shift design and SBS digital beamforming optimization problem to maximize the system sum-rate. Then, we decouple the resulting non-convex optimization problem into two subproblems.

2.5.4.1 Problem Formulation

As shown in (2.125) and (2.134), the achievable rate of an MU is determined by the phase shifts of the IOS $\boldsymbol{s}$ and by the digital beamformer of the SBS $\mathbf{V}_D$. With an appropriate design of the IOS phase shifts and SBS digital beamformer, the signal-to-interference + noise ratio (SINR) at the MUs can be improved, and the data rate of the MUs can be enhanced. To study the impact of the IOS on the MUs in terms of both reflected and transmitted signals, we aim to maximize the sum-rate of these N MUs in the system by jointly optimizing the IOS phase shifts $\boldsymbol{s}$ and the SBS digital beamformer $\mathbf{V}_D$. In particular, this optimization problem can be formulated as

$$\max_{\boldsymbol{s}, \mathbf{V}_D} \sum_{i=1}^{N} R_i, \tag{2.135a}$$

$$s.t. \ \mathrm{Tr}\left(\mathbf{V}_D \mathbf{V}_D^H\right) \leq P_B, \tag{2.135b}$$

$$s_m \in \mathscr{S}_a, m = 1, 2, \ldots, M. \tag{2.135c}$$

In (2.135), constraint (2.135b) accounts for the maximum power budget of the SBS, and constraint (2.135c) denotes the feasible set for the phase shift of each IOS reconfigurable element.

2.5.4.2 Problem Decomposition

Problem (2.135) is a mixed-integer non-convex optimization problem that consists of the discrete-valued variables $\boldsymbol{s}$ and the continuous-valued variables $\mathbf{V}_D$. Therefore, (2.135) is known to be a complex optimization problem to solve. To tackle it, we decouple (2.135) into two subproblems: (i) the optimization of the digital beamforming at the SBS and (ii) the optimization of the analog beamforming (phase shifts) at the IOS.

1. Digital Beamforming Optimization at the SBS: To optimize the digital beamforming at the SBS, we set phase shifts of the IOS, i.e. $\boldsymbol{s}$, to fixed values. Therefore, (15) reduces to

$$\max_{\mathbf{V}_D} \sum_{i=1}^{N} R_i, \tag{2.136a}$$

$$s.t. \ \mathrm{Tr}\left(\mathbf{V}_D \mathbf{V}_D^H\right) \leq P_B. \tag{2.136b}$$

2. Analog Beamforming Optimization at the IOS: To optimize the phase shifts of the IOS, we assume that the digital beamforming matrix $\mathbf{V}_D$ is given. Therefore, (15) reduces to

$$\max_{s} \sum_{i=1}^{N} R_i, \tag{2.137a}$$

$$s.t.\ s_m \in \mathcal{S}_a, m = 1, 2, \ldots, M. \tag{2.137b}$$

2.5.5 Sum-Rate Maximization: Algorithm Design

In this section, we first design two algorithms to solve subproblems (2.136) and (2.137) individually, and we then devise an iterative algorithm for solving (2.135).

2.5.5.1 Digital Beamforming Optimization at the SBS

In this section, we solve the digital beamforming optimization at the SBS stated in (2.136). In particular, we consider zero-forcing (ZF) beamforming and optimal transmit power optimization, in order to alleviate the interference among the MUs. As introduced in [20], the ZF-based digital beamforming can be formulated as

$$\mathbf{V}_D = \mathbf{H}^H(\mathbf{H}\mathbf{H}^H)^{-1}\mathbf{P}^{1/2} = \widetilde{\mathbf{V}}_D\mathbf{P}^{1/2}, \tag{2.138}$$

where $\widetilde{\mathbf{V}}_D = \mathbf{H}^H(\mathbf{H}\mathbf{H}^H)^{-1}$ and $\mathbf{P}$ is a diagonal matrix whose ith diagonal element, which is denoted by p_i, is the received power at the ith MU.

Based on the ZF beamforming design in (2.138), problem (2.136) can be simplified as

$$\max_{p_i \geq 0} \sum_{i=1}^{N} W_B \log_2\left(1 + \frac{p_i}{\sigma^2}\right), \tag{2.139a}$$

$$s.t.\ \mathrm{Tr}\left(\mathbf{P}^{1/2}\widetilde{\mathbf{V}}_D^H\widetilde{\mathbf{V}}_D\mathbf{P}^{1/2}\right) \leq P_B. \tag{2.139b}$$

The optimal solution of problem (2.139) is the well-known water-filling power allocation [26]

$$p_i = \frac{1}{\nu_i}\max\left(\frac{1}{\mu} - \nu_i\sigma^2, 0\right), \tag{2.140}$$

where v_i is the ith diagonal element of $\widetilde{\mathbf{V}}_D^H\widetilde{\mathbf{V}}_D$ and μ is a normalization factor that fulfills the constraint $\sum_{i=1}^{N} \max\left(\frac{1}{\mu} - v_i\sigma^2, 0\right) = P_B$. After obtaining $\mathbf{P}$ through the water-filling algorithm, the digital beamforming matrix is directly obtained from (2.138).

2.5.5.2 Analog Beamforming Optimization at the IOS

In this section, we solve the analog beamforming optimization at the IOS stated in (2.137). Since the digital beamforming is assumed to be fixed and to be given by the ZF-based precoding matrix in (2.139), the problem in (2.137) can be simplified as

$$\max_{s} \sum_{i=1}^{N} W_B \log_2\left(1 + \frac{|(\mathbf{H}^i)^H \mathbf{V}_D^i|^2}{\sigma^2}\right), \tag{2.141a}$$

$$s.t.\ 0 \leq s_m < 2\pi, m = 1, 2, \ldots, M. \tag{2.141b}$$

Problem (2.141) is tackled in two steps: (i) first, a relaxed problem that assumes continuous phase shifts is considered, and (ii) then, a branch-and-bound based algorithm is proposed to account for the discrete phase shifts.

1. Continuous IOS Phase Shift Design: To tackle this problem efficiently, problem (2.141) can be solved by optimizing phase shifts of these M reconfigurable elements iteratively. More precisely, the downlink data rate of each MU, i.e., R_i, is a convex function of each variable in $\boldsymbol{s}$. Therefore, the objective function in (2.141a), i.e., $\sum_{i=1}^{N} R_i$, is a convex function of each variable in $\boldsymbol{s}$ while keeping the others fixed. To optimize the phase shifts of all the reconfigurable elements of the IOS, we first set a random initial solution, which is denoted by $\boldsymbol{s}^0 = (s_1^0, \ldots, s_M^0)$. Then, we iteratively optimize the phase shift of each reconfigurable element. Without loss of generality, let us consider the optimization of s_m at the rth iteration. We fix the phase shift of all the other reconfigurable elements at the newly solved value, i.e., $s_1 = s_1^r, \ldots, s_{m-1} = s_{m-1}^r, s_{m+1} = s_{m+1}^{r-1}, \ldots, s_M = s_M^{r-1}$, and maximize $\sum_{i=1}^{N} W_B \log_2\left(1 + \frac{|(\mathbf{H}^i)^H \mathbf{V}_D^i|^2}{\sigma^2}\right)$ as a function of only s_m. Since $|(\mathbf{H}^i)^H \mathbf{V}_D^i|^2\ \forall i \in \mathcal{N}$ is a convex function with respect to s_m, the maximizer can be obtained efficiently by using existing convex optimization techniques. The obtained solution s_m^r is then updated as a temporary solution for the phase shift of the mth reconfigurable element, and the corresponding sum-rate increment is given by

Algorithm 3: Continuous IOS phase shift design

begin
  Initialization: Set an initial solution $s^0 = (s_1^0, \ldots, s_M^0)$ to problem (2.141);
  while $\sum_{m=1}^{M} \Delta R_m^r \geq R_{th}$ **do**
    for $m = 1 : M$ **do**
      Solve s_m in problem (2.141) while keeping the other variables fixed;
      Compute ΔR_m^r in (2.142);
      Update the IOS phase shifts to
      $s = (s_1 = s_i^r, \cdots, s_m = s_m^r, s_{m+1} = s_{m+1}^{r-1}, \cdots, s_M = s_M^{r-1})$;
    end
  end
end

$$\Delta R_m^r = \sum_{i=1}^{N} \log_2\left(1 + \frac{|(\mathbf{H}^i)^H \mathbf{V}_D^i|^2}{\sigma^2}\right)|_{s_m^r} - \sum_{i=1}^{N} \log_2\left(1 + \frac{|(\mathbf{H}^i)^H \mathbf{V}_D^i|^2}{\sigma^2}\right)|_{s_m^{r-1}}. \tag{2.142}$$

The algorithm terminates when the increment of the downlink sum-rate between two consecutive iterations is below a specified threshold, i.e., $\sum_{m=1}^{M} \Delta R_m^r < R_{th}$. Since problem (2.141) is non-convex with respect to all the variables in $\boldsymbol{s}$ jointly, the proposed solution converges to a locally optimal solution. The convergence of the proposed algorithm is proved in Sect. 2.5.5.3. The proposed continuous IOS phase shift design is summarized in Algorithm 3.

2. Discrete IOS Phase Shift Design: By applying Algorithm 3, the continuous phase shifts of the M reconfigurable elements are obtained, which are denoted by $s_m^{opt}, m = 1, 2, \ldots, M$. However, s_m^{opt} may not correspond to any of the finite and discrete phase shifts in $\mathcal{S}_a$. In general, in fact, the continuous phase shift of the mth reconfigurable element of the IOS that is solution of problem (2.141) lies in the range determined by the two consecutive phase shifts equal to $l_m \Delta\phi_m$ and $(l_m + 1)\Delta\phi_m$, i.e., $l_m \Delta\phi_m \leq s_m^{opt} \leq (l_m + 1)\Delta\phi_m$. Therefore, after computing the continuous phase shifts of the IOS with the aid of Algorithm 3, the search space for the M discrete phase shifts still encompasses 2^M possibilities. To overcome the associated computational complexity, we propose an efficient branch-and-bound algorithm that yields the optimal discrete phase shifts of the IOS that belong to the finite set $\mathcal{S}_a$.

The solution space of the IOS phase shift vector $\boldsymbol{s}$ can be viewed as a binary tree structure, as shown in Fig. 2.30. Each node of the tree contains the phase shift information of all the M reconfigurable elements, i.e., $\boldsymbol{s} = (s_1, \ldots, s_M)$. At the root node, all the variables in $\boldsymbol{s}$ are unfixed. After applying Algorithm 3, the value of an unfixed variable s_m at a father node can be one of the two phase shifts $l_m \Delta\phi_m$ or $(l_m + 1)\Delta\phi_m$. As illustrated in Fig. 2.30, this branches the father node into two child nodes. This operation can be repeated for each parent node. Our objective is to devise an efficient algorithm that allows us to solve problem (2.137) based on the tree structure illustrated in Fig. 2.30, which is determined by Algorithm 3.

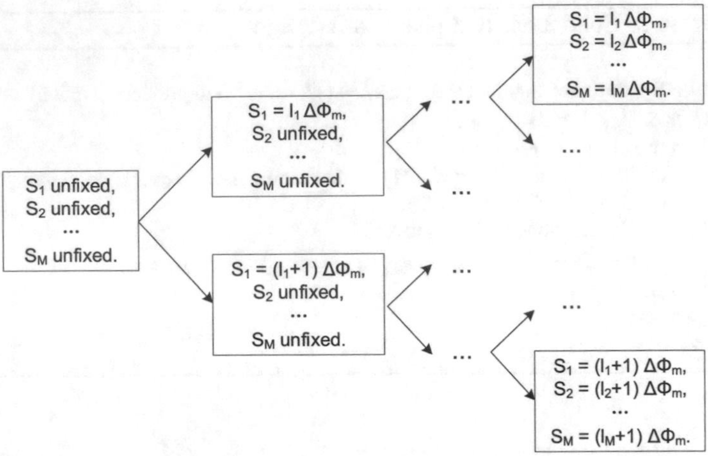

Fig. 2.30 Tree structure of the proposed branch-and-bound algorithm

The proposed algorithms starts by randomly setting a feasible solution for the phase shift s_m to any of the two possible solutions $l_m \Delta\phi_m$ or $(l_m + 1)\Delta\phi_m$. The corresponding value of the objective function yields a lower bound to problem (2.141). Then, the proposed branch-and-bound algorithm recursively splits the search space of problem (2.137) into smaller spaces based on the tree structure illustrated in Fig. 2.30. The random initialization at the end of Algorithm 3 corresponds to one node in the tree structure illustrated in Fig. 2.30. This node is characterized by fixed and unfixed ($\mathbf{s}^{uf}$) variables, as illustrated in Fig. 2.30. An upper bound for the sum rate is obtained solving problem (2.141) by relaxing the unfixed discrete variables $\mathbf{s}^{uf}$ into continuous variables. The obtained upper bound for the rate is compared with the current lower bound, and, depending on their relation, the branch that corresponds to the node under analysis is either kept or pruned. The process is iterated until the current lower bound for the rate is improved and all the nodes of the tree in Fig. 2.30 are either visited or pruned. The details of the branch-and-bound method that solves the IOS phase shift design problem (2.137) is summarized in Algorithm 4.

2.5.5.3 Joint SBS Digital Beamforming and IOS Phase Shift Optimization

Based on the proposed solutions for the ZF-based digital beamforming at the SBS and the branch-and-bound algorithm for obtaining the discrete phase shifts of the IOS by leveraging Algorithm 3 and Algorithm 4, problem (2.135) can be solved by using alternating optimization as summarized in Algorithm 5. The convergence and complexity of Algorithm 5 are analyzed in the following two propositions.

Algorithm 4: IOS phase shift design

---

1 **begin**
2   **Initialization:** Set an initial feasible solution s to problem (2.137) based on Algorithm 3 and set the corresponding sum-rate R_{lb} as the lower bound;
3   **while** *All nodes in Fig. 2.30 are not visited or pruned* **do**
4   Calculate the upper bound of the current
      $$\text{node } R_{ub} = \max_{s^{uf}} \sum_{i=1}^{N} \log_2 \left(1 + \frac{|(\mathbf{H}^i)^H \mathbf{V}_D^i|^2}{\sigma^2} \right);$$
5   **if** $R_{ub} < R_{lb}$ **then**
6   Prune the corresponding branch;
7   Return;
8   **else**
9   **if** *The current node has any child nodes* **then**
10   Move to one of its two child nodes;
11   Continue;
12   **end**
13   **else**
14   Calculate the corresponding sum-rate R_{curr};
15   **end**
16   **if** $R_{curr} > R_{lb}$ **then**
17                  $R_{lb} = R_{curr}$;
18   Return;
19   **end**
20   **end**
21   **end**
22 **end**

Algorithm 5: Joint SBS digital beamforming and IOS phase shift optimization

---

1 **begin**
2   **while** *The sum-rate difference between two consecutive iterations is below a threshold ω* **do**
3   Perform SBS digital beamforming as introduced in Sect. 2.5.5.1;
4   Compute the continuous IOS phase shifts by leveraging Algorithm 3;
5   Compute the discrete IOS phase shifts by leveraging Algorithm 4;
6   **end**
7   Obtain the maximum sun-rate of the IOS-assisted communication system;
8 **end**

Proposition 2.13 *The proposed joint SBS digital beamforming and IOS analog beamforming optimization algorithm is convergent to a local optimum solution.*

Proof We denote the sum-rate at the rth iteration by $\mathscr{R}(s^r, \mathbf{V}_D^r)$. At the $(r + 1)$th iteration, the solution of the optimal SBS digital beamforming given $s = s^r$ yields a sum-rate $\mathscr{R}(s^r, \mathbf{V}_D^{r+1}) \geq \mathscr{R}(s^r, \mathbf{V}_D^r)$. Similarly, the solution of the optimal IOS analog beamforming given $\mathbf{V}_D = \mathbf{V}_D^{r+1}$ yields a sum-rate $\mathscr{R}(s^{r+1}, \mathbf{V}_D^{r+1}) \geq \mathscr{R}(s^r, \mathbf{V}_D^{r+1})$. Therefore, we obtain the following inequalities $\mathscr{R}(s^{r+1}, \mathbf{V}_D^{r+1}) \geq$

$\mathscr{R}(s^r, \mathbf{V}_D^{r+1}) \geq \mathscr{R}(s^r, \mathbf{V}_D^r)$, i.e., the sum-rate does not decrease at each iteration of Algorithm 5. Since the sum-rate is upper bounded thanks to the constraint on the total transmit power, Algorithm 5 converges in a finite number of iterations to a local optimum solution.

Proposition 2.14 *The complexity of each iteration of the proposed joint SBS digital beamforming and IOS analog beamforming optimization algorithm is* $O(2^M + N)$.

Proof At each iteration, the SBS digital beamforming problem for the N MUs can be solved by using convex optimization methods, whose complexity is $O(N)$. The complexity of the IOS analog beamforming problem is the sum of the complexities of Algorithm 3 and Algorithm 4. The complexity of Algorithm 3 is $O(N)$, and the complexity of the branch-and-bound algorithm is, in the worst case, $O(2^M)$. Therefore, the complexity of each iteration of Algorithm 5 is $O(2^M + N)$.

2.5.6 Performance Analysis of the IOS-Assisted Communication System

In this section, we first discuss the impact of the phase shift design on the reflected and transmitted signals, and we then analyze the downlink sum-rate as a function of the power ratio of the reflected and transmitted signals. In this section, for ease of understanding, the impact of small-scale fading is ignored, and only the impact of the distances, the radiation pattern of the reconfigurable elements of the IOS, and the power allocation ratio between the reflected and transmitted signals are considered.

2.5.6.1 Analysis of the Phase Shift Design

Our objective is to study the relation between the optimal phase shifts of the IOS when it is used as a transmissive and reflective surface, and to understand the differences between these two cases. For analytical convenience, we assume that the power ratio $|\gamma_m|^2$ between the power of the signal re-emitted by the IOS and the power of the incident signal is a constant value.

In order to understand the differences and similarities between an IRS (only reflections are allowed) and an IOS (both reflections and transmissions are allowed simultaneously), the following proposition considers the case study with two MUs, when one MU (the ith MU) lies in the reflective side of the IOS and the other MU (the jth MU) lies in the transmissive side of the IOS.

Proposition 2.15 *Consider an IOS-assisted communication system with two MUs, which are denoted by the indices* i *and* j*. If the two users are located symmetrically with respect to the IOS, as shown in Fig. 2.31a, and the impact of small-scale fading is ignored, the optimal SBS digital beamforming and IOS analog beamforming are the same for both MUs.*

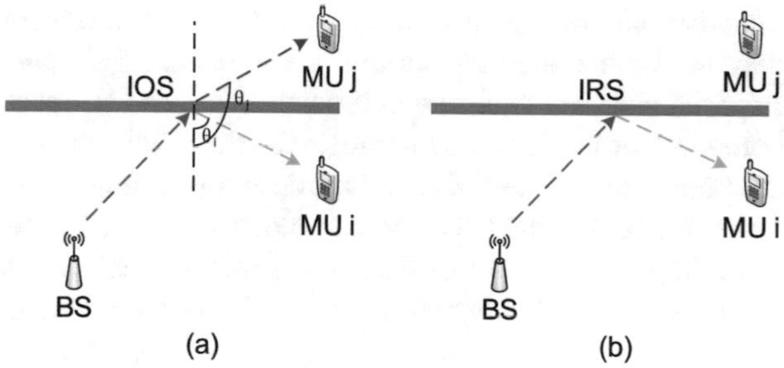

Fig. 2.31 Examples of MU distributions for the performance analysis of an IOS-assisted communication system. (**a**) IOS setup in which two MUs are located symmetrically with respect to the IOS; (**b**) IRS setup in which only reflections are allowed

Proof For a pair of MUs that are located symmetrically with respect to the IOS (e.g., the ith and jth MUs in Fig. 2.31a), the distance from each reconfigurable element to the two MUs is the same, i.e.,

$$d_{m,i} = d_{m,j}, \forall m \in \mathcal{M}. \tag{2.143}$$

According to (2.123), the power radiation pattern of the reflected signals $K^D(m)|_r$ and that of the transmitted signals $K^D(m)|_t$ of the MUs that are located symmetrically with respect to the IOS satisfies the relation

$$\frac{K^D(m)|_r}{K^D(m)|_t} = \epsilon. \tag{2.144}$$

As shown in Fig. 2.31a, in addition, the angles with respect to the normal to the surface satisfy the relation $\theta_i^D(m) + \theta_j^D(m) = \pi$, and then we have

$$|\cos(\theta_i^D(m))| = |\cos(\theta_j^D(m))|. \tag{2.145}$$

When substituting (2.143), (2.144), and (2.145) into (2.121), we have $\frac{g_m(\xi^A(m),\xi_j^D(m),s_m)}{g_m(\xi^A(m),\xi_i^D(m),s_m)} = \epsilon$ for any IOS phase shifts s. If we assume that the antenna gains of the MUs are the same, and substitute (2.121) into (2.125), we have $\frac{h_{j,k}^{m,LoS}}{h_{i,k}^{m,LoS}} = \epsilon$, $\forall m \in \mathcal{M}, k \in \mathcal{K}$, and for any IOS phase shifts s. Under these assumptions, and if the small-scale fading is not considered, we evince that the optimal SBS digital beamforming and IOS analog beamforming of the ith and jth MUs coincide. This concludes the proof.

In an IOS-assisted communication system, IOS phase shifts are designed for the MUs on both sides of the IOS jointly, and the optimal solution for the MUs in $\mathcal{N}_r$

(i.e., in the reflective side of the surface) and $\mathcal{N}_t$ (i.e., in the transmissive side of the surface) may not be obtained concurrently. Proposition 3, in particular, can be regarded as a special case in which the optimal solution can be guaranteed for the two users in both sides of the IOS. To maximize the downlink sum-rate for multiple MUs located in both sides of the IOS, we introduce a priority-based approach that gives higher priority to either the MUs on the reflective or transmissive sides of the IOS. In particular, the *priority index* of the ith MU that describes the strength of the reflective/transmissive channel gain with respect to its location is defined as

$$\mathscr{P}_i = \begin{cases} \sum_{m=1}^{M} \frac{|\cos^3 \theta_i^D(m)|}{(1+\epsilon)d_{m,i}^{\alpha}}, \ i \in \mathcal{N}_r, \\ \sum_{m=1}^{M} \frac{\epsilon |\cos^3 \theta_i^D(m)|}{(1+\epsilon)d_{m,i}^{\alpha}}, \ i \in \mathcal{N}_t \end{cases} \tag{2.146}$$

Proposition 2.16 *Let us assume that the distance between the SBS and the IOS is fixed and only the distances of the MUs change, as well as that the small-scale fading is not considered. Thanks to the IOS, the MU with the highest priority index in (2.146) obtains the largest data rate gain.*

Proof The improvement of data rate in the presence of an IOS is determined by the sum of the channel gains of the M SBS-IOS-MU links. Let us assume that the small scale fading can be ignored (long-term performance). From (2.121) and (2.125), the LoS component of the SBS-IOS-MU link from the kth antenna of the SBS to the ith MU via the mth reconfigurable element of the IOS is

$$h_{i,k}^{m,LoS} = \frac{\lambda K^A(m) K_i^D(m) \sqrt{G_k^{tx} G_m G_i^{rx} \delta_x \delta_y |\gamma_m|^2} \exp\left(\frac{-j2\pi(d_{k,m}+d_{m,i})}{\lambda} - j\psi_m\right)}{(4\pi)^{\frac{3}{2}} d_{k,m}^{\alpha} d_{m,i}^{\alpha}}, \tag{2.147}$$

The hardware-related parameters G_k^{tx}, G_m, G_i^{rx}, δ_x, δ_y, and γ_m can be considered as constants in a given IOS-assisted communication system. The phase shift of the SBS-IOS-MU link $\frac{-j2\pi(d_{k,m}+d_{m,i})}{\lambda} - j\psi_m$ is optimized with the algorithm proposed in Sect. 2.5.5.2. Assuming that the locations of the SBS and IOS are given, the MU-related term that affects the channel gain of the LoS component is $\frac{K_i^D(m)}{d_{m,i}^{\alpha}}$, $\forall i \in \mathcal{N}, k \in \mathcal{K}, m \in \mathcal{M}$. Therefore, the impact of the M reconfigurable elements can be expressed as

$$\mathscr{P}_i = \sum_{m=1}^{M} \frac{K_i^D(m)}{d_{m,i}^{\alpha}} = \begin{cases} \sum_{m=1}^{M} \frac{|\cos^3\theta_i^D(m)|}{(1+\epsilon)d_{m,i}^{\alpha}}, & i \in \mathcal{N}_r, \\ \sum_{m=1}^{M} \frac{\epsilon|\cos^3\theta_i^D(m)|}{(1+\epsilon)d_{m,i}^{\alpha}}, & i \in \mathcal{N}_t. \end{cases} \tag{2.148}$$

This concludes the proof.

Based on Proposition 2.16, the following remarks for some asymptotic regimes of the IOS-assisted communication system can be made.

Remark 2.8 When $\epsilon \to 0$, the IOS boils down to an IRS, and $\sum_{i\in\mathcal{N}_r} \mathscr{P}_i \gg \sum_{i\in\mathcal{N}_t} \mathscr{P}_i$ is satisfied. The IOS phase shift design only considers the MUs that belong to the set $\mathcal{N}_r$.

Remark 2.9 When $\epsilon \to \infty$, the IOS only transmits the signals to the opposite side of the SBS, and $\sum_{i\in\mathcal{N}_r} \mathscr{P}_i \ll \sum_{i\in\mathcal{N}_t} \mathscr{P}_i$ is satisfied. The IOS phase shift design only considers the MUs that belong to the set $\mathcal{N}_t$.

2.5.6.2 Analysis of the Transmission/Reflection Power Ratio

In this section, we analyze the impact of the power ratio of the reflected and transmitted signals ϵ on the sum-rate of an IOS-assisted communication system.

Proposition 2.17 *Given the average distance between the IOS and the MUs, the power ratio of the reflected and transmitted signals ϵ is positively correlated with the ratio of the number of MUs on the two sides of the IOS, i.e., $\epsilon \propto \mathcal{N}_t/\mathcal{N}_r$.*

Proof Assume that the SNR of the direct link from the SBS to the ith MU is α_i and that the SNR of the link from the IOS to the ith MU is β_i. The rate of the ith MU is

$$R_i = \log_2(1 + \alpha_i + \beta_i).^{12} \tag{2.149}$$

Therefore, the sum-rate of all the MUs in the set $\mathcal{N}$ is given by $\sum_{k\in\mathcal{N}_r} R_k + \sum_{j\in\mathcal{N}_r} R_j$. Given the total power of the transmitted and reflected signals, we aim to maximize the normalized sum-rate of the MUs, which is given as

$$\sum_{i\in\mathcal{N}} R_i = \sum_{k\in\mathcal{N}_r} \log_2(1+\alpha_k+\frac{1}{1+\epsilon}\beta_k) + \sum_{j\in\mathcal{N}_t} \log_2(1+\alpha_j+\frac{\epsilon}{1+\epsilon}\beta_j). \tag{2.150}$$

The first-order derivative of (2.150) with respect to ϵ is

[12] For simplicity, we do not consider the specific impact of the phase shift optimization, and assume that the SNRs of the SBS-to-MU link and the IOS-to-MU link can just be added directly.

$$\sum_{i\in\mathcal{N}}\frac{dR_i}{d\epsilon}=\frac{1}{\ln 2}\left(\sum_{j\in\mathcal{N}_t}\frac{\beta_j/(1+\epsilon)^2}{1+\alpha_j+\frac{\beta_j\epsilon}{1+\epsilon}}-\sum_{k\in\mathcal{N}_r}\frac{\beta_k/(1+\epsilon)^2}{1+\alpha_k+\frac{\beta_k}{1+\epsilon}}\right). \tag{2.151}$$

We denote the optimal value of ϵ that maximizes (2.150) by ϵ^{opt}. When the sum-rate is maximized, we have $\sum_{i\in\mathcal{N}}\frac{d\Delta R_i}{d\epsilon}|_{\epsilon=\epsilon^{opt}}=0$.

If the MU j' is added to the set $\mathcal{N}_t$, a positive value is added to (2.151), and we have $\sum_{i\in\mathcal{N}}\frac{d\Delta R_i}{d\epsilon}|_{\epsilon=\epsilon^{opt}}>0$. Thus, the sum-rate can be further improved by increasing ϵ. On the contrary, if the MU k' is added to the set $\mathcal{N}_r$, a negative value is added to (2.151). In this case a smaller value of ϵ improves the sum-rate of the system. This concludes the proof.

Proposition 2.18 *Given a pair of symmetrically located MUs as illustrated in Fig. 2.31a, a larger proportion of the available power is allocated to the MUs with a weak direct link (i.e., low received power). More specifically, $\epsilon>1$ when the sum of the received power of the direct links from the SBS to the MUs in $\mathcal{N}_r$ is larger than that of the MUs in $\mathcal{N}_t$. Otherwise, $\epsilon<1$ holds.*

Proof We denote a pair of symmetrically-located MUs that receive the reflected signals and transmitted signals by MU i and MU j, respectively, as shown in Fig. 2.31a. Based on the notation and formulas introduced in Appendix C, the normalized sum-rate of the MUs can be given as ($\beta_j=\beta_i$ since the MUs are located symmetrically with respect to the IOS)

$$R_i+R_j=\log_2\left(1+\alpha_i+\frac{1}{1+\epsilon}\beta_i\right)+\log_2\left(1+\alpha_j+\frac{\epsilon}{1+\epsilon}\beta_i\right). \tag{2.152}$$

The derivative function of (2.152) with respect to ϵ is

$$\frac{d(R_i+R_j)}{d\epsilon}=\frac{1}{\ln 2}\frac{\beta_i^2(1-\epsilon)(1+\epsilon)+\beta_i(1+\epsilon)^2(\alpha_i-\alpha_j)}{(\epsilon\beta_i(1+\epsilon)+(1+\epsilon)^2)(\beta_i(1+\epsilon)+(1+\epsilon)^2)}. \tag{2.153}$$

The optimal value of the power ratio of the reflected and transmitted signals satisfies $\epsilon=\frac{\beta_i+\alpha_i-\alpha_j}{\beta_i-\alpha_i+\alpha_j}$. When $\alpha_i>\alpha_j$, we have $\epsilon>1$. Otherwise, we have $\epsilon<1$. This concludes the proof.

In particular, when the distance from the SBS to the IOS is much larger than the distance from the IOS to the MUs, the distance from the SBS to each MU is approximately the same. Therefore, (2.151) in Appendix C can be simplified as

$$\frac{d(R_i+R_j)}{d\epsilon}\simeq\frac{1}{\ln 2}\frac{\beta_i^2(1-\epsilon)(1+\epsilon)}{(\epsilon\beta_i(1+\epsilon)+(1+\epsilon)^2)(\beta_i(1+\epsilon)+(1+\epsilon)^2)}. \tag{2.154}$$

From (2.154), we evince that the maximum value of $\sum_{i\in\mathcal{N}}\Delta R_i$ is obtained for $\epsilon=1$. Therefore, the following remark follows.

Remark 2.10 When the distance from the SBS to the IOS is much larger than the distance from the IOS to the MUs, the IOS maximizes the throughput of the system for $\epsilon = 1$, i.e., the power of the transmitted and reflected signals is the same.

The assumptions in Proposition 2.18 and Remark 2.10 are usually satisfied when the IOS is deployed at the cell edge for coverage extension. Therefore, an IOS with $\epsilon = 1$ is capable of maximizing the sum-rate of the MUs at the cell edge of an IOS-assisted communication system. Finally, the following proposition yields the largest theoretical gain that an IOS-assisted system provides with respect to the benchmark IRS-assisted system.

Proposition 2.19 *The ratio of the downlink sum-rate of an IOS-assisted system and an IRS-assisted system is upper-bounded by two.*

Proof Consider the two case studies illustrated in Fig. 2.31. In the IRS-assisted communication system, the total power that impinging upon the IRS is reflected towards the ith MU, and the downlink rate of the corresponding link can be expressed as

$$R_i^{IRS} = \log_2(1 + \beta_i). \tag{2.155}$$

As far as the IOS-assisted communication system is concerned, on the other hand, the downlink rate can be expressed as

$$R_i^{IOS} = \log_2(1 + \frac{\beta_i}{\epsilon + 1}) + \log_2(1 + \frac{\epsilon\beta_i}{\epsilon + 1}). \tag{2.156}$$

The ratio between the IOS-MU downlink rate and the IRS-MU downlink rate is

$$\begin{aligned} \frac{R_i^{IOS}}{R_i^{IRS}} &= \frac{\log_2(1 + \frac{\beta_i}{\epsilon+1}) + \log_2(1 + \frac{\epsilon\beta_i}{\epsilon+1})}{\log_2(1 + \beta_i)} \\ &= \frac{\log_2(1 + \beta_i + \frac{\epsilon\beta_i^2}{(\epsilon+1)^2})}{\log_2(1 + \beta_i)}. \end{aligned} \tag{2.157}$$

Equation (2.157) is maximized for $\epsilon = 1$, and the maximum value is

$$\frac{R_i^{IOS}}{R_i^{IRS}} = \frac{2\log_2(1 + \frac{\beta_i}{2})}{\log_2(1 + \beta_i)}. \tag{2.158}$$

When $\beta_i \to \infty$, the ratio in (2.158) is maximized and attains its maximum equal to two. This concludes the proof.

Proposition 2.19 unveils that an IOS may double the downlink sum-rate when compared to an IRS. This upper-bound is, however, difficult to be attained because it requires that the signal-to-interference-plus-noise ratio is infinite.

2.5.7 Simulation Results

In this section, we evaluate the performance of the considered IOS-assisted system based on the proposed algorithm, and compare it with and IRS-assisted system [62] and a conventional cellular system in the absence of IRS or IOS. In the IRS-assisted system, the IRS only reflects the signals from the SBS to the MUs, and the surface does not work in transmission mode. In the conventional cellular system, the MUs receive only the direct links from the SBS without the assistance of a reconfigurable surface.

In the simulations, we set the height of the SBS and the center of the IOS at 2 m, and the distance between the SBS and the IOS is 100 m. The MUs are randomly deployed within a disk of radius 2 m centered at the IOS. The maximum transmit power of the SBS is $P_B = 40$ dBm, the carrier frequency is 5.9 GHz, the noise power is −96 dBm, the antenna separation at the SBS is 0.2 m, and the inter-distance between the reconfigurable elements of the IOS is 0.025 m (i.e., half of the wavelength). The numbers of MUs (N) and SBS antennas (K) are 5, and the power ratio of the reflected and transmitted signals is $\epsilon = 1$ (unless stated otherwise). The path-loss exponent of the direct link is 3, and the Rician factor is $\kappa = 4$.

Figure 2.32a illustrates the average sum-rate of the MUs as a function of the size of the IOS. The average sum-rate is defined as the average of the sum-rate in (2.135a) as a function of the spatial distribution of the MUs. The IOS is modeled as a square array with $\sqrt{M}$ elements on each line and each row. The average sum-rate increases with the number of IOS reconfigurable elements, and the growth rate gradually decreases with the IOS size. An IOS with 30 × 30 elements improves the average sum-rate of about 2.7 times when compared to a conventional cellular system. On the other hand, an IRS of the same size improves the average sum-rate of about 2.2 times only. The IOS provides a higher average sum-rate since it is

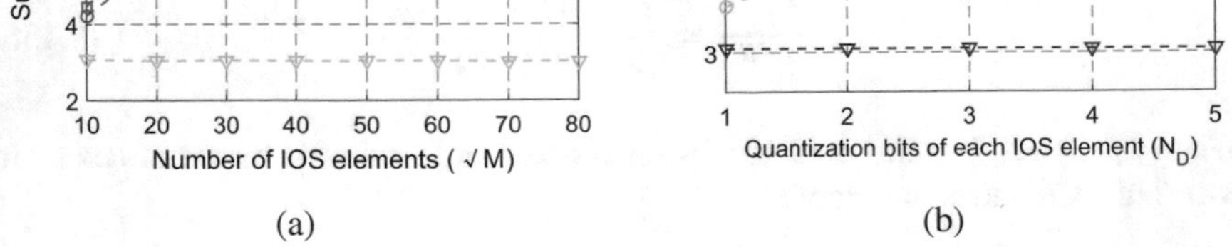

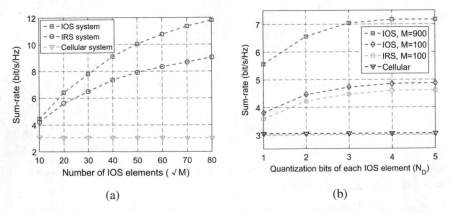

Fig. 2.32 Impact of the surface size and quantization bits on the sum-rate. (**a**) Number of IOS reconfigurable elements ($\sqrt{M}$) vs. sum-rate ($S_a = 4$). (**b**) Quantization bits of each IOS element (N_D) vs. sum-rate

capable of supporting the transmission of the MUs that are located on both sides of the surface. The rate improvement offered by an IOS over an IRS is less than two, which is in agreement with Proposition 2.19.

Figure 2.32b shows the average sum-rate as a function of the number of quantization bits N_D of each reconfigurable element of the IOS, with $N_D = \log_2(S_a)$. The average sum-rate increases with N_D and converges to a stable value as N_D increases. We observe, in particular, that few quantization bits are sufficient to achieve most of the average sum-rate and that the convergence rate increases with the size M of the IOS.

In Fig. 2.33, we analyze the relation between the average sum-rate and the power ratio of the reflected and transmitted signals ϵ. In Fig. 2.33a, the MUs are distributed around the IOS within a radius of 2 m, and the distance from the IOS to the MUs is much shorter than the distance from the SBS to the MUs. The average sum-rate of all the MUs is maximized for $\epsilon = 1$, which is in agreement with Remark 3. In Fig. 2.33b, the MUs are distributed around the IOS within a radius of 20 m. In this configuration, the MUs in $\mathcal{N}_r$ receive a higher power through the direct links when compared to the MUs in $\mathcal{N}_t$. The average sum-rate is maximized if $\epsilon > 1$, which agrees with Proposition 2.18.

The impact of the distribution of the MU with respect to the location of the IOS is illustrated in Fig. 2.34. In the IOS-assisted communication system, the average sum-rate is maximized, in the considered setup, when the MUs are equally distributed on the two sides of the IOS (i.e., $N_r/N = 0.5$ in the figure). When the MUs are mostly located on one side of the IOS, i.e., $N_t/N \to 0$ or $N_t/N \to 1$, the power of the signals re-emitted by the IOS on both sides of the surface cannot be optimized for all the MUs concurrently. In an IRS-assisted communication system, on the other

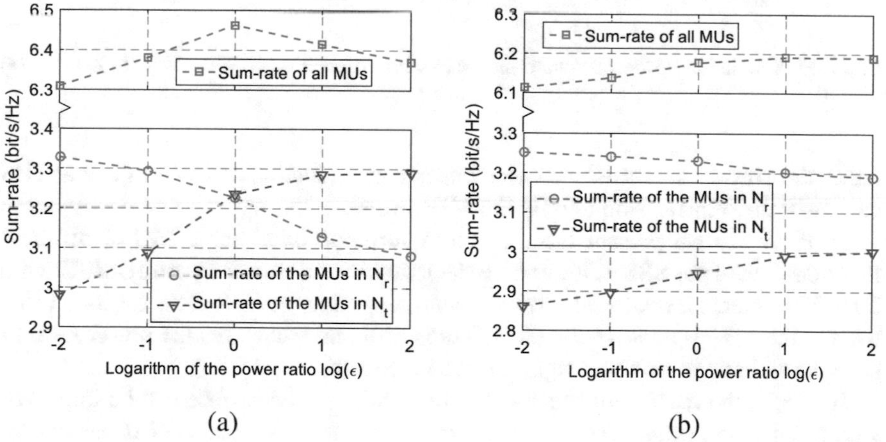

Fig. 2.33 Power ratio of the reflected and transmitted signals (ϵ) vs. the sum-rate. (**a**) MU distribution radius: 2 m. (**b**) MU distribution radius: 20 m

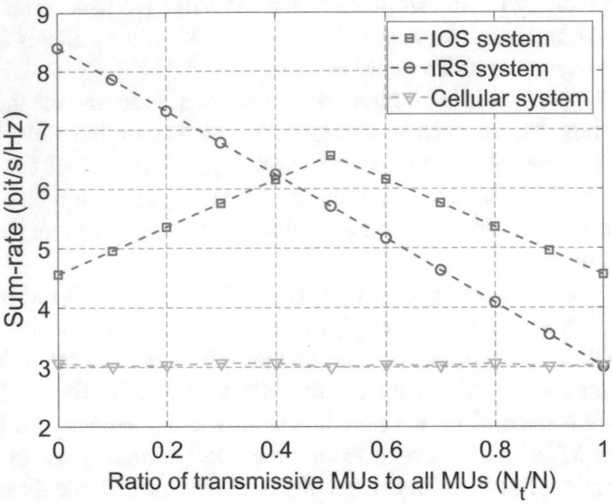

Fig. 2.34 Ratio of MUs receiving transmitted signals to all MUs (N_t/N) vs. sum-rate

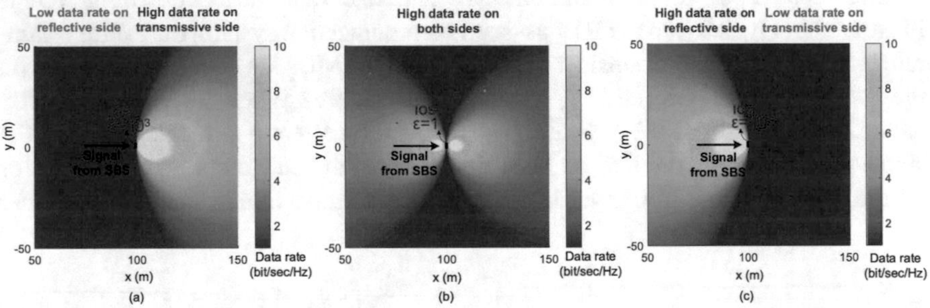

Fig. 2.35 Simulation of the maximum data rate with different ϵ. (**a**) $\epsilon = 10^3$. (**b**) $\epsilon = 1$. (**c**) $\epsilon = 10^{-3}$

hand, the average sum-rate decreases linearly as a function of N_t/N, since an RIS can assist the transmission of only the MUs in $\mathcal{N}_r$.

In Fig. 2.35, we present the maximum sum-rate of a single MU at different locations, when the SBS is located on the left side of the surface. An IOS/IRS with 20×20 elements is deployed vertically with respect to the SBS at the location (100, 0), and the SBS is deployed at (0,0). Three different values for the power ratio of the reflected and transmitted signals ϵ are illustrated.

If $\epsilon = 1$, the maximum sum-rate on both sides of the surface can be improved significantly. An omni-service extension is provided since the IOS is capable of enhancing the received power of the MUs on both sides of the surface. The MU can obtain a higher data rate when it is closer to the center of the IOS, where the reflective-transmissive channel has a better quality. If $\epsilon = 10^3$ and $\epsilon = 10^{-3}$, most of the power is re-emitted to the right-hand and left-hand side of the IOS, respectively. This is in agreement with Remarks 2.8 and 2.9.

2.5.8 Summary

In this section, we have introduced the IOS, which is a reconfigurable surface capable of reflecting and transmitting the impinging signals towards both sides of the surface concurrently. We have studied the utilization of the IOS in an indoor multi-user downlink communication system. The IOS is capable of enhancing the received signals of the MUs on both sides of the surface through an appropriate design of the IOS phase shifts. We have formulated a joint IOS analog beamforming and SBS digital beamforming optimization problem to maximize the sum-rate of the system, and have proposed an iterative algorithm to solve it efficiently. From the obtained numerical simulations and analysis, three main conclusions can be drawn:

1. The sum-rate of an IOS-assisted communication system is higher than that of an IRS-assisted communication system. An IOS enhances the data rate of the MUs located on both sides of the surface concurrently, which asymptotically double the achievable sum-rate and the service coverage.
2. The optimal power ratio of the reflected and transmitted signals of the IOS is positively correlated with the ratio of the number of MUs on the two sides of the IOS, and is negatively correlated with the received power on the direct links from the SBS to the MUs.
3. The sum-rate of an IOS-assisted communication system increases with the geometric size of the surface and with the number of quantization bits of the phase shifts, but converges to a stable value when the IOS size and the quantization bits are sufficiently large.

References

1. E.W. Paper, More than 50 billion connected devices. Utah Education Network, Salt Lake City. Techical Report 284 23–3149 (2011)
2. A. Gatherer, What will 6G be? (2018). [Online] Available: https://www.comsoc.org/publications/ctn/what-will-6g-be
3. Huawei, Asia-Pacific leads 5G innovation, Huawei enables sustainable development of a digital economy (2019). [Online] Available: https://www.huawei.com/en/press-events/news/2019/9/huawei-5th-asia-pacific-innovation-day
4. P. Yang, M. Di Renzo, Y. Xiao, S. Li, L. Hanzo, Design guidelines for spatial modulation. IEEE Commun. Surveys Tuts. **17**(1), 6–26, 1st Quart (2015)
5. E. Larsson, O. Edfors, F. Tufvesson, T. Marzetta, Massive MIMO for next generation wireless systems. IEEE Commun. Mag. **52**(2), 186–195 (2014)

6. S. Gong, X. Lu, D.T. Hoang, D. Niyato, L. Shu, D.I. Kim, Y.C. Liang, Towards smart wireless communications via intelligent reflecting surfaces: a contemporary survey. IEEE Commun. Surveys Tuts. **22**(4), 2283–2314, 4th Quart (2020)
7. J.G. Andrews, S. Buzzi, W. Choi, S.V. Hanly, A. Lozano, A.C.K. Soong, J.C. Zhang, What will 5G be? IEEE J. Sel. Areas Commun. **32**(6), 1065–1082 (2014)
8. M. Di Renzo, M. Debbah, D. Phan-Huy, A. Zappone, M. Alouini, C. Yuen, V. Sciancalepore, G.C. Alexandropoulos, J. Hoydis, H. Gacanin, J. Rosny, A. Bounceur, G. Lerosey, M. Fink, Smart radio environments empowered by reconfigurable AI meta-surfaces: an idea whose time has come. EURASIP J. Wirel. Commun. Netw. **2019**(129), 1–20 (2019)
9. H. Hashida, Y. Kawamoto, N. Kato, Intelligent reflecting surface placement optimization in air-ground communication networks towards 6G. IEEE Wirel. Commun. **27**(6), 146–151 (2020)
10. M. Di Renzo, K. Ntontin, J. Song, F.H. Danufane, X. Qian, F. Lazarakis, J. De Rosny, D. Phan-Huy, O. Simeone, R. Zhang, M. Debbah, G. Lerosey, M. Fink, S. Tretyakov, S. Shamai, Reconfigurable intelligent surfaces vs. relaying: differences, similarities, and performance comparison. IEEE Open J. Commun. Soc. **1**, 798–807 (2020)
11. M.A. Elmossallamy, H. Zhang, L. Song, K. Seddik, Z. Han, G.Y. Li, Reconfigurable intelligent surfaces for wireless communications: Principles, challenges, and opportunities. IEEE Trans. Cognitive Commun. Netw. **6**(3), 990–1002 (2020)
12. M. Di Renzo, A. Zappone, M. Debbah, M.S. Alouini, C. Yuen, J. De Rosny, S. Tretyakov, Smart radio environments empowered by reconfigurable intelligent surfaces: how it works, state of research, and road ahead. IEEE J. Sel. Areas Commun. **38**(11), 2450–2525 (2020)
13. W. Yan, X. Yuan, Z.Q. He, X. Kuai, Passive beamforming and information transfer design for reconfigurable intelligent surfaces aided multiuser MIMO systems. IEEE J. Sel. Areas Commun. **38**(8), 1793–1808 (2020)
14. L. Li, H. Ruan, C. Liu, Y. Li, Y. Shuang, A. Alu, C. Qiu, T. Cui, Machine-learning reprogrammable metasurface imager. Nat. Comm. **10**(1), 1–8 (2019)
15. H. Zhang, H. Zhang, B. Di, K. Bian, Z. Han, L. Song, Towards ubiquitous positioning by leveraging reconfigurable intelligent surface. IEEE Commun. Lett. **25**(1), 284–288 (2021)
16. J. Hu, H. Zhang, B. Di, L. Li, L. Song, Y. Li, Z. Han, H. Vincent Poor, Reconfigurable intelligent surfaces based radio-frequency sensing: Design, optimization, and implementation. IEEE J. Sel. Areas Commun. **38**(11), 2700–2716 (2020)
17. X. Yu, D. Xu, R. Schober, MISO wireless communication systems via intelligent reflecting surfaces, in *Proceedings of the IEEE/CIC ICCC*, Changchun (2019)
18. C. Huang, A. Zappone, G.C. Alexandropoulos, M. Debbah, C. Yuen, Reconfigurable intelligent surfaces for energy efficiency in wireless communication. IEEE Trans. Wirel. Commun. **18**(8), 4157–4170 (2019)
19. T. Cui, D. Smith, R. Liu, *Metamaterials: Theory, Design, and Applications* (Springer, Berlin, 2010)
20. H. Zhang, B. Di, L. Song, Z. Han, Reconfigurable intelligent surfaces assisted communications with limited phase shifts: how many phase shifts are enough? IEEE Trans. Veh. Technol. **69**(4), 4498–4502 (2020)
21. 3GPP TR 38.901, Study on channel model for frequencies from 0.5 to 100 GHz (Release 14) (2018)
22. Y. Han, W. Tang, S. Jin, C.-K. Wen, X. Ma, Large intelligent surface-assisted wireless communication exploiting statistical CSI. IEEE Trans. Veh. Technol. **68**(8), 8238–8242 (2019)
23. H. Zhang, B. Di, Z. Han, H.V. Poor, L. Song, Reconfigurable intelligent surface assisted multi-user communications: how many reflective elements do we need? IEEE Wirel. Commun. Lett. **10**(5), 1098–1102 (2021)
24. Ö. Özdogan, E. Björnson, E.G. Larsson, Intelligent reflecting surfaces: physics, propagation, and pathloss modeling. IEEE Wirel. Commun. Lett. **9**(5), 581–585 (2020)
25. B. Di, H. Zhang, L. Song, Y. Li, Z. Han, H.V. Poor, Hybrid beamforming for reconfigurable intelligent surface based multi-user communications: Achievable rates with limited discrete phase shifts. IEEE J. Sel. Areas Commun. **38**(8), 1809–1822 (2020)
26. D. Tse, P. Viswanath, *Fundamentals of Wireless Communications* (Cambridge University Press, Cambridge, 2005)

27. H.Q. Ngo, E.G. Larsson, T.L. Marzetta, Energy and spectral efficiency of very large multiuser MIMO systems. IEEE Trans. Commun. **61**(4), 14361449 (2013)
28. A.M. Tulino, S. Verdú, Random matrix theory and wireless communications. Found. Trends Commun. Inf. Theory **1**(1), 1–182 (2004)
29. M.A. Elmossallamy, H. Zhang, R. Sultan, K. Seddik, L. Song, G.Y. Li, Z. Han, On spatial multiplexing using reconfigurable intelligent surfaces. IEEE Wirel. Commun. Lett. **10**(2), 226–230 (2021)
30. S. Zeng, H. Zhang, B. Di, Z. Han, L. Song, Reconfigurable intelligent surface (RIS) assisted wireless coverage extension: RIS orientation and location optimization. IEEE Commun. Lett. **25**(1), 269–273 (2021)
31. L. Zhang, Z. Wang, R. Shao, J. Shen, X. Chen, X. Wan, Q. Cheng, T. Cui, Dynamically realizing arbitrary multi-bit programmable phases using a 2-bit time-domain coding metasurface. IEEE Trans. Antennas Propag. **68**(4), 2984–2992 (2020)
32. W. Tang, M. Chen, X. Chen, J. Dai, Y. Han, M. Renzo, Y. Zeng, S. Jin, Q. Cheng, T. Cui, Wireless communications with reconfigurable intelligent surface: path loss modeling and experimental measurement. IEEE Trans. Wirel. Commun. **20**(1), 421–439 (2021)
33. J. Thornton, D. Grace, M.H. Capstick, T.C. Tozer, Optimizing an array of antennas for cellular coverage from a high altitude platform. IEEE Trans. Wirel. Commun. **2**(3), 484–492 (2003)
34. H. Wang, P. Zhang, J. Li, X. You, Radio propagation and wireless coverage of LSAA-based 5G millimeter-wave mobile communication systems. China Comm. **16**(5), 1–18 (2019)
35. S. Boyd, L. Vandenberghe, *Convex Optimization* (Cambridge University Press, Cambridge, 2004)
36. E. Basar, M. Renzo, J. Rosny, M. Debbah, M. Alouini, R. Zhang, Wireless communications through reconfigurable intelligent surfaces. IEEE Access **7**, 116753–116773 (2019)
37. C. Huang, G. Alexandropoulos, A. Zappone, M. Debbah, C. Yuen, Energy efficient multi-user MISO communication using low resolution large intelligent surfaces, in *IEEE Global Commmunications Workshops*, Abu Dhabi (2018)
38. S. Nie, J. Jornet, I. Akyildiz, Intelligent environments based on ultra-massive MIMO platforms for wireless communication in millimeter wave and Terahertz bands, in *IEEE International Conference on Acoustics Speech and Signal Processing*, Brighton (2019)
39. A. Goldsmith, *Wireless Communications* (Cambridge Press, Cambridge, 2005)
40. F. Sohrabi, W. Yu, Hybrid digital and analog beamforming design for large-scale antenna arrays. IEEE J. Sel. Topics Signal Process. **10**(3), 501–513 (2016)
41. N. Shlezinger, O. Dicker, Y. Eldar, I. Yoo, M. Imani, D. Smith, Dynamic metasurface antennas for Uplink massive MIMO systems. IEEE Trans. Commun. **67**(10), 6829–6843 (2019)
42. D. Smith, O. Yurduseven, L. Pulido-Mancera, P. Bowen, N. Kundtz, Analysis of a waveguide-fed metasurface antenna. Phys. Rev. Appl. **8**(5), 1–16 (2017)
43. H. Chen, A. Taylor, N. Yu, A review of metasurfaces: physics and applications. Rep. Prog. Phys. **79**(7) (2016)
44. X. Zhang, A.F. Molisch, S. Kung, Variable-phase-shift-based RF-baseband codesign for MIMO antenna selection. IEEE Trans. Signal Process. **53**(11), 4091–4103 (2005)
45. F. Rusek, D. Persson, B.K. Lau, E.G. Larsson, T.L. Marzetta, O. Edfors, F. Tufvesson, Scaling up MIMO: opportunities and challenges with very large arrays. IEEE Signal Process. Mag. **30**(1), 40–60 (2013)
46. C.B. Peel, B.M. Hochwald, A.L. Swindlehurst, A vector-perturbation technique for near-capacity multiantenna multiuser communication-Part I: channel inversion and regularization. IEEE Trans. Commun. **53**(1), 195–202 (2005)
47. J. Gallier, *The Schur Complement and Symmetric Positive Semidefinite (and Definite) Matrices* (Penn Engineering, Danboro, 2010)
48. T. Gally, M.E. Pfetsch, S. Ulbrich, A framework for solving mixed-integer semidefinite programs. Optim. Methods Softw. **33**(3), 594–632 (2017)
49. S. Zhang, H. Zhang, B. Di, L. Song, Cellular UAV-to-X communications: design and optimization for multi-UAV networks. IEEE Trans. Wirel. Commun. **18**(2), 1346–1359 (2019)

50. A. Alkhateeb, J. Mo, N. Gonzalez-Prelcic, R. Heath, MIMO precoding and combining solutions for millimeter-wave systems. IEEE Commun. Mag. **52**(12), 122–131 (2014)
51. M. Shehata, A. Mokh, M. Crussire, M. Helard, P. Pajusco, On the equivalence of hybrid beamforming to full digital zero forcing in mmWave MIMO, in *International Conference on Telecommunication (ICT)*, Hanoi (2019)
52. B. Di, L. Song, Y. Li, Sub-channel assignment, power allocation, and user scheduling for non-orthogonal multiple access networks. IEEE Trans. Wirel. Commun. **15**(11), 7686–7698 (2016)
53. H. El-Sallabi, M. Abdallah, K. Qaraqe, Modelling of parameters of Rician fading distribution as a function of polarization parameter in reconfigurable antenna, in *IEEE International Conference on Communication in China (ICCC)*, Shanghai (2014)
54. C. Pan, H. Ren, K. Wang, W. Xu, M. Elkashlan, A. Nallanathan, L. Hanzo, Multicell MIMO communications relying on intelligent reflecting surfaces. IEEE Trans. Wirel. Commun. **19**(8), 5218–5233 (2020)
55. NTT DOCOMO, DOCOMO conducts world's first successful trial of transparent dynamic metasurface (2020) [Online]. Available: https://www.nttdocomo.co.jp/english/info/media_center/pr/2020/0117_00.html
56. F.H. Danufane, M. Di Renzo, J. de Rosny, S. Tretyakov, On the path-loss of reconfigurable intelligent surfaces: an approach based on Green's theorem applied to vector fields [Online]. Available: https://arxiv.org/abs/2007.13158.pdf
57. S. Zhang, H. Zhang, B. Di, Y. Tan, Z. Han, L. Song, Beyond intelligent reflecting surfaces: reflective-transmissive metasurface aided communications for full-dimensional coverage extension. IEEE Trans. Veh. Technol. **69**(11), 13905–13909 (2020)
58. T. Cai, G. Wang, J. Liang, Y. Zhuang, T. Li, High-performance transmissive meta-surface for C-/X-band lens antenna application. IEEE Trans. Antennas Propag. **65**(7), 3598–3606 (2017)
59. M.L.N. Chen, L.J. Jiang, W.E.I. Sha, Detection of orbital angular momentum with metasurface at microwave band. IEEE Antennas Wirel. Propag. Lett. **17**(1), 110–113 (2018)
60. V. Arun, H. Balakrishnan, RFocus: beamforming using thousands of passive antennas, in *Proceedings of the USENIX Symposium on Netwroked System Design and Implementation (NSDI)*, Santa Clara (2020)
61. C. Pfeiffer, A. Grbic, Metamaterial Huygens' surfaces: tailoring wave fronts with reflectionless sheets. Phys. Rev. Lett. **110**(197401), 1–5 (2013)
62. B. Di, H. Zhang, L. Li, L. Song, Y. Li, Z. Han, Practical hybrid beamforming with finite-resolution phase shifters for reconfigurable intelligent surface based multi-user communications. IEEE Trans. Veh. Technol. **69**(4), 4565–4570 (2020)

Chapter 3
Convergences of RISs with Existing Wireless Technologies

In addition to realize low-cost MIMO systems, reconfigurable intelligent surfaces (RISs) have shown its potential to improve the performance of existing wireless techniques. In this chapter, we will show how to integrate RISs into existing wireless communication systems and give some case studies. This chapter is organized as follows: We give some introductions of RIS aided Device-to-Device (D2D) communications in Sect. 3.1. In Sect. 3.2, we show how to incorporate RISs into cell-free MIMO systems. In Sect. 3.3, a novel application of RIS, i.e., spatial equalizer is presented.

3.1 RIS Aided Device-to-Device Communications

3.1.1 Motivations

In recent years, with the popularity of mobile devices and smart terminals, data traffic demands in wireless networks have increased dramatically. In order to meet the rapid growth of data traffic demands and achieve seamless communications, D2D technology is considered to be a promising solution [1]. In D2D communications, users that are physically close to each other can communicate directly without forwarding through a base station (BS). Due to the short distance transmission, D2D communications can reduce energy consumption, provide early warning in emergencies, and upgrade users' quality of service (QoS) demands [2, 3]. Typically, D2D links are allowed to share the uplink spectrum with cellular links, which can alleviate the shortage of spectrum. However, the inevitable interference requires to protect cellular networks from harmful interference under the strict requirement of communication quality [4]. Fortunately, an innovative and revolutionary technology is developed, namely RIS, which can be utilized to effectively eliminate D2D interference and fulfill demanding data rates [5]. This can be achieved by config-

H. Zhang et al., *Reconfigurable Intelligent Surface Empowered 6G*, Wireless Networks, https://doi.org/10.1007/978-3-030-73499-2_3

uring the RIS to control the reflection, refraction and scattering of electromagnetic waves that impinge on the surface. With this reconfigurable property, the incident signals are superposed and reflected by adjusting phase shifts to create a favorable beam steering towards users, in order to effectively control the multi-path effects [6]. Therefore, by adjusting the amplitude and/or phase shift of RIS elements, the purpose of boosting the received intended power gain and destructively mitigating interference can be achieved, and consequently the spectrum and energy efficiency are improved [5, 6].

However, the integration of the RIS into D2D communications is very challenging and the research on this topic has not yet been fully carried out. *First*, in addition to direct paths, the RIS brings reflective paths which also carry intended signals. The reflective paths will not only provide the diversity gain but also be possible to enhance the interference among users. How to well design phase shifts of the RIS to enhance the received signals and reduce the interference is important but difficult. *Second*, power control is another necessary scheme to mitigate the interference. Due to the introduction of reflective paths by the RIS, the power optimization is coupled together with phase shifts design at the RIS, which makes it more complicated and existing schemes impossible to be applied directly.

In this section, we study an uplink RIS-assisted heterogeneous network, where a cellular link shares the same frequency band with multiple D2D links and an RIS is deployed to eliminate the interference. The goal is to maximize the system transmission rate subjected to QoS constraints and a transmission power budget. Optimization variables include limited discrete phase shifts and continuous transmission power, both of which are coupled and not independent in the numerator and denominator of the signal to interference plus noise ratio (SINR) term in the objective function and constraints. The problem is a mixed integer non-convex non-linear problem, thereby challenging to be solved. In particular, due to co-channel interference and the fact that D2D transmitters are not capable of precoding, the optimal scheme is very hard to obtain [7]. To address this issue, we use an alternating optimization method, which iteratively updates two separated sub-problems, i.e., power allocation and discrete phase shift optimizations. The power allocation sub-problem is indeed a difference of concave/convex functions (DC) program [8], where the multivariate Taylor expansion linearization with the gradient descent method is utilized to solve it efficiently. Another computationally affordable approach, namely local search, is adopted to optimize phase shifts.

The rest of this section is organized as follows. In Sect. 3.1.2, we introduce the investigated system. Then, we formulate the sum rate maximization problem and decompose it into a power allocation sub-problem and a discrete phase shift optimization sub-problem in Sect. 3.1.3. Next, two algorithms are designed, and the sum rate maximization algorithm is obtained by alternately iterating these two sub-problems in Sect. 3.1.4. Meanwhile, theoretical analysis of the proposed algorithm is also provided. In Sect. 3.1.5, we conduct a performance evaluation and compare the proposed algorithm with other benchmark algorithms. Finally, a summary is given in Sect. 3.1.6.

3.1.2 *System Model*

3.1.2.1 System Description

We consider an uplink single-cell RIS-aided heterogeneous network with one cellular user and multiple D2D pairs involved. As illustrated in Fig. 3.1, an RIS is deployed to mitigate the interference, which is composed of a great quantity of built-in programmable elements [9]. It reflects signals that impinge on the surface and maps towards intended receivers with directional beams. With the help of RIS, a reflective channel between the source and the destination is established. Therefore, for the cellular communications, the BS can not only receive the direct signals from the cellular user, but also receive the reflected signals through the RIS. Besides, we also assume that there exists D D2D links, and they share the same frequency band with the cellular link. For D2D links, similar to the transmission process in the cellular communications, the receiver can receive both direct signals and reflected ones. Note that these links share the same spectrum, and thus the interference among these links is inevitable, which will limit the system performance. To sum up, there

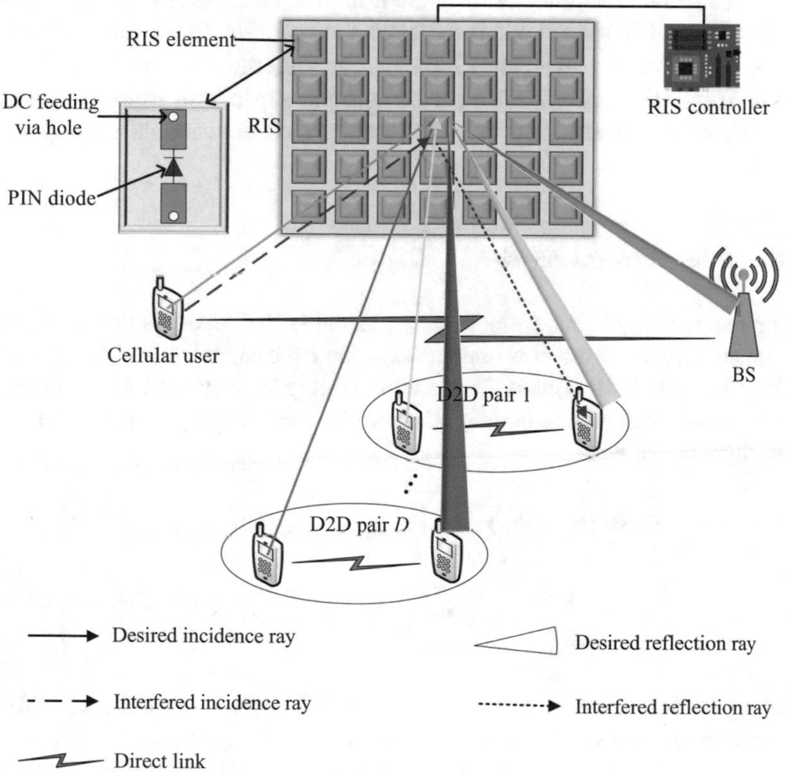

Fig. 3.1 System model for an uplink RIS-aided heterogeneous network

are a total of $D + 1$ links denoted as the set $\mathbf{L} = \{1, 2, \ldots, D, D + 1\}$. For a link $i \in \mathbf{L}$, we denote the corresponding transmitter and receiver as t_i and r_i, respectively.

In our investigated system, the RIS is a uniform planar array consisting of $N \times N$ elements. When a wave is incident to the RIS, the phase shift of each element can be tuned in a real-time manner by voltage-controlled PIN diodes with ON/OFF states. According to the regulation voltage, a certain phase shift will be provided by the RIS element, as shown in Fig. 3.1. With reconfigurable characteristics, the propagation environment can be manipulated to alleviate the interference among links and enhance the system performance. For simplicity, the range of phase shift of each element is constrained and we just take finite discrete values with equal quantization intervals between $[0, 2\pi]$. Assuming the number of quantization bits is e, there are 2^e phase shift values generated. The coefficient of the response induced by the RIS element in the l_z-th row and the l_y-th column is $q_{l_z,l_y} = e^{j\theta_{l_z,l_y}}$ with phase shift $\theta_{l_z,l_y} = \frac{2m_{l_z,l_y}\pi}{2^e-1}$, $m_{l_z,l_y} = \{0, \ldots, 2^e - 1\}$, $1 \le l_z, l_y \le N$, and j denotes the imaginary unit.

For the communication channels, as we have introduced before, there exists two types of channels. The first one is the reflective channel through the RIS. To be specific, we denote the reflective channel from t_i to r_i associated with RIS element $\{l_z, l_y\}$ as $h_{l_z,l_y}^{r_i,t_i}$. The second one is the direct channel. We denote the direct channel between transmitter t_i and receiver r_i as h_{r_i,t_i}. It is worthwhile to point out that since we assume that the CSI is known perfectly to the transmitter, expressions of these channels will not influence the power allocation and the phase shift design in what follows.

3.1.2.2 Interference Analysis

For the receiver r_i of link i, the received signal is the superposition of the signal transmitted directly from the transmitter t_i and the signal reflected by the RIS. On the other hand, all signals from the remaining co-channel links comprise the superposed interference. Summing all above interference together, we can express the received signal s_{r_i} as

$$\begin{aligned} s_{r_i} &= (h_{r_i,t_i} + \sum_{l_z,l_y} h_{l_z,l_y}^{r_i,t_i} q_{l_z,l_y})\sqrt{p_i} s_{t_i} \\ &+ \sum_{j \in \mathbf{L}, j \neq i} (h_{r_i,t_j} + \sum_{l_z,l_y} h_{l_z,l_y}^{r_i,t_j} q_{l_z,l_y})\sqrt{p_j} s_{t_j} + w_{r_i}, \end{aligned} \tag{3.1}$$

where s_{t_i} is the unit-power transmitted symbol from transmitter t_i, and p_i is the transmission power of link i. To facilitate subsequent presentations, $\mathbf{P} = [p_1, \ldots, p_{D+1}]^T$ is a $(D+1) \times 1$ matrix representing the transmission power of all links $i = 1, \ldots, D + 1$. w_{r_i} is thermal Gaussian white noise with the components independently drawn from $\mathcal{CN}(0, \sigma^2)$. In order to simplify the expression of the formula, we have some matrices defined as

$$\mathbf{F} = \sum_{l_z, l_y} q_{l_z, l_y} \mathbf{H}_{l_z, l_y}. \tag{3.2}$$

Here, $\mathbf{H}_{l_z, l_y}$ is a $(D+1) \times (D+1)$ channel vector composed of reflection channel coefficients between all transmitters and receivers, $\{h_{l_z, l_y}^{r_i, t_j}, 1 \leq i, j \leq D+1\}$. F_{r_i, t_j} represents the element in the j-th row and i-th column of the matrix $\mathbf{F}$. Similarly, $\mathbf{H}^L$ is a matrix of direct channel coefficients with the same dimension of $(D+1) \times (D+1)$. It includes h_{r_i, t_j}, $1 \leq i, j \leq D+1$. H_{r_i, t_j}^L denotes the element in the j-th row and i-th column of the matrix $\mathbf{H}^L$. With the definition of these matrices, s_{r_i} can be rewritten as

$$s_{r_i} = (H_{r_i, t_i}^L + F_{r_i, t_i}) \sqrt{p_i} s_{t_i} + \sum_{j \in \mathbf{L}, j \neq i} (H_{r_i, t_j}^L + F_{r_i, t_j}) \sqrt{p_j} s_{t_j} + w_{r_i}. \tag{3.3}$$

Accordingly, the SINR received by the link $i \in \mathbf{L}$ is given as

$$\Gamma_{r_i} = \frac{|H_{r_i, t_i}^L + F_{r_i, t_i}|^2 p_i}{\sum_{j \in \mathbf{L}, j \neq i} |H_{r_i, t_j}^L + F_{r_i, t_j}|^2 p_j + \sigma^2}. \tag{3.4}$$

According to the Shannon's capacity formula, the corresponding achievable transmission rate can be written as

$$R_{r_i} = \log_2 \left(1 + \frac{|H_{r_i, t_i}^L + F_{r_i, t_i}|^2 p_i}{\sum_{j \in \mathbf{L}, j \neq i} |H_{r_i, t_j}^L + F_{r_i, t_j}|^2 p_j + \sigma^2} \right). \tag{3.5}$$

3.1.3 Problem Formulation

In this subsection, we first formulate our optimization problem based on the above system model, and then decompose the complex optimization problem into two sub-problems in order to obtain a sub-optimal solution efficiently.

3.1.3.1 Sum Rate Maximization Problem Formulation

The optimization objective is to maximize the system sum rate, where optimization variables are phase shift values of all RIS elements and the transmission power of all links. Moreover, we record phase shift values applied to all elements as a set

$\boldsymbol{\Theta} = \{\theta_{l_z,l_y}, 1 \le l_z, l_y \le N\}$, and the optimization problem can be written as follows.

$$
\begin{aligned}
&\max_{\mathbf{P},\boldsymbol{\Theta}} \sum_{i=1}^{D+1} R_{r_i} \\
&s.t.\ (a)\ \Gamma_{r_i} \ge \gamma_{min},\ \forall i = 1, 2, \ldots, D+1, \\
&\quad (b)\ 0 \le p_i \le P_{max},\ \forall i = 1, 2, \ldots, D+1, \\
&\quad (c)\ q_{l_z,l_y} = e^{j\theta_{l_z,l_y}},\ \theta_{l_z,l_y} = \frac{2m_{l_z,l_y}\pi}{2^e - 1}, \\
&\quad m_{l_z,l_y} = \{0, 1, \ldots, 2^e - 1\}, 1 \le l_z, l_y \le N,
\end{aligned} \tag{3.6}
$$

where constraint (a) indicates the minimum SINR requirements for the cellular link and D2D links to ensure the QoS. Then, we limit the transmission power as in constraint (b) to effectively manage the interference. In constraint (c), the amplitude reflection coefficient of each element is 1, and the phase shift is a discrete variable. The above formulated problem is a mixed integer non-convex optimization problem. Both the objective function and constraint (a) involve the non-convex SINR formula, in which two variables, $\mathbf{P}$ and $\boldsymbol{\Theta}$, are coupled. However, this coupling relationship is hard to eliminate, which makes the problem challenging and difficult to solve.

3.1.3.2 Problem Decomposition

To solve the optimization problem in (3.6) efficiently, we decouple it into two easy-to-solve sub-problems.

1. Power Allocation: This sub-problem is to allocate the appropriate transmission power within the power constraint to maximize the objective function on the basis that SINR constraints are satisfied. When the other variable $\boldsymbol{\Theta}$ is fixed, the problem in (3.6) about transmission power $\mathbf{P}$ can be written as

$$
\begin{aligned}
&\max_{\mathbf{P}} \sum_{i=1}^{D+1} R_{r_i} \\
&s.t.\ (a)\ \Gamma_{r_i} \ge \gamma_{min},\ \forall i = 1, 2, \ldots, D+1, \\
&\quad (b)\ 0 \le p_i \le P_{max},\ \forall i = 1, 2, \ldots, D+1.
\end{aligned} \tag{3.7}
$$

2. Discrete Phase Shift Optimization: We fix $\mathbf{P}$ and this sub-problem is to provide an efficient phase shift from 2^e values for each RIS element. After that, the optimization problem is transformed to

$$
\begin{aligned}
&\max_{\boldsymbol{\Theta}} \sum_{i=1}^{D+1} R_{r_i} \\
&s.t.\ (a)\ \Gamma_{r_i} \geq \gamma_{min},\ \forall i = 1, 2, \ldots, D+1, \\
&\quad (b)\ q_{l_z,l_y} = e^{j\theta_{l_z,l_y}},\ \theta_{l_z,l_y} = \frac{2m_{l_z,l_y}\pi}{2^e - 1}, \\
&\quad m_{l_z,l_y} = \{0, 1, \ldots, 2^e - 1\},\ 1 \leq l_z, l_y \leq N.
\end{aligned} \tag{3.8}
$$

In view of these two sub-problems, we will design efficient algorithms to solve them in the next section, and then achieve the goal of solving the original optimization problem.

3.1.4 *Sum Rate Maximization Algorithm*

In this subsection, we first solve the aforementioned two sub-problems, and then we propose a sum rate maximization algorithm, which adopts the alternating method to solve these two sub-problems iteratively until the algorithm converges and a sub-optimal solution is obtained.

3.1.4.1 Power Allocation Sub-problem Algorithm Design

Although the integer discrete variables are removed, the sub-problem is still non-convex with respect to **P**. Motivated by the special structure, we use some simple mathematical transformations. In the objective function, both the cellular transmission rate and D2D transmission rate can be written as the difference of two concave functions, and the objective function of this sub-problem is equivalent to

$$
\begin{aligned}
&\max_{\mathbf{P}} \sum_{i=1}^{D+1} R_{r_i} \\
&= \max_{\mathbf{P}} \sum_{i=1}^{D+1} \log_2 \left(1 + \frac{|H_{r_i,t_i}^L + F_{r_i,t_i}|^2 p_i}{\sum\limits_{j \in \mathbf{L}, j \neq i} |H_{r_i,t_j}^L + F_{r_i,t_j}|^2 p_j + \sigma^2} \right) \\
&= -\min_{\mathbf{P}} \sum_{i=1}^{D+1} \Bigg[\log_2 \Bigg(\sum_{j \in \mathbf{L}, j \neq i} |H_{r_i,t_j}^L + F_{r_i,t_j}|^2 p_j + \sigma^2 \Bigg) \\
&\quad - \log_2 \Bigg(|H_{r_i,t_i}^L + F_{r_i,t_i}|^2 p_i + \sum_{j \in \mathbf{L}, j \neq i} |H_{r_i,t_j}^L + F_{r_i,t_j}|^2 p_j + \sigma^2 \Bigg) \Bigg].
\end{aligned} \tag{3.9}
$$

Based on the above mathematical decomposition, for link i, we regard these two logarithmic function terms in the objective as two functions, denoted as $g_i(\mathbf{P})$ and

$\varphi_i(\mathbf{P})$, which is known as the difference of two concave functions.

$$g_i(\mathbf{P}) = \log_2\left(\sum_{j\in\mathbf{L}, j\neq i} |H^L_{r_i,t_j} + F_{r_i,t_j}|^2 p_j + \sigma^2\right). \tag{3.10}$$

$$\varphi_i(\mathbf{P}) = \log_2\left(|H^L_{r_i,t_i} + F_{r_i,t_i}|^2 p_i + \sum_{j\in\mathbf{L}, j\neq i} |H^L_{r_i,t_j} + F_{r_i,t_j}|^2 p_j + \sigma^2\right). \tag{3.11}$$

Then, problem (3.7) is a DC problem, and is equivalently written as

$$\begin{aligned} &\min_{\mathbf{P}} \sum_{i=1}^{D+1} f_i(\mathbf{P}) \triangleq g_i(\mathbf{P}) - \varphi_i(\mathbf{P}) \\ &s.t.\ (a)\ g_i(\mathbf{P}) - \varphi_i(\mathbf{P}) \leq -\log_2(\gamma_{min}+1),\ \forall i = 1,\ldots,D+1, \\ &\quad\ (b)\ 0 \leq p_i \leq P_{max},\ \forall i = 1,\ldots,D+1. \end{aligned} \tag{3.12}$$

So far, this DC problem is still not easy to solve. Thus, we use the first-order Taylor expansion to approximate it as a convex function, i.e., to provide an upper bound for the objective function that needs to be minimized, and then gradually approach the optimal solution from the upper bound. This transformation has been mathematically proven that it can achieve a local optimal solution with a fast convergence rate. With this approximation, both the objective function and constraints are convex with respect to optimization variable $\mathbf{P}$. The Taylor expansion of $g_i(\mathbf{P})$ in the n-th iteration is written as follows.

$$g_i(\mathbf{P}) = g_i(\mathbf{P}^{(n)}) + \sum_{k=1}^{D+1} \frac{\partial g_i(\mathbf{P})}{\partial p_k}\bigg|_{\mathbf{P}=\mathbf{P}^{(n)}} (p_k - p_k^{(n)}). \tag{3.13}$$

Substituting (3.13) into problem (3.12) and let

$$f_i^{(n)}(\mathbf{P}) = g_i(\mathbf{P}^{(n)}) + \sum_{k=1}^{D+1} \frac{\partial g_i(\mathbf{P})}{\partial p_k}\bigg|_{\mathbf{P}=\mathbf{P}^{(n)}} (p_k - p_k^{(n)}) - \varphi_i(\mathbf{P}), \tag{3.14}$$

the problem can be further simplified into

$$\begin{aligned} &\min_{\mathbf{P}} \sum_{i=1}^{D+1} f_i^{(n)}(\mathbf{P}) \\ &s.t.\ (a)\ f_i^{(n)}(\mathbf{P}) \leq -\log_2(\gamma_{min}+1),\ \forall i = 1,\ldots,D+1, \\ &\quad\ (b)\ 0 \leq p_i \leq P_{max},\ \forall i = 1,\ldots,D+1. \end{aligned} \tag{3.15}$$

Therefore, this problem is converted into a convex problem, which can be solved by the Lagrangian method. Specifically, we first construct a Lagrangian unconstrained function for the problem. Then, by utilizing an iterative gradient descent method, the objective function is monotonically decreasing, so that it can converge to a static point. At the n-th iteration, the Lagrangian function with respect to optimization variable $\mathbf{P}$ can be written as,

$$L^{(n)}(\mathbf{P}, \boldsymbol{\lambda}^{(n)}) = \sum_{i=1}^{D+1} f_i^{(n)}(\mathbf{P}) + \lambda_i^{(n)}[f_i^{(n)}(\mathbf{P}) + \log_2(\gamma_{min} + 1)], \tag{3.16}$$

where $\lambda_i^{(n)}, i = 1, \ldots, D+1$ refer to the Lagrangian multiplier corresponding to the constraint (a) in problem (3.15) at the n-th iteration. By solving the Lagrangian function in (3.16), we obtain the transmission power $\mathbf{P}^{(n+1)}$ and substitute it into (3.16). Then, the Lagrangian function for Lagrangian multiplier $\boldsymbol{\lambda}$ is given as,

---

Algorithm 6: DC programming for power allocation

Input : $n = 0, \delta^{(0)} = 50, \mu^{(0)} = 100, \lambda_i^{(0)} = 100, \forall i = 1, \ldots, D+1$.
Output: $\mathbf{P}^*$.

1 Construct an auxiliary optimization problem (3.15) using DC programming theory;
2 Define Lagrangian unconstrained function $L^{(n)}(\mathbf{P}, \boldsymbol{\lambda}^{(n)})$ as (3.16);
3 Solve the optimization problem about $p_i, i = 1, 2, \ldots, D+1$;
$p_i^{(n+1)} = \left(p_i^{(n)} - \delta^{(n)} \frac{\partial L^{(n)}(\mathbf{P},\boldsymbol{\lambda}^{(n)})}{\partial p_i}\Big|_{p_i = p_i^{(n)}}\right)^+$; **if** $p_i^{(n+1)} > P_{max}$ **then**
4 | $p_i^{(n+1)} = P_{max}$;
5 **end**
6 **if** $\delta^{(n)} > 1$ **then**
7 | $\delta^{(n+1)} = \frac{\delta^{(n)}}{2}$;
8 **end**
9 $\lambda_i^{(n+1)} = \left(\lambda_i^{(n)} - \mu^{(n)} \left(\frac{\partial L^{(n)}(\mathbf{P}^{(n+1)},\boldsymbol{\lambda})}{\partial \lambda_i}\Big|_{\lambda_i = \lambda_i^{(n)}}\right)^+\right)^+, i = 1, \ldots, D+1$;
10 **if** $\mu^{(n)} > 1$ **then**
11 | $\mu^{(n+1)} = \frac{\mu^{(n)}}{2}$;
12 **end**
13 **if** $|R(\mathbf{P}^{(n+1)}) - R(\mathbf{P}^{(n)})| < \epsilon$ **then**
14 | $\mathbf{P}^* = \mathbf{P}^{(n+1)}$;
15 **else**
16 | $n = n + 1$, and go to line 1;
17 **end**

$$
\begin{aligned}
L^{(n)}(\mathbf{P}^{(n+1)}, \boldsymbol{\lambda}) = &\sum_{i=1}^{D+1} \left[g_i(\mathbf{P}^{(n)}) + \sum_{k=1}^{D+1} \frac{\partial g_i(\mathbf{P})}{\partial p_k}\Big|_{\mathbf{P}=\mathbf{P}^{(n)}} (p_k^{(n+1)} - p_k^{(n)}) - \varphi_i(\mathbf{P}^{(n+1)}) \right] \\
&+ \sum_{i=1}^{D+1} \lambda_i \Bigg[g_i(\mathbf{P}^{(n)}) + \sum_{k=1}^{D+1} \frac{\partial g_i(\mathbf{P})}{\partial p_k}\Big|_{\mathbf{P}=\mathbf{P}^{(n)}} (p_k^{(n+1)} - p_k^{(n)}) \\
&- \varphi_i(\mathbf{P}^{(n+1)}) + \log_2(\gamma_{min} + 1) \Bigg].
\end{aligned} \tag{3.17}
$$

The details of the DC programming algorithm for the power allocation are shown in Algorithm 6. We initialize Lagrangian multipliers $\lambda_i, \forall i = 1, 2, \ldots, D+1$. Besides, initial values of corresponding gradient descent step sizes $\delta^{(0)}$ and $\mu^{(0)}$ for transmission power and λ_i are also given, respectively. Then, we construct an optimization problem in (3.15) using the DC programming and multivariate Taylor expansion method introduced before is used in order to approximate the non-convex original problem. In step 2, the Lagrangian unconstrained function is defined for the transformed convex problem with $\boldsymbol{\lambda}^{(n)}$ fixed. As for optimization variable p_i, the gradient descent method computes the optimal gradient descent direction as the opposite direction of the first-order partial derivative, and the suitable step size δ is selected to gradually decrease with the number of iterations based on the given initial value. Combined with the basic constraint of transmission power, $0 \leq p_i \leq P_{max}, \forall i = 1, 2, \ldots, D+1$, the optimal values for this iteration or required initial values for the next iteration are generated. From step 11, we turn to optimize λ_i. It should be noted that optimizing λ_i requires substituting the optimal value of p_i solved from steps 4–7. The algorithm terminates when the difference between two results of the objective in two successive iterations is less than a preset threshold. Then, we output the optimal solution $\mathbf{P}^*$. Otherwise, we proceed to the next iteration with the power $\mathbf{P}^{(n+1)}$ obtained.

Algorithm 7: Local search for phase shift

Input : The number of quantization bits e.
Output: $\boldsymbol{\Theta}^*$.
```
for l_z = 1 : N do
    for l_y = 1 : N do
        Assign all possible values to θ_{l_z,l_y}, and select the value maximizing the sum rate on
        the premise that constraint (3.8a) is satisfied, denoted as θ*_{l_z,l_y};
        θ_{l_z,l_y} = θ*_{l_z,l_y};
    end
end
```

3.1.4.2 Discrete Phase Shift Optimization Sub-problem Algorithm Design

When fixing the transmission power **P**, we can remove the power range constraint. The objective function and constraints with respect to $\boldsymbol{\Theta}$ are still non-convex. Besides, $\boldsymbol{\Theta}$ contains a series of discrete variables, and the range available for each phase shift depends on the RIS quantization bits. Considering the complexity, we use the local search method as shown in Algorithm 7 to solve this problem. Specifically, keeping the other $N^2 - 1$ phase shift values fixed, for each element θ_{l_z,l_y}, we traverse all possible values and choose the optimal one which satisfies SINR constraints. Then, use this optimal solution $\theta^*_{l_z,l_y}$ as the new value of θ_{l_z,l_y} for the optimization of another phase shift, until all phase shifts in the set $\boldsymbol{\Theta}$ are fully optimized.

3.1.4.3 Sum Rate Maximization

We summarize the above power allocation sub-problem algorithm and discrete phase shift optimization sub-problem algorithm, and propose the sum rate maximization algorithm. As shown in Algorithm 8, we set all links to transmit at a maximum power budget in the initial state, and randomly generate a phase shift matrix. Then, we update the transmission power and phase shift in an alternating manner until the algorithm converges, i.e., the system transmission rate difference between two successive iterations is less than a certain threshold, $|R^{(\varrho+1)} - R^{(\varrho)}| < \epsilon$.

Algorithm 8: Sum rate maximization

Input: $\epsilon = 10^{-3}$, $\varrho = 0$, $p_i^{(0)} = P_{max}, \forall i = 1, 2, \ldots, D+1$;
The number of quantization bits e;
Randomly generate $\boldsymbol{\Theta}$, $\boldsymbol{\Theta}^* = \boldsymbol{\Theta}$.
1 Given $\boldsymbol{\Theta}^*$, update **P** using Algorithm 6;
2 Given $\mathbf{P}^*$, update $\boldsymbol{\Theta}$ using Algorithm 7;
3 **if** $|R^{(\varrho+1)} - R^{(\varrho)}| < \epsilon$ **then**
4     $R^* = R^{(\varrho+1)}$;
5   **Output** $\mathbf{P}^*, \boldsymbol{\Theta}^*, R^*$;
6 **else**
7     $\varrho = \varrho + 1$, and go to line 1;
8 **end**

3.1.4.4 Convergence, Feasibility and Complexity Analysis

1. Convergence: In the sum rate maximization algorithm, iterative calculations of two sub-problems are involved. When $\boldsymbol{\Theta}$ is fixed, **P** is solved using Algorithm 6. For Algorithm 6, first, for the objective function $f_i(\mathbf{P})$, we construct an auxiliary function using DC programming theory and multivariate Taylor expansion at the n-th iteration as $f_i^{(n)}(\mathbf{P})$. Then, $f_i(\mathbf{P}^{(n)}) = f_i^{(n)}(\mathbf{P}^{(n)})$ holds. In each iteration, we can get the optimal solution $\mathbf{P}^{(n+1)}$ of optimization problem (3.15) using the gradient descent method. Thus, $f_i^{(n)}(\mathbf{P}^{(n)}) \geq f_i^{(n)}(\mathbf{P}^{(n+1)})$, that is, the system transmission rate has increased. Since $\log_2\left(\sum_{j\in\mathbf{L}, j\neq i} |H_{r_i,t_j}^L + F_{r_i,t_j}|^2 p_j + \sigma^2\right)$ is concave,

$$\log_2\left(\sum_{j\in\mathbf{L}, j\neq i} |H_{r_i,t_j}^L + F_{r_i,t_j}|^2 p_j + \sigma^2\right) \leq g_i(\mathbf{P}^{(n)}) + \sum_{k=1}^{D+1} \frac{\partial g_i(\mathbf{P})}{\partial p_k}\bigg|_{\mathbf{P}=\mathbf{P}^{(n)}} (p_k - p_k^{(n)}), \tag{3.18}$$

and the inequality

$$f_i^{(n)}(\mathbf{P}) \geq f_i(\mathbf{P}), \tag{3.19}$$

is also satisfied. Based on the above analysis, we can have

$$f_i(\mathbf{P}^{(n)}) = f_i^{(n)}(\mathbf{P}^{(n)}) \geq f_i^{(n)}(\mathbf{P}^{(n+1)}) \geq f_i(\mathbf{P}^{(n+1)}). \tag{3.20}$$

Thus, as the number of iterations increases, the original objective function $f_i(\mathbf{P})$ also decreases monotonically, and the convergence of the original problem is proved. Next, we need to show in the following theorem that the solution obtained by the approximated problem is also a feasible solution to the original one.

Theorem 3.1 *$g_i(\mathbf{P})$ and $\varphi_i(\mathbf{P})$ are continuously differentiable in the domain of transmission power. Then, solving the auxiliary problem can get a stationary point of the original problem.*

Proof The first thing to explain is that in any problem domain, the stationary point of a function $f_i(\mathbf{P})$ is a point $\bar{\mathbf{P}}$ that satisfies the following conditions, $\frac{\partial f_i(\mathbf{P})}{\partial p_k}\big|_{\mathbf{P}=\bar{\mathbf{P}}}(p_k - \bar{p}_k) \geq 0, \forall k = 1, 2, \ldots, D+1$. Since the $f_i(\mathbf{P})$ can be converged, when the number of iterations approaches infinity, $\mathbf{P}^{(n)} = \mathbf{P}^{(n+1)}$, and both of them are optimal solutions. Furthermore, $f_i(\mathbf{P}^{(n)}) = f_i^{(n)}(\mathbf{P}^{(n)}) \geq f_i^{(n)}(\mathbf{P}^{(n+1)}) \geq f_i(\mathbf{P}^{(n+1)})$ are changed to $f_i(\mathbf{P}^{(n)}) = f_i^{(n)}(\mathbf{P}^{(n)}) = f_i^{(n)}(\mathbf{P}^{(n+1)}) = f_i(\mathbf{P}^{(n+1)})$. $\mathbf{P}^{(n)}$ and $\mathbf{P}^{(n+1)}$ are optimal solutions of $f_i^{(n)}(\mathbf{P})$. Let $\mathbf{P}^{(n)}$ be the final solution, $f_i^{(n)}(\mathbf{P})$ is convex, and $\mathbf{P}^{(n)}$ is the stationary point of $f_i^{(n)}(\mathbf{P})$. Thus, $\frac{\partial f_i^{(n)}(\mathbf{P})}{\partial p_k}\big|_{\mathbf{P}=\mathbf{P}^{(n)}}(p_k - p_k^{(n)}) \geq 0$, that is, $\frac{\partial f_i(\mathbf{P})}{\partial p_k}\big|_{\mathbf{P}=\mathbf{P}^{(n)}}(p_k - p_k^{(n)}) \geq 0$. Thus, $\mathbf{P}^{(n)}$ is also the stationary point of function $f_i(\mathbf{P})$.

The goal is to maximize the system transmission rate with solving the Lagrangian unconstrained function obtained by mathematical transformations. In the line 1 of Algorithm 8, with $\boldsymbol{\Theta}^*$ given at the ϱ-th iteration, $R(\mathbf{P}^{*(\varrho+1)}, \boldsymbol{\Theta}^*) \geq R(\mathbf{P}^{*(\varrho)}, \boldsymbol{\Theta}^*)$

is satisfied. Then, we turn to the second step of Algorithm 8. When $\mathbf{P}^*$ is fixed, $\boldsymbol{\Theta}$ is updated using the Algorithm 7. The local search algorithm aims at maximizing the sum rate, and will eventually get $\boldsymbol{\Theta}^{*(\varrho+1)}$ corresponding to fixed $\mathbf{P}^*$. In other words, the optimal phase shift $\boldsymbol{\Theta}^{*(\varrho+1)}$ must be greater than or equal to the initial value of $\boldsymbol{\Theta}$, which is also the value of $\boldsymbol{\Theta}^*$ obtained from the ϱ-th iteration of Algorithm 8. Thus, $R(\mathbf{P}^*, \boldsymbol{\Theta}^{*(\varrho+1)}) \geq R(\mathbf{P}^*, \boldsymbol{\Theta}^{*(\varrho)})$. In summary, $R(\mathbf{P}^{*(\varrho+1)}, \boldsymbol{\Theta}^{*(\varrho+1)}) \geq R(\mathbf{P}^{*(\varrho)}, \boldsymbol{\Theta}^{*(\varrho)})$. We can see the objective function of original optimization problem is non-decreasing. In addition, the number of discrete phase shifts is limited, and the continuous transmission power is also constrained by the upper and lower bounds, which make the problem of maximizing the sum rate bounded and the output solutions guaranteed. Therefore, we have completed the proof of the convergence of the sum rate maximization algorithm.

2. *Feasibility*: The proposed optimization problem in (3.6) is not always feasible, and it is only feasible under certain conditions. For the constraint (a) of the optimization problem written as follows,

$$|H^L_{r_i,t_i} + F_{r_i,t_i}|^2 p_i - \gamma_{min} \left(\sum_{j \in \mathbf{L}, j \neq i} |H^L_{r_i,t_j} + F_{r_i,t_j}|^2 p_j \right) \geq \gamma_{min} \sigma^2, \tag{3.21}$$

when the γ_{min} is given, it can be written as $\mathbf{AP} \geq b$ and then we consider $\mathbf{AP} = b$. $\mathbf{A}$ is a $1 \times (D+1)$ matrix including all coefficients of $p_i, i = 1, \ldots, D+1$, and $b = \gamma_{min}\sigma^2$. Thus, the sufficient conditions for a feasible solution of $\mathbf{P}$ are that: (1) there is a phase shift matrix $\boldsymbol{\Theta}$ that satisfies the constraint (c) of the problem in (3.6) such that $rank(\mathbf{A}) = rank(\mathbf{A}, b) \leq D+1$. (2) $0 \leq p_i \leq P_{max}, \forall i = 1, \ldots, D+1$.

3. *Complexity*: The complexity of the sum rate maximization algorithm is not only related to the number of iterations, which has been set to N_{outer} to achieve the convergence condition $|R^{(\varrho+1)} - R^{(\varrho)}| < c$, but also related to the complexity of power allocation sub-problem and phase shift optimization sub-problem. For the former one, the number of gradient updates and the optimization of transmission power of all links in each gradient iteration create complexity. The number of cellular link and D2D links is $D + 1$, and we denote the number of iterations of the gradient descent method as N_{inner}. Consequently, the complexity of power allocation sub-problem is $O(N_{inner} * (D+1))$. For the latter one, for each element $\{l_z, l_y\}$, the local search algorithm keeps the remaining phase shifts unchanged, selects the best one among 2^e phase shifts, and updates a value for θ_{l_z,l_y}. Since the RIS is an $N \times N$ planar array, the complexity of this part is $O(N^2 * 2^e)$. Therefore, we get the complexity of the proposed sum rate maximization algorithm as $O(N_{outer} * (N_{inner} * (D+1) + N^2 * 2^e))$.

3.1.5 Performance Evaluation

In this section, in comparison with other benchmark algorithms, we obtain numerical results under various representative parameters to verify the effectiveness of our proposed scheme, and investigate the impact of different parameters on system performance.

3.1.5.1 Simulation Setup

In the simulation, since the channel model is highly related to the positions of the RIS, transmitters, and receivers, we first establish a three dimensional Cartesian coordinate system to represent positions of communication nodes. As shown in Fig. 3.2, the RIS is placed on the Y-Z plane with the origin located at the bottom left corner, and the Y-axis and Z-axis as the alignment edges. $d_{ye} = 0.005$ m and $d_{ze} = 0.005$ m are the spacing between adjacent elements along Y and Z axes, respectively. It should be noted that, for each RIS element, we record its position of the top right corner, i.e., $L_{\{l_z,l_y\}} = (0, l_y d_{ye}, l_z d_{ze})$. In addition, we assume that the BS, the cellular user, and D2D pairs are deployed in a rectangular area on the X-Y plane with four points $(0, -100, 0)$, $(0, 100, 0)$, $(100, 100, 0)$ and $(100, -100, 0)$ as vertices. The cellular user and BS are distributed on both sides of the positive semi-axis of X with a distance of 10 m. Since the size of the RIS is very small, it is equivalent to that the cellular user and BS are distributed on both sides of the RIS. All D2D users, regardless of the transmitter or the receiver, are uniformly and randomly scattered in the rectangular area. In accordance with the

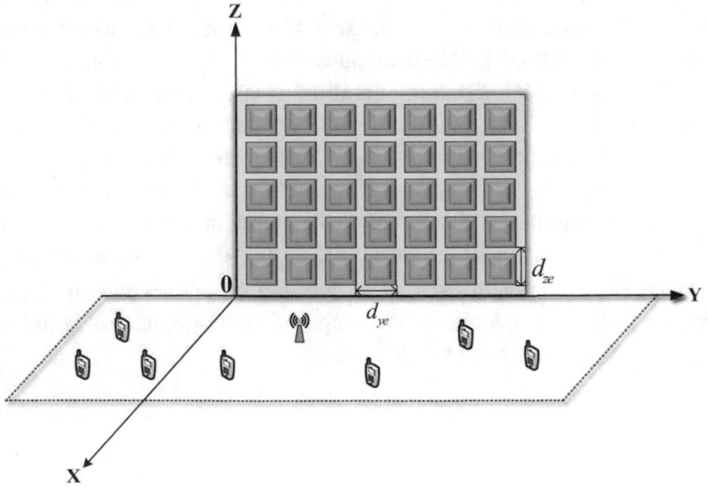

Fig. 3.2 The Cartesian coordinate for the RIS-assisted D2D network

realistic case, we set the maximum distance between two D2D users to 10 m. Take a link i as an example, the corresponding coordinates for t_i and r_i are denoted as $L_{t_i} = (t_{ix}, t_{iy}, 0)$ and $L_{r_i} = (r_{ix}, r_{iy}, 0)$. Based on given positions, the distance from t_i to the element $\{l_z, l_y\}$, and from the element $\{l_z, l_y\}$ to r_i, denoted as $DS_{t_i}^{l_z,l_y}$ and $DS_{l_z,l_y}^{r_i}$, respectively, can be easily obtained as follows,

$$DS_{t_i}^{l_z,l_y} = \sqrt{(t_{ix})^2 + (t_{iy} - l_y d_{ye})^2 + (-l_z d_{ze})^2}, \tag{3.22}$$

$$DS_{l_z,l_y}^{r_i} = \sqrt{(r_{ix})^2 + (r_{iy} - l_y d_{ye})^2 + (-l_z d_{ze})^2}. \tag{3.23}$$

Based on the constructed coordinate system, we will introduce the adopted channel model. For the reflective channel, it is modeled as a Rician channel, which consists of Line-of-Sight (LoS) and Non-LoS (NLoS) components. More precisely, the channel model is written as

$$h_{l_z,l_y}^{r_i,t_i} = \sqrt{\frac{\beta}{1+\beta}} \tilde{h}_{l_z,l_y}^{r_i,t_i} + \sqrt{\frac{1}{1+\beta}} \hat{h}_{l_z,l_y}^{r_i,t_i}, \tag{3.24}$$

where $\beta = 4$ is the Rician factor, $\tilde{h}_{l_z,l_y}^{r_i,t_i}$ is the LoS component, and $\hat{h}_{l_z,l_y}^{r_i,t_i}$ is the NLoS component of the reflective channel.

$$\tilde{h}_{l_z,l_y}^{r_i,t_i} = \sqrt{\beta_0 \left(DS_{t_i}^{l_z,l_y} \cdot DS_{l_z,l_y}^{r_i} \right)^{-\alpha}} e^{-j\theta'}, \tag{3.25}$$

where $\beta_0 = -61.3849$ dB denotes the channel gain at the reference distance of 1 m, θ' is the phase which is a random variable in $[0, 2\pi]$, and α is the path loss exponent in the LoS case.

$$\hat{h}_{l_z,l_y}^{r_i,t_i} = \sqrt{\beta_0 \left(DS_{t_i}^{l_z,l_y} \cdot DS_{l_z,l_y}^{r_i} \right)^{-\alpha'}} \hat{h}_{NLoS,l_z,l_y}^{r_i,t_i}, \tag{3.26}$$

where α' is the path loss exponent in the NLoS case, and $\hat{h}_{NLoS,l_z,l_y}^{r_i,t_i} \sim \mathcal{CN}(0, 1)$ is the small-scale fading.

Besides, the direct channel is

$$h_{r_i,t_i} = h_i \sqrt{\beta_0 \left(DS_{t_i}^{r_i} \right)^{-\alpha}}, \tag{3.27}$$

where h_i is the small-scale fading that obeys Nakagami-m_i distribution with parameters $\{m_i, \omega_i\}$, where $m_i = 3$ is the fading depth parameter and $\omega_i = 1/3$ is the average power in the fading signal. $\left(DS_{t_i}^{r_i} \right)^{-\alpha}$ is the large-scale path loss and the distance between t_i and r_i is also calculated based on the coordinates.

Furthermore, we give some parameter settings. For large-scale fading, the path loss exponent is $\alpha = 2.5$ in the LoS case, and it is $\alpha' = 3.6$ in the NLoS case [10]. In this section, we use the millimeter-wave (mm-wave) band with a center frequency of 28 GHz for simulation, and the mm-wave noise power spectral density is −134 dBm/MHz. Besides, the maximum transmission power P_{max} is 23 dBm, and all individuals have the same minimum SINR $\gamma_{min} = 5$ dB. Finally, not only for the power allocation algorithm, but also for the sum rate maximization algorithm, we set the termination iteration threshold as 10^{-3}. Basic system parameters are set as above, unless specified later.

To show the system performance of the **Proposed-algorithm**, we compare it with the following algorithms:

- **Maximum Power Transmission (MPT)**: in this algorithm, all cellular and D2D links are operated with the maximum transmission power. Phase shifts of the RIS are adjusted by the local search algorithm to mitigate the interference and improve the system performance. Compared with the proposed scheme, the MPT algorithm omits the power allocation.
- **Random Phase Shift (RPS)**: this algorithm randomly selects a feasible phase shift for each RIS element and keeps these phase shifts unchanged. Then, the maximum power is used as the initial transmission power for all links. By using the power allocation sub-problem algorithm proposed in this section, the goal of maximizing the sum rate is approached.
- **Without-RIS**: this scheme does not use RIS for signal reflection, and receivers can only receive signals through direct channels. The same power allocation algorithm is used to mitigate the interference.
- **Continuous Phase Shift (CPS)**: the only difference between this scheme and the proposed algorithm is the optimization of the phase shift. Instead of directly processing the discrete phase shift, the CPS algorithm relaxes the phase shift of each RIS element into a continuous variable from 0 to 2π. With this relaxation, the phase shift sub-problem is also a DC program, which can use the algorithm introduced in the power allocation sub-problem to solve it, but it is necessary to apply the multivariate Taylor expansion for all logarithmic function terms to get the convex auxiliary problem. Phase shift optimization and power allocation sub-problems are also solved iteratively.

3.1.5.2 Performance Evaluation

In Fig. 3.3, we set $N = 4$, $e = 3$, and provide the sum rate performance curves of these four schemes as the number of D2D links changes from 1 to 6. It is observed that the proposed algorithm achieves the highest system sum rate, and the sum rates of all schemes monotonically increase with the number of D2D links. As more D2D links are deployed, the gap between the proposed scheme and the MPT algorithm is increasing, which shows that the effect of the power control is becoming more significant. Then, a noticeable difference is observed between the proposed one and

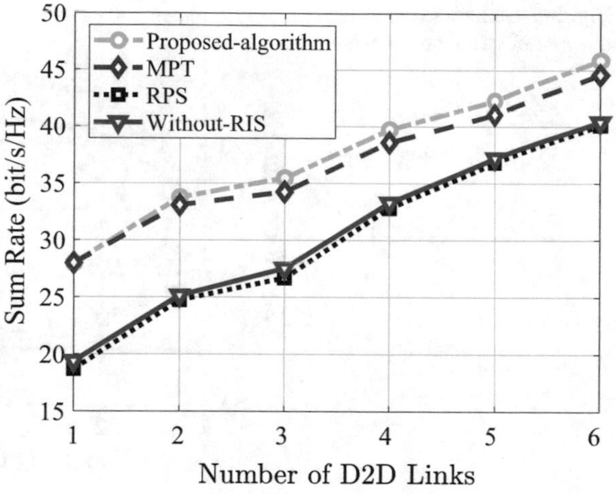

Fig. 3.3 Sum rate versus the number of D2D links

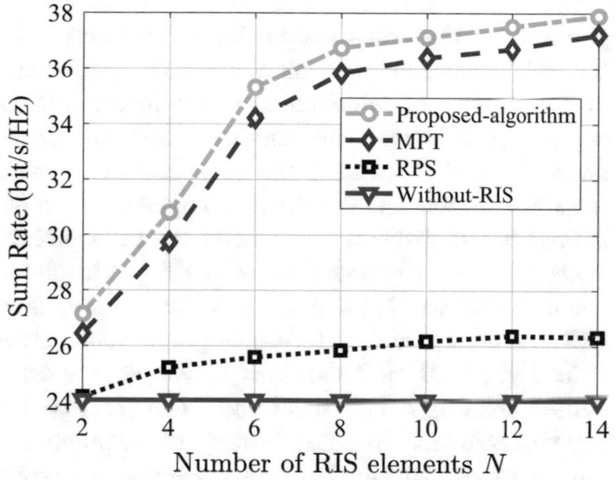

Fig. 3.4 Sum rate versus the number of RIS elements N

other two schemes, i.e., Without-RIS and RPS. This can be attributed to the adoption of the RIS. Without using the RIS to mitigate the interference or using the RIS without optimized phase shifts will suffer a significant performance loss. Besides, for RPS and Without-RIS schemes, under different setups, it is unclear which one is better, but a narrow gap is provided. In general, when the number of D2D links is 6, the performance gap between the proposed algorithm and MPT, Without-RIS, RPS is 2.9%, 13.5% and 14.4%, respectively.

In Fig. 3.4, both the number of D2D links and e are set to be 3. We plot the sum rates of these four schemes versus N. As seen from the given results, the proposed algorithm outperforms others with the number of RIS elements varied from 2^2 to 14^2. Then, the proposed one and the MPT show an increasing trend as the number of RIS elements increases, which show that by using the RIS for beamforming, undesired co-channel interference can be more effectively suppressed. It worths

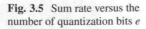

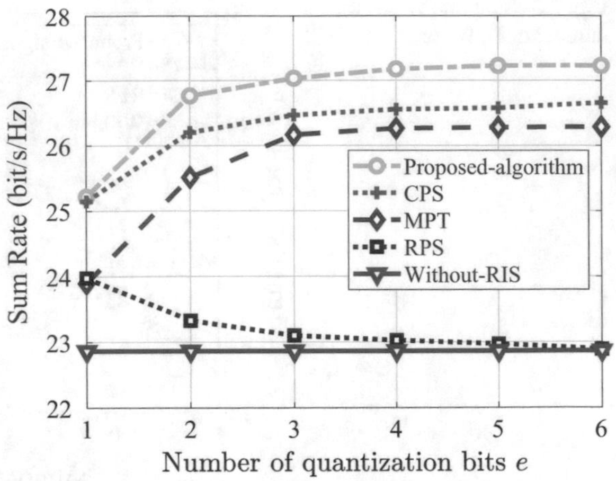

Fig. 3.5 Sum rate versus the number of quantization bits e

pointing out that when the number of elements is large, the growth of these two schemes becomes slightly slower. This is because the RIS contributes an increase in the number of interference signal paths. Nevertheless, it is undeniable that continuing to increase the number of RIS elements and adaptively adjusting the phase shifts of elements is still considerable for these two schemes. However, for the RPS, without the constructive alignment of reflecting beams, the increase in the number of elements has caused the interference to increase significantly, which leads to a slower growth from 2 to 12, and results in a slightly decreasing curve when N exceeds 12. Finally, comparing the proposed algorithm with the Without-RIS scheme, a 57.6% performance gain is obtained for $N = 14$.

In Fig. 3.5, $D = 3$ and $N = 3$. We plot the achievable sum rate performance while increasing the bit-quantization number from 1 to 6. There are several important observations. First, the advantage of the proposed algorithm is pronounced over the other four benchmark schemes. Second, the proposed, CPS and MPT schemes gradually increase with e grows from 1 to 4, and then basically remain unchanged from 5 to 6. This reveals that the system performance tends to be saturated when the number of quantization bits exceeds 4. In addition, the RPS scheme exhibits a slight sum rate degradation. The reason behind this phenomenon is that the distance of the reflection link that generates the useful signal and the distance of the reflection link that generates the interference signal are both random, so the useful reflection signal is not necessarily stronger than the interfering reflection signal. Under these parameter settings, the effect of changing e on the superposition of multiple reflected interference links is more prominent. Last but not least, comparing the proposed scheme and the CPS scheme, we can see that the system performance obtained by relaxing discrete phase shifts into continuous ones, and then recovering them is worse than that obtained by directly optimizing discrete phase shifts. This is because, unlike the continuous power allocation sub-problem, we cannot guarantee that the original problem is convergent from the fact that the auxiliary problem can converge. At the same time, the consistency of the static point of auxiliary

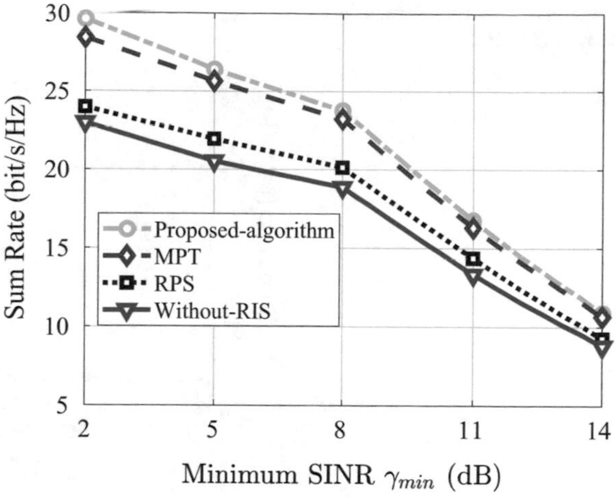

Fig. 3.6 Sum rate versus minimum SINR γ_{min}

problem and original problem can not be proved. Above two points greatly damage the system performance. From the perspective of quantitative analysis, the sum rate of the proposed algorithm is 2.3% higher than that of the CPS scheme when $e = 4$.

In Fig. 3.6, we set $D = 3$, $N = 4$ and $e = 3$. We vary the γ_{min} from 2 dB to 14 dB and depict the comparison of four schemes in terms of the sum rate. It should be noted that under different γ_{min}, randomly generated user locations, and channel coefficients will cause the optimization problem to be infeasible. If the problem is not feasible, that is, the obtained sum rate is 0. From this figure, we can observe that, curves of all schemes show a downward trend with the increase of the minimum value of individual SINR, and the gap between these schemes is getting smaller. This is because by improving the lower bound of the network QoS constraint, the feasible region of the optimization problem becomes smaller and the average sum rate will drop. When γ_{min} increases to infinity, the sum rates of all schemes are 0. From the simulation results, we can see the proposed algorithm has a highest system sum rate than other schemes, while the RPS and Without-RIS are the worst two of the four schemes.

In Fig. 3.7, we set $D = 3$, $N = 3$ and $e = 3$. We study the impact of the deployment location of the RIS on the performance of the proposed algorithm. First, we choose nine positions as the bottom left corner of the RIS, which are $(0, -100, 0)$, $(0, -75, 0)$, $(0, -50, 0)$, $(0, -25, 0)$, $(0, 0, 0)$, $(0, 25, 0)$, $(0, 50, 0)$, $(0, 75, 0)$ and $(0, 100, 0)$. In other words, RIS is always in the Y-Z plane and moves along the Y axis. For convenience, we denote the position of RIS as pos, and different positions are distinguished by the distance between the bottom left corner and the reference point $(0, 0, 0)$. Based on these settings, we plot the performance curves with the pos from -100 to 100 for the proposed algorithm and the Without-RIS scheme. Interestingly, when the RIS position pos is equal to -100 or 100, i.e., at the edge of the investigated area, the system performance is the worst. While pos is equal to 0, which is the center of the long side of the rectangular area, the system

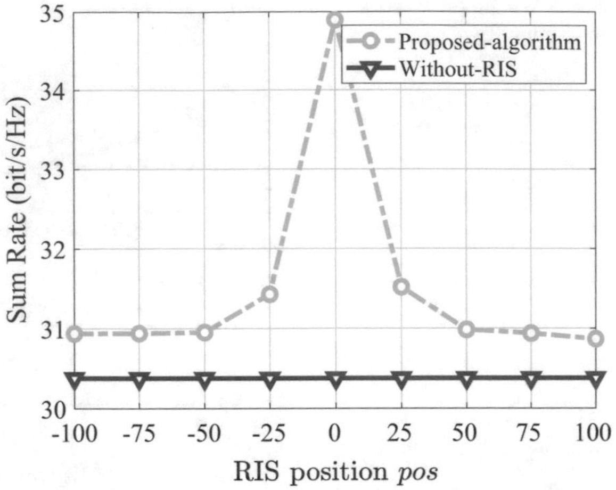

Fig. 3.7 Sum rate versus RIS position *pos*

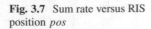

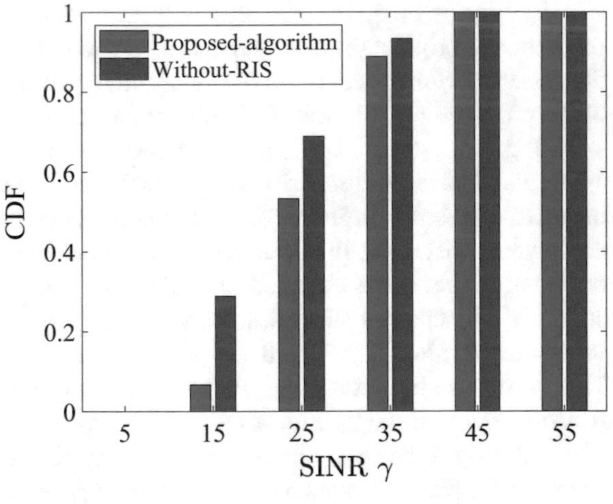

Fig. 3.8 CDF histograms for SINR γ

performance is the best. This is owing to the fact that when the RIS moves closer to the center of the area, the path loss of the reflection link will change accordingly, which results in an enhanced reflected signal and that the benefits of RIS are fully utilized. In general, the system performance is sensitive to the deployment location of RIS.

In Fig. 3.8, we set $D = 4$, $N = 4$ and $e = 3$. We plot the cumulative distribution function (CDF) of the SINR received by all links, which is denoted by γ, varying from 5 dB to 55 dB. Without loss of generality, all receivers are assumed to have the same SINR target, i.e., 5 dB. Comparing the proposed algorithm and the Without-RIS scheme, we can see that the performance gain brought by the deployment of RIS is completely non-negligible, and the effect of RIS on reducing interference of D2D networks is very considerable. This is because that, by adding more controllable reflection paths, RIS can boost the signal quality at the intended receiver.

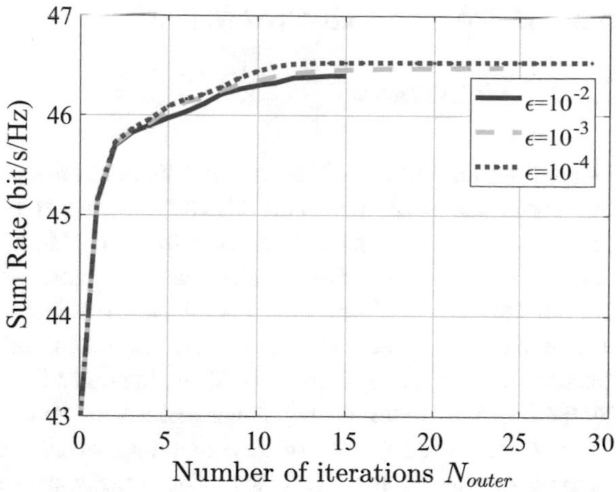

Fig. 3.9 Convergence behaviour of the proposed algorithm

In Fig. 3.9, we set $D = 4$, $N = 4$ and $e = 3$. We simulate the complexity of the proposed algorithm under different stopping threshold settings, i.e., the number of iterations of the sum rate maximization algorithm. When the values of ϵ are 10^{-2}, 10^{-3}, and 10^{-4}, respectively, the corresponding values of N_{outer} required for convergence are 15, 24, and 29. First, we can see that, no matter the pre-defined threshold is large or small, the overall complexity of the sum rate maximization algorithm is still acceptable. In addition, the smaller the threshold is, the slower the proposed algorithm reaches the stationary solution, but the obtained system performance is better.

3.1.6 *Summary*

In this section, we have investigated uplink D2D-enabled cellular networks assisted by an RIS. A joint power allocation design and RIS phase shift optimization problem has been studied to maximize the system sum rate under the individual QoS, power and practical discrete phase shift constraints. Due to the mixed integer non-convex nature of the formulated problem, we have proposed an alternative optimization algorithm to obtain a sub-optimal solution. Finally, numcrical results have validated the superiority of the proposed algorithm and we have observed that, by providing multiple controllable reflected signals based on the number of RIS elements, RIS can convert a channel in poor condition into a well-conditioned channel.

3.2 RIS Aided Cell-Free MIMO

3.2.1 Motivations

During the past few decades, MIMO has drawn a great attention [11]. However, the performance of multi-cell MIMO systems is typically limited by inter-cell interference due to the *cell-centric* implementation. To address this issue, the concept of cell-free networks has been proposed as a *user-centric* paradigm to enable multiple randomly-distributed BSs without cell boundaries to coordinate with each other to serve all users in the network simultaneously [12, 13]. Benefited from the distributed deployment of BSs, the diversity and coverage can be improved by the cell-free networks [14]. Nevertheless, the traditional cell-free system requires a large-scale deployment of BSs, leading to an unsatisfying energy efficiency performance due to the high costs of both hardware and power sources.

To tackle this issue, the RIS has emerged as a potential cost-efficient technique by creating favorable propagation conditions from BSs and users [15]. Benefited from a large number of RIS elements whose phase shifts are controlled by simple programmable PIN diodes, RISs can reflect signals and generate directional beams from BSs to users [16]. Unlike large-scale phased array antennas in the cell-free system enabled by phase shifters with inevitable power consumption, RIS requires no extra hardware implementation, such as complex digital phase shift circuits, thus greatly saving the energy consumption and complexity for signal processing [17]. Hence, compared with the conventional cell-free systems, a lower level of power consumption is required to achieve the same QoS. In other words, RIS provides a new dimension for cell-free systems to enhance the energy efficiency of the cell-free systems.

In this section, we consider an *RIS aided cell-free MIMO* system where multiple BSs and RISs are coordinated to serve various users. Benefited from the programmable characteristic of RIS elements which mold the wavefronts into desired shapes, the propagation environment can be reconfigurable in a low-cost way, thereby enhancing the cell-free MIMO communications. To achieve favorable propagation, it is vitally important to determine the phase shifts of all RIS elements, each of which can be viewed as an antenna, inherently capable of realizing analog beamforming. Since the RISs do not have any digital processing capability, we consider a hybrid beamforming (HBF) scheme [18] where the digital beamforming is performed at BSs, and the RIS-based analog beamforming is conducted at RISs. In such an RIS-aided cell-free system, RISs act as antenna arrays far away from the BS to achieve a broader coverage in a low-cost way. Moreover, the spatial resource can be better utilized benefited from the multipath effects, i.e., the direct and RIS-reflected paths, for each transmitted signal to improve the energy efficiency performance.

To fully explore how RISs as a cost-efficient technique influence the energy efficiency of the cell-free system, we aim to optimize the digital beamformer at BSs and phase shifts of RISs. Challenges have arisen in such a system. *First*, unlike conventional cell-free systems, the propagation environment is more complicated due to the extra reflected links via RISs. It is hard to obtain the optimal solution to the joint digital beamforming at BSs and phase shift configuration of RISs. *Second*, compared to single-RIS systems, the dispersed distribution of multiple RISs brings a new degree of freedom for energy efficiency maximization. It remains to be explored how the number of RISs as well as the RIS size influences the performance of the multi-RIS aided cell-free system.

To address the above challenges, we first propose an RIS aided cell-free MIMO system where various users are served by multiple BSs both directly and assisted by RISs. An HBF scheme consisting of the digital beamforming at BSs and RIS-based analog beamforming is designed. Then, an energy efficiency maximization problem for the HBF scheme is formulated and decomposed into two subproblems, i.e., the digital beamforming subproblem and the RIS-based analog beamforming subproblem. An iterative energy efficiency maximization (EEM) algorithm is proposed to solve them. Moreover, the impact of the transmit power, the number of RISs, and the size of each RIS on energy efficiency is analyzed theoretically. Finally, simulation results indicate that the RIS aided cell-free system has a better energy efficiency performance compared to traditional ones including conventional distributed antenna system (DAS), conventional cell-free system, and the no-RIS case. The impact of the number of RISs, the size of each RIS, the quantization bits of discrete phase shifts on energy efficiency is also shown numerically, which verify our theoretical analysis.

The rest of this section is organized as follows. In Sect. 3.2.2, we provide the system model of the RIS aided cell-free MIMO system. The RIS reflection model and the channel model are presented. In Sect. 3.2.3, the HBF scheme is proposed for the RIS aided cell-free system. An energy efficiency maximization problem is formulated and then decomposed into two subproblems, i.e., the digital beamforming subproblem and RIS-based analog beamforming subproblem. An EEM algorithm is designed in Sect. 3.2.4 to solve the above subproblems iteratively. In Sect. 3.2.5, we discuss the influence of the number of RISs, the size of each RIS, and the transmit power on energy efficiency theoretically. The complexity and convergence of the proposed EEM algorithm are also analyzed. The simulation results are given in Sect. 3.2.6 to evaluate the energy efficiency and sum rate performance to validate our analysis. Finally, we draw our conclusions in Sect. 3.2.7.

3.2.2 System Model

In this section, we first introduce an RIS-aided downlink cell-free MIMO communication system in which multi-antenna BSs and multiple RISs coordinate to serve various single-antenna users. The RIS reflection model and the channel model are then given.

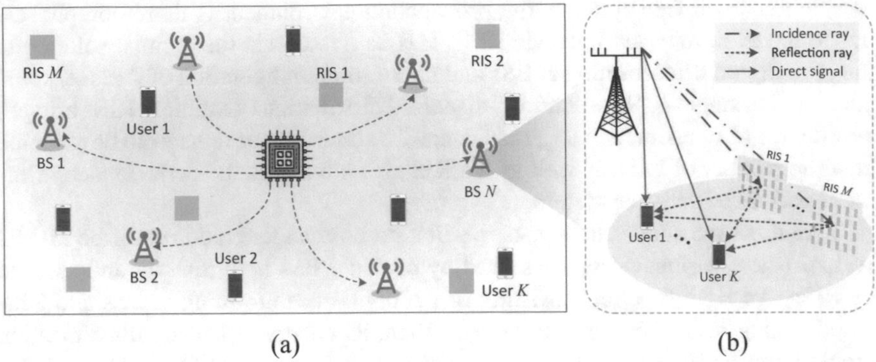

Fig. 3.10 (**a**) RIS aided cell-free MIMO system; (**b**) Illustration of one RIS-aided cell in the MIMO system

3.2.2.1 Scenario Description

Consider a downlink multi-user system in Fig. 3.10a where K single-antenna users are served by N BSs each of which is equipped with N_a antennas [19]. To improve the energy efficiency performance, we further consider to deploy M RISs around these BSs, which passively reflect the signals from BSs and directly project to users, thereby forming an RIS aided cell-free MIMO system. These RISs act as antenna arrays far away from the BS to improve the capacity and achieve a broader coverage in a low-cost way. Moreover, the RIS technique enables signals to be transmitted via both the direct and reflected links, and thus, a more diverse transmission is obtained to improve the performance, and thus enhancing the cell-free communications.

In such a system, all BSs and RISs are coordinated to serve all users. A central processing unit (CPU) is deployed for network control and planning, and decides the transmission scheduling based on the locations of BSs, RISs, and users. To characterize the optimal performance of the RIS aided cell-free MIMO system with discrete phase shifts, without loss of generality [20, 21], we assume that the channel state information (CSI) of all channels, i.e., direct channel between BSs and users and reflected channel via multiple RISs, is perfectly known at the CPU. As shown in Fig. 3.10b, after receiving the signals transmitted from BSs, multiple RISs operating at the same frequency are employed to reflect these signals to the users synchronously. Specifically, users receive both the signals transmitted from BSs directly and various reflection signals via multiple RISs.

3.2.2.2 RIS Reflection Model

RIS is an artificial thin film of electromagnetic and reconfigurable materials, consisting of L *RIS elements*, each of which is a sub-wavelength meta-material particle connected by multiple electrically controllable PIN diodes [22]. Each PIN diode can be switched ON/OFF based on the applied bias voltages, thereby manipulating the electromagnetic response of RIS elements towards incident waves. Benefit from these built-in programmable elements, RISs serve as low-cost reconfigurable phased arrays without extra active power sources which only rely on the combination of multiple programmable radiating elements to realize a desired transformation on the transmitted, received, or reflected waves.

By a b-bit re-programmable meta-material, each RIS element can be configured into 2^b possible amplitudes/phase shifts to reflect the radio wave. Without loss of generality, we assume that each RIS element is configured to maximize the reflected signal, i.e., the amplitude of each element of each RIS is set as 1 [23, 24]. Let $\theta_{m,l}$ denote the phase shift of the l-th RIS element of the m-th RIS, assuming that signals are reflected ideally without the hardware imperfections such as non-linearity and noise, the corresponding frequency response of each RIS element $q_{m,l}$ $(1 \leq m \leq M, 1 \leq l \leq L)$ can be expressed by

$$q_{m,l} = e^{j\theta_{m,l}}, \tag{3.28}$$

$$\theta_{m,l} = \frac{i_{m,l}\pi}{2^{b-1}}, i_{m,l} \in \{0, 1, \ldots, 2^b - 1\}. \tag{3.29}$$

3.2.2.3 Channel Model

As shown in Fig. 3.10, each user receives signals from both N BSs directly and M RISs which reflect the signals sent by BSs. The channel between each antenna of BS n and each user k consists of a **direct link** from BS n to user k and $M \times L$ **reflected links** since there are M RISs each of which relies on L elements to reflect the received signals.

Therefore, the equivalent channel $\mathbf{H} \in \mathbb{C}^{K \times NN_a}$ can be given by

$$\mathbf{H} = \underbrace{\mathbf{H}_D}_{\text{Direct link}} + \underbrace{\mathbf{H}_{RU}^H \mathbf{Q} \mathbf{H}_{BR}}_{\text{Reflected links}}, \tag{3.30}$$

where $\mathbf{H}_D \in \mathbb{C}^{K \times NN_a}$ denotes the direct component between BSs and users, the $\mathbf{H}_{RU}^H \mathbf{Q} \mathbf{H}_{BR}$ denotes the reflected links via M RISs. Specifically, each RIS m reflects incident signal based on an equivalent $ML \times ML$ diagonal phase shift matrix $\mathbf{Q}$ consisting of the phase shifts $\{q_{m,l}\}$. $\mathbf{H}_{BR} \in \mathbb{C}^{ML \times NN_a}$ and $\mathbf{H}_{RU}^H \in \mathbb{C}^{K \times ML}$ represent the reflected channel between BSs and RISs and that between RISs and users, respectively. We assume that all channels are statistically independent and follow Rician distribution.

3.2.3 Hybrid Beamforming and Problem Formulation

In this section, we first present an HBF scheme based on the system model introduced in Sect. 3.2.2. An energy efficiency maximization problem is formulated and decomposed for the proposed HBF scheme.

3.2.3.1 Hybrid Beamforming Scheme

For each RIS m, L RIS elements can be viewed as antenna elements far away from the BSs, inherently capable of realizing analog beamforming via determining the phase shifts of all RIS elements. Moreover, BSs are responsible for the signal processing since RIS elements are passive devices without any digital processing capacity. To create reflected waves towards preferable directions to serve multiple users, we present an HBF scheme for the proposed RIS aided cell-free MIMO system given the RIS reflection model and the channel model. Specifically, the digital beamforming is performed at BSs while the analog beamforming is achieved by these RISs.

Digital Beamforming: For K users, BSs encode K different data streams via a digital beamformer $\mathbf{V}_D \in \mathbb{C}^{NN_a \times K}$, and then up-converts the encoded signals over the carrier frequency and allocates the transmit powers. We assume that $K \leq NN_a$. The users' signals are sent directly through N BSs each equipped with N_a antennas, which can be given by

$$\mathbf{x} = \mathbf{V}_D\mathbf{s}, \tag{3.31}$$

where $\mathbf{s} \in \mathbb{C}^{K\times 1}$ denotes the incident signal vector for K users.

RIS-based Analog Beamforming: In the RIS aided cell-free MIMO system, the analog beamforming is achieved by determining the phase shifts of all RIS elements, i.e., $\mathbf{Q}$. Specifically, RIS m with L elements performs a linear mapping from the incident signal vector to a reflected signal vector. Therefore, the received signal at user k can be expressed by

$$z_k = \mathbf{H}_k\mathbf{V}_{D,k}s_k + \underbrace{\sum_{k'\neq k}\mathbf{H}_k\mathbf{V}_{D,k'}s_{k'}}_{\text{inter-user interference}} + \omega_k, \tag{3.32}$$

where $\omega_k \sim \mathscr{CN}(0, \sigma^2)$ is the additive white Gaussian noise. $\mathbf{H}_k$ and $\mathbf{V}_{D,k}$ denote the k-th row and the k-th column of matrices $\mathbf{H}$ and $\mathbf{V}_D$, respectively.

3.2.3.2 Energy Efficiency Maximization Problem Formulation

Based on the RIS reflection model, the channel model, and the HBF scheme, the data rate of user k can be given by (3.33),

$$\begin{aligned} R_k &= \log_2\left(1+\frac{|\mathbf{H}_k\mathbf{V}_{D,k}|^2}{\sum_{k'\neq k}|\mathbf{H}_k\mathbf{V}_{D,k'}|^2+\sigma^2}\right) \\ &= \log_2\left(1+\frac{|\mathbf{H}_{D,k}\mathbf{V}_{D,k}+\mathbf{H}^H_{RU,k}\mathbf{Q}\mathbf{H}_{BR}\mathbf{V}_{D,k}|^2}{\sum_{k'\neq k}|\mathbf{H}_{D,k}\mathbf{V}_{D,k'}+\mathbf{H}^H_{RU,k}\mathbf{Q}\mathbf{H}_{BR}\mathbf{V}_{D,k'}|^2+\sigma^2}\right). \end{aligned} \tag{3.33}$$

where $\mathbf{H}_{D,k}$ and $\mathbf{H}^H_{RU,k}$ denote the k-th rows of matrices $\mathbf{H}_D$ and $\mathbf{H}^H_{RU}$, respectively. The sum rate of all the users can be given by

$$R = \sum_{1\leq k\leq K} R_k. \tag{3.34}$$

Since the signals are *passively* reflected by RISs via controlling phase shifts of its elements, the power consumption at RISs only relies on the hardware cost which depends on the number of quantization bits b with no transmit power consumption. Therefore, the total power consumption $\mathscr{P}$ consists of the transmit power of BSs, the hardware static power consumption of each BS $P_B^{(n)}$, each element l of RIS m with b-bit phase shifts $P_R^{(m,b)}$, and each user k $P_U^{(k)}$, expressed by

$$\mathscr{P} = \sum_{n}^{N}(\omega_n p_t^{(n)} + P_B^{(n)}) + \sum_{m,l}^{M,L} P_R^{(m,b)} + \sum_{k}^{K} P_U^{(k)}, \tag{3.35}$$

where $\omega_n > 0$ is a constant reflecting the power amplifier efficiency, feeder loss and other loss factors due to power supply and cooling for BS n. $p_t^{(n)} = \mathrm{Tr}(\mathbf{V}^H_{D,n}\mathbf{V}_{D,n})$ is the transmit power at BS n, where $\mathbf{V}_{D,n}$ denote the n-th block of $\mathbf{V}_D$ which represents digital beamformer of the n-th BS.

Therefore, the energy efficiency of the RIS aided cell-free system can be expressed by [25]

$$\eta = \frac{BR}{\mathscr{P}}, \tag{3.36}$$

where B denotes the transmission bandwidth.

The energy efficiency maximization problem can be formulated as

$$\max_{\mathbf{V}_D,\mathbf{Q}} \quad \eta, \tag{3.37a}$$

$$s.t.\ p_t^{(n)} \leq P_T^{(n)}, \forall 1\leq n\leq N \tag{3.37b}$$

$$q_{m,l} = e^{j\theta_{m,l}}, \tag{3.37c}$$

$$\theta_{m,l} = \frac{i_{m,l}\pi}{2^{b-1}}, i_{m,l} \in \{0, 1, \ldots, 2^b - 1\}, \tag{3.37d}$$

where $P_T^{(n)}$ is the transmit power budget available at each BS n, the digital beamformer $\mathbf{V}_D$ and phase shift matrix $\mathbf{Q}$ are required to be optimized.

3.2.3.3 Problem Decomposition

The energy efficiency maximization problem (3.37) is difficult to solve due to the complicated interference items. Moreover, it needs to jointly optimize both the digital beamformer and the RIS-based beamformer where the latter involves a large number of discrete variables, i.e., the phase shifts of each RIS element $\theta_{m,l}$. To solve this problem efficiently, we decouple it into two subproblems as shown below.

Digital Beamforming Subproblem: Given the fixed RIS-based beamformer $\mathbf{Q}$, the digital beamforming subproblem can be formulated by

$$\max_{\mathbf{V}_D} \quad \eta, \tag{3.38a}$$

$$s.t. \; p_t^{(n)} \le P_T^{(n)}, \forall 1 \le n \le N. \tag{3.38b}$$

RIS-based Analog Beamforming Subproblem: Similarly, the RIS-based analog beamforming subproblem with fixed digital beamformer $\mathbf{V}_D$ can be written by

$$\max_{\mathbf{Q}} \quad \eta, \tag{3.39a}$$

$$s.t. \; q_{m,l} = e^{j\theta_{m,l}}, \tag{3.39b}$$

$$\theta_{m,l} = \frac{i_{m,l}\pi}{2^{b-1}}, i_{m,l} \in \{0, 1, \ldots, 2^b - 1\}. \tag{3.39c}$$

3.2.4 *Energy Efficiency Maximization Algorithm Design*

In this section, we develop an EEM algorithm to obtain a suboptimal solution of the energy efficiency maximization problem (3.37) via solving the digital beamforming subproblem and RIS-based analog beamforming subproblem in an iterative manner. Finally, we summarize the overall algorithm.

3.2.4.1 Digital Beamforming Design

For the interference management among users, without loss of generality, we consider zero-forcing (ZF) beamforming with power allocation as the digital beamformer to achieve a near-optimal solution in MIMO systems [26]. According to the results in [27], the digital beamformer with a fixed RIS based analog beamformer can be given as

$$\mathbf{V}_D = \mathbf{H}^H(\mathbf{H}\mathbf{H}^H)^{-1}\mathbf{P}^{\frac{1}{2}} = \widetilde{\mathbf{V}}_D\mathbf{P}^{\frac{1}{2}}, \tag{3.40}$$

where $\widetilde{\mathbf{V}}_D = \mathbf{H}^H(\mathbf{H}\mathbf{H}^H)^{-1}$ and $\mathbf{P} \triangleq \text{diag}(p_1, \ldots, p_K)$ is the power matrix in which the k-th diagonal element denotes the transmit power allocated to the signal intended for the k-th user from all the BSs.

Note that the ZF beamformer has the following properties:

$$\begin{aligned} |\mathbf{H}_k\mathbf{V}_{D,k}| &= \sqrt{p_k}, \\ |\mathbf{H}_k\mathbf{V}_{D,k'}| &= 0, \forall k' \neq k. \end{aligned} \tag{3.41}$$

For convenience, we assume that each BS has the same transmit power amplifier efficiency, i.e., $\omega_1 = \ldots = \omega_n = \omega$. Based on these properties of ZF beamforming, subproblem (3.38a) can be reduced to a power allocation problem which can be given by

$$\max_{\{p_k \geqslant 0\}} \quad \frac{\sum_{k=1}^K \log_2(1 + p_k\sigma^{-2})}{\omega\sum_{k=1}^K p_k + \mathscr{P}_s}, \tag{3.42a}$$

$$s.t. \ \text{Tr}(\mathbf{P}^{\frac{1}{2}}\widetilde{\mathbf{V}}_{D,n}^H\widetilde{\mathbf{V}}_{D,n}\mathbf{P}^{\frac{1}{2}}) \leq P_T^{(n)}, \forall 1 \leq n \leq N, . \tag{3.42b}$$

where $\mathscr{P}_s = \sum_n^N P_B^{(n)} + \sum_{m,l}^{M,L} P_R^{(m,b)} + \sum_k^K P_U^{(k)}$ is the hardware static power consumption which is independent of the optimization variables in problem (3.37), i.e., $\mathbf{V}_D$ and $\mathbf{Q}$. $\widetilde{\mathbf{V}}_{D,n}$ denote the n-th block of $\widetilde{\mathbf{V}}_D$.

Lemma 3.1 *By adopting Benson's transform method [28], the energy efficiency maximization problem in (3.37) is equivalent to*

$$\max_{\{p_k \geqslant 0\}} \quad \eta(\mathbf{P}, y) = 2y[\sum_{k=1}^K \log_2(1 + p_k\sigma^{-2})]^{\frac{1}{2}} - y^2(\omega\sum_{k=1}^K p_k + \mathscr{P}_s), \tag{3.43a}$$

$$s.t. \ Tr(\mathbf{P}^{\frac{1}{2}}\widetilde{\mathbf{V}}_{D,n}^H\widetilde{\mathbf{V}}_{D,n}\mathbf{P}^{\frac{1}{2}}) \leq P_T^{(n)}, \forall 1 \leq n \leq N, \tag{3.43b}$$

where the optimal y^* can be obtained based on the results in [29], expressed by

$$y^* = \frac{\sum_{k=1}^{k} \log_2(1 + p_k \sigma^{-2})}{\omega \sum_{k=1}^{K} p_k + \mathscr{P}_s} \tag{3.44}$$

As such, problem (3.42) is transformed into a convex problem of **P**, and thus, the digital beamforming subproblem can be solved by existing convex optimization techniques. The algorithm can be summarized in Algorithm 9.

Algorithm 9: Digital beamforming design algorithm

Input: RIS-based beamformer matrix $\mathbf{Q}$
Output: Digital beamformer matrix $\mathbf{V}_D$
1 Initialize $\mathbf{P}$ to a feasible value;
2 **for** *each iteration* **do**
3 | Update y by (3.44);
4 | Update $\mathbf{P}$ by solving the convex optimization problem (3.43) with fixed y.
5 **end**
6 Derive the digital beamforming matrix from the optimal power allocation solution based on (3.40).

3.2.4.2 RIS-Based Analog Beamforming Design

Note that we optimize the energy efficiency via performing the digital beamforming and RIS-based analog beamforming iteratively. Based on the designed digital beamforming as shown in (3.40), the energy efficiency in (3.42) depends on the RIS-based analog beamformer **Q** only through the power constraint (3.42b). Therefore, the RIS-based analog beamforming subproblem can be reformulated as a power minimization problem, expressed by

$$\min_{\theta_{m,l}} \quad f(\mathbf{Q}), \tag{3.45a}$$

$$s.t. \quad q_{m,l} = e^{j\theta_{m,l}}, \tag{3.45b}$$

$$\theta_{m,l} = \frac{i_{m,l}\pi}{2^{b-1}}, i_{m,l} \in \{0, 1, \ldots, 2^b - 1\}. \tag{3.45c}$$

where

$$f(\mathbf{Q}) = \mathrm{Tr}(\mathbf{V}_D^H \mathbf{V}_D). \tag{3.46}$$

However, problem (3.45a) is still difficult to solve due to the large-scale of **Q**. For simplicity, we ignore the items about direct link $\mathbf{H}_D$ in $f(\mathbf{Q})$ which is unrelated to the optimization objective **Q**. For $K = NN_a$, $f(\mathbf{Q})$ can be rewritten by

$$
\begin{aligned}
f(\mathbf{Q}) &= \mathrm{Tr}(\mathbf{V}_D^H \mathbf{V}_D) \\
&= \mathrm{Tr}[(\mathbf{P}^{-\frac{1}{2}}\mathbf{H}_{RU}^H \mathbf{Q}\mathbf{H}_{BR}\mathbf{H}_{BR}^H \mathbf{Q}^H \mathbf{H}_{RU}\mathbf{P}^{-\frac{1}{2}})^{-1}].
\end{aligned}
\tag{3.47}
$$

To obtain the optimal RIS based analog beamforming matrix $\mathbf{Q}$, the phase shift of each RIS element (m, l) is updated sequentially in an iterative manner. Therefore, we first separate the optimization variable $q_{m,l}$ from other fixed elements. For brevity, we define the index $j = (m-1)L + l$, and thus $q_{m,l}$ is abbreviated to q_j. We set the j-th diagonal element of $\mathbf{Q}$ as 0 and use $\mathbf{Q}^{(-j)}$ to denote the new matrix which is a constant. Denote $\widetilde{\mathbf{P}} = \mathbf{P}^{-\frac{1}{2}}\mathbf{H}_{RU}^H$, $\mathbf{P}^{-\frac{1}{2}}\mathbf{H}_{RU}^H\mathbf{Q}\mathbf{H}_{BR}$ in $f(\mathbf{Q})$ can be rewritten as $\mathbf{P}^{-\frac{1}{2}}\mathbf{H}_{RU}^H\mathbf{Q}\mathbf{H}_{BR} = \widetilde{\mathbf{P}}\mathbf{Q}^{(-j)}\mathbf{H}_{BR} + \widetilde{\mathbf{P}}^{(j)}q_j\mathbf{H}_{BR}^{(j)} = \mathbf{A}_j + \mathbf{B}_j q_j$, where $\mathbf{A}_j = \widetilde{\mathbf{P}}\mathbf{Q}^{(-j)}\mathbf{H}_{BR}$ and $\mathbf{B}_j = \widetilde{\mathbf{P}}^{(j)}\mathbf{H}_{BR}^{(j)}$. The j-th column and j-th row of $\widetilde{\mathbf{P}}$ and $\mathbf{H}_{BR}$ are represented by $\widetilde{\mathbf{P}}^{(j)}$ and $\mathbf{H}_{BR}^{(j)}$, respectively. Hence, $f(\mathbf{Q})$ can be rewritten as

$$
\begin{aligned}
f(\mathbf{Q}) &= \mathrm{Tr}[((\mathbf{A}_j + \mathbf{B}_j q_j)(\mathbf{A}_j^H + \mathbf{B}_j^H \overline{q}_j))^{-1}] \\
&= \mathrm{Tr}[(\mathbf{A}_j\mathbf{A}_j^H + \mathbf{A}_j\mathbf{B}_j^H\overline{q}_j + \mathbf{B}_j q_j \mathbf{A}_j^H + \mathbf{B}_j q_j \mathbf{B}_j^H \overline{q}_j)^{-1}] \\
&= \mathrm{Tr}[(\mathbf{D}_j + \mathbf{C}_j\overline{q}_j + \mathbf{C}_j^H q_j)^{-1}],
\end{aligned}
\tag{3.48}
$$

where $\mathbf{C}_j = \mathbf{A}_j\mathbf{B}_j^H$ and $\mathbf{D}_j = \mathbf{A}_j\mathbf{A}_j^H + \mathbf{B}_j\mathbf{B}_j^H$.

Note that $\mathbf{C}_j\overline{q}_j$ and $\mathbf{C}_j^H q_j$ are one-rank matrices and $\mathbf{D}_j$ is a full-rank matrix. Therefore, $f(\mathbf{Q})$ can be rewritten according to the Sherman Morrison formula [30], i.e., $(\mathbf{A}+\mathbf{B})^{-1} = \mathbf{A}^{-1} - \frac{\mathbf{A}^{-1}\mathbf{B}\mathbf{A}^{-1}}{1+\mathrm{Tr}(\mathbf{A}^{-1}\mathbf{B})}$, expressed by

$$
\begin{aligned}
f(\mathbf{Q}) &= \mathrm{Tr}\left[(\mathbf{D}_j + \mathbf{C}_j\overline{q}_j)^{-1} - \frac{(\mathbf{D}_j + \mathbf{C}_j\overline{q}_j)^{-1}\mathbf{C}_j^H q_j (\mathbf{D}_j + \mathbf{C}_j\overline{q}_j)^{-1}}{1 + \mathrm{Tr}[(\mathbf{D}_j + \mathbf{C}_j\overline{q}_j)^{-1}\mathbf{C}_j^H q_j]}\right] \\
&\overset{(a)}{=} \left(\frac{a_1 e^{3j\theta_{m,l}} + a_2 e^{2j\theta_{m,l}} + a_3 e^{j\theta_{m,l}} + a_4}{a_5 e^{3j\theta_{m,l}} + a_6 e^{2j\theta_{m,l}} + a_7 e^{j\theta_{m,l}} + a_8}\right).
\end{aligned}
\tag{3.49}
$$

where (a) is obtained by expanding $(\mathbf{D}_j + \mathbf{C}_j\overline{q}_j)^{-1}$ by adopting the Sherman Morrison formula again. The details of $a_1 \sim a_8$ are defined in Table 3.1, where $E_1 = \mathrm{eig}(\mathbf{A}_j)$, $E_2 = \mathrm{Tr}(\mathbf{D}_j^{-2}\mathbf{C}_j)$, $E_3 = \mathrm{Tr}(\mathbf{D}_j^{-1}\mathbf{C}_j)$, $E_4 = \mathrm{Tr}(\mathbf{D}_j^{-2}\mathbf{C}_j^H)$, $E_5 = \mathrm{Tr}(\mathbf{D}_j^{-2}\mathbf{C}_j\mathbf{D}_j^{-1}\mathbf{C}_j^H + \mathbf{D}_j^{-1}\mathbf{C}_j\mathbf{D}_j^{-2}\mathbf{C}_j^H)$, $E_6 = \mathrm{Tr}[(\mathbf{D}_j^{-1}\mathbf{C}_j\mathbf{D}_j^{-1})^2\mathbf{C}_j^H]$, $E_7 = \mathrm{Tr}(\mathbf{D}_j^{-1}\mathbf{C}_j\mathbf{D}_j^{-1}\mathbf{C}_j^H)$, and $E_8 = \mathrm{Tr}(\mathbf{D}_j^{-1}\mathbf{C}_j^H)$.

The minimum value of $f(\mathbf{Q})$ can be obtained with respect to $\theta_{m,l}$ which satisfies

$$
\frac{\partial f(\mathbf{Q})}{\partial \theta_{m,l}} = 0. \tag{3.50}
$$

Table 3.1 Definition of $a_1 \sim a_8$

Parameter	Value
a_1	$E_1E_8 - E_4$
a_2	$(2E_1E_3 - E_2)E_8$ $-E_1E_7 + E_5 - 2E_3E_4 + E_1$
a_3	$(E_1E_3^2 - E_2E_3)E_8+$ $(E_2 - E_1E_3)E_7 - E_6 + E_3E_5 - E_3^2E_4$
a_4	$E_1E_3^2 - E_2E_3$
a_5	E_8
a_6	$2E_3E_8 - E_7 + 1$
a_7	$E_3^2E_8 - E_3E_7 + 2E_3$
a_8	E_3^2

Algorithm 10: RIS-based analog beamforming algorithm

for *each iteration* **do**
  for $\mathscr{J} = 1 \rightarrow M \times L$ **do**
    Calculate χ according to Appendix B;
    Obtain $\theta_{m,l}^* = 2\arctan\chi$;
    Quantize $\theta_{m,l}^*$ to the nearest points in the feasible set $\mathscr{F}$;
    Set $q_{m,l} = e^{j\theta_{m,l}^*}$.
  end
end

Based on the results in Sect. 3.2.8, the optimal value of $\theta_{m,l}$ can be obtained by

$$\theta_{m,l}^* = 2\arctan\chi, \tag{3.51}$$

where χ is defined as in Sect. 3.2.8. Since $0 \leq \theta_{m,l} \leq 2\pi$, only the smaller of the two solutions of $\arctan\chi$, i.e., the solution which is less than π, is taken into account. Moreover, since only a limited number of discrete phase shifts are available at the RIS-based analog beamforming, the optimal value $\theta_{m,l}^*$ should be quantized to the nearest points in the set $\mathscr{F} = \{\frac{i_{m,l}\pi}{2^{b-1}}\}, i_{m,l} \in \{0, 1, \ldots, 2^b - 1\}$.

Therefore, starting from a randomly initiated RIS-based analog beamformer, the optimal $\mathbf{Q}$ can be obtained by sequentially updating the phase shift of each RIS element (m, l) in an iterative manner, until the algorithm converges to a local minimum of $f(\mathbf{Q})$. The algorithm is summarized in Algorithm 10.

3.2.4.3 Overall Algorithm Description

In this section, we propose the EEM algorithm to solve the energy efficiency maximization problem (3.37) in an iterative manner. We first design the digital beamforming $\mathbf{V}_D$ by Algorithm 9 with fixed RIS-based analog beamforming. Based on the achieved digital beamforming, the RIS-based analog beamforming is optimized by Algorithm 10. Let η denote the objective function of the energy efficiency maximization problem (3.37). The EEM algorithm converges if, in the ζ-th iteration, the value difference of the objective functions between two adjacent iterations is less than a predefined threshold ε, i.e., $\eta^{(\zeta)} - \eta^{(\zeta-1)} \leq \varepsilon$.

3.2.5 Theoretical Analysis of RIS Aided Cell-Free System

In this section, we first analyze the convergence and computational complexity of the proposed EEM algorithm, then the impact of the transmit power, number of RISs, and size of each RIS on the energy efficiency of the RIS aided cell-free system are explore theoretically.

3.2.5.1 Properties of the Energy Efficiency Maximization Algorithm

We now analyze the convergence, computational complexity, and feasibility of the EEM algorithm.

1. Convergence: In the ζ-th iteration, by performing Algorithm 9, a better digital beamforming $\mathbf{V}_D^{(\zeta)}$ is achieved with a fixed RIS-based analog beamforming $\mathbf{Q}^{(\zeta-1)}$. Hence, we have

$$\eta(\mathbf{V}_D^{(\zeta)}, \mathbf{Q}^{(\zeta-1)}) \geqslant \eta(\mathbf{V}_D^{(\zeta-1)}, \mathbf{Q}^{(\zeta-1)}). \tag{3.52}$$

Similarly, given digital beamforming $\mathbf{V}_D^{(\zeta)}$, the RIS-based analog beamforming is optimized to maximize the energy efficiency by Algorithm 10, expressed by

$$\eta(\mathbf{V}_D^{(\zeta)}, \mathbf{Q}^{(\zeta)}) \geqslant \eta(\mathbf{V}_D^{(\zeta)}, \mathbf{Q}^{(\zeta-1)}). \tag{3.53}$$

Based on the above inequalities, we can obtain that

$$\eta(\mathbf{V}_D^{(\zeta)}, \mathbf{Q}^{(\zeta)}) \geqslant \eta(\mathbf{V}_D^{(\zeta-1)}, \mathbf{Q}^{(\zeta-1)}). \tag{3.54}$$

This indicates that the objective value of problem (3.37) is non-decreasing after each iteration, and thus, the proposed EEM algorithm is guaranteed to converge.

2. Computational Complexity: We now analyze the computational complexity of the proposed EEM algorithm for two subproblems separately.

- Digital beamforming: According to Algorithm 9, the received power for each user should be optimized by solving a convex problem which has a polynomial complexity in the number of optimization variables, i.e., the number of users K. Therefore, its computational complexity is $\mathcal{O}(K^t)$, where $1 \leq t \leq 4$ [31].
- RIS-based analog beamforming: In each iteration, the phase shift of each RIS element (m, l) is updated sequentially to obtain the optimal $\mathbf{Q}$ according to the

closed-form expression in Algorithm 10 which is *unrelated* to the quantization bits for discrete phase shifts b. Therefore, the complexity of Algorithm 10 is $\mathcal{O}(ML)$ in each iteration while that of exhaustive search method is $\mathcal{O}(2^{bML})$.

3. Feasibility: We now analyze the feasibility of the proposed EEM algorithm.

Lemma 3.2 *The proposed scheme is feasible if* $rank(\mathbf{H}_D + \mathbf{H}_{RU}^H \mathbf{H}_{BR}) = K$.

Proof To guarantee the proposed hybrid beamforming is feasible, the (right) pseudo inverse of $\mathbf{H}$ must exist to perform zero-forcing (ZF) beamforming with power allocation as the digital beamformer, i.e., $\mathbf{V}_D = \mathbf{H}^H(\mathbf{H}\mathbf{H}^H)^{-1}\mathbf{P}^{\frac{1}{2}}$. Based on the results in [32], the channel matrix should satisfy that $rank(\mathbf{H}_D + \mathbf{H}_{RU}^H \mathbf{H}_{BR}) = K$.

3.2.5.2 Performance Analysis of RIS Aided Cell-Free System

In this part, the impact of the transmit power, number of RISs, and size of each RIS on the energy efficiency of the RIS aided cell-free system is analyzed, respectively.

1. Impact of Transmit Power: As given in the energy efficiency maximization problem (3.37), the transmit power is one of the key optimization variables. In this part, we analyze the impact of transmit power on energy efficiency in the RIS aided cell-free system.

Proposition 3.1 *The energy efficiency grows rapidly with a low transmit power budget available at each BS n, i.e.,* $P_T^{(n)}$, *and gradually flattens as* $P_T^{(n)}$ *is large enough.*

Proof Based on problem (3.42), the energy efficiency η at *high-SNR* can be rewritten as

$$\begin{aligned}\eta &\approx \frac{B\sum_{k=1}^{K}\log_2(p_k\sigma^{-2})}{\omega\sum_{k=1}^{K}p_k + \mathscr{P}_s}\\ &\overset{(a)}{\approx} \frac{B\log_2(\sum_{k=1}^{K}p_k\sigma^{-2})}{\omega\sum_{k=1}^{K}p_k + \mathscr{P}_s} = \frac{\log_2(B\sum_{k=1}^{K}p_k) - B\log_2(\sigma^2)}{\omega\sum_{k=1}^{K}p_k + \mathscr{P}_s},\end{aligned} \tag{3.55}$$

where (a) is obtained by the property of the logarithmic function [33]. Hence, the derivative of energy efficiency η with respect to $\sum_{k=1}^{K} p_k$ can be given by (3.56).

$$\begin{aligned}&\frac{\partial\eta}{\partial(\sum_{k=1}^{K}p_k)}\\ &= \frac{\frac{\omega B}{\ln 2}\mathscr{P}_s + \omega B\log_2\sigma^2\sum_{k=1}^{K}p_k + [\sum_{k=1}^{K}p_k - \sum_{k=1}^{K}p_k\ln(\sum_{k=1}^{K}p_k)]\frac{\omega B}{\ln 2}}{\mathscr{P}_s^2\sum_{k=1}^{K}p_k + 2\omega\mathscr{P}_s(\sum_{k=1}^{K}p_k)^2 + \omega^3(\sum_{k=1}^{K}p_k)^3}.\end{aligned} \tag{3.56}$$

It is easy to find that the denominator is positive. We define the molecule as $h(p_k)$, when the transmit power tends to zero, $h(p_k)$ is also positive, expressed by

$$\lim_{\sum_{k=1}^{K} p_k \to 0} h(p_k) = \frac{\omega B}{\ln 2}\mathscr{P}_s > 0. \tag{3.57}$$

The derivative of $h(p_k)$ with respect to $\sum_{k=1}^{K} p_k$ can be given by

$$h'(p_k) = -\omega B \log_2 \frac{\sum_{k=1}^{K} p_k}{\sigma^2}, \tag{3.58}$$

where $\log_2 \frac{\sum_{k=1}^{K} p_k}{\sigma^2}$ is approximately equal to the sum rate which is positive, thus, $h'(p_k) < 0$.

Therefore, the equation $\frac{\partial \eta}{\partial(\sum_{k=1}^{K} p_k)} = 0$ has an unique solution $\sum_{k=1}^{K} p_k^*$, and satisfies that

$$\frac{\partial \eta}{\partial(\sum_{k=1}^{K} p_k)} \begin{cases} > 0, & \text{if } \sum_{k=1}^{K} p_k < \sum_{k=1}^{K} p_k^* \\ < 0, & \text{if } \sum_{k=1}^{K} p_k > \sum_{k=1}^{K} p_k^*. \end{cases} \tag{3.59}$$

Hence, with a low transmit power budget available at each BS n, i.e., $P_T^{(n)}$, the energy efficiency grows with $P_T^{(n)}$. However, when $P_T^{(n)}$ is large enough to achieve $\sum_{k=1}^{K} p_k = \sum_{k=1}^{K} p_k^*$, each p_k is no longer growing as $P_T^{(n)}$ continues to increase, and thus, the energy efficiency gradually flattens. This completes the proof.

2. *Impact of The Number of RISs:* To improve the energy efficiency performance, *multiple* RISs are deployed to create favorable propagation conditions via configurable reflection from BSs to users to enhance the cell-free communication. Therefore, it is important to analyze the impact of the number of RISs on energy efficiency given a fixed total number of BSs and RISs.

By ignoring the items about direct link $\mathbf{H}_D$ which is unrelated to the number of RISs M, the transmit power allocated to the signals intended for the user k can be expressed by

$$\begin{aligned} p_k &= |\mathbf{H}_k \mathbf{V}_{D,k}|^2 = \mathbf{V}_{D,k}^H \mathbf{H}_{BR}^H \mathbf{Q}^H \mathbf{H}_{RU,k} \mathbf{H}_{RU,k}^H \mathbf{Q} \mathbf{H}_{BR} \mathbf{V}_{D,k} \\ &= \mathbf{V}_{D,k}^H \mathbf{H}_{BR}^H \mathbf{Q}^H \mathrm{Tr}(\mathbf{H}_{RU,k}^H \mathbf{H}_{RU,k}) \mathbf{Q} \mathbf{H}_{BR} \mathbf{V}_{D,k} \\ &\overset{(a)}{\approx} M^2 L^2 \mathbf{V}_{D,k}^H \mathbf{V}_{D,k}, \end{aligned} \tag{3.60}$$

where (a) is obtained by the well-known *channel hardening* effect in MIMO communication systems [34]. Specifically, as the number of receive (or transmit) antennas grows large while keeping the number of transmit (or receive) antennas constant, the column-vectors (or row-vectors) of the propagation matrix are asymptotically orthogonal. Hence, we have

$$\begin{aligned}\lim_{M\times L\to\infty} \mathbf{H}_{BR}^H\mathbf{H}_{BR} \approx (M \times L)\mathbf{I},\\ \lim_{M\times L\to\infty} \mathbf{H}_{RU}^H\mathbf{H}_{RU} \approx (M \times L)\mathbf{I},\end{aligned} \tag{3.61}$$

where $\mathbf{I}$ denotes the identity matrix. Moreover, the phase shift matrix $\mathbf{Q}$ satisfy $\mathbf{Q}\mathbf{Q}^H = \mathbf{I}$ since $\mathbf{Q}$ is a diagonal matrix where the module of each element is 1, i.e., $|q_{m,l}| = 1, \forall m, l$. Therefore, approximate formula (a) in (3.60) can be obtained. Based on (3.60), we present the following proposition.

Proposition 3.2 *When a large amount of RIS elements are deployed to assist the cell-free system, i.e., $M \times L \to \infty$, the energy efficiency decreases as the number of RISs grows.*

Proof Based on (3.42), the energy efficiency η at **high SNR** can be rewritten as

$$\begin{aligned}\eta &\approx \frac{\sum_{k=1}^K \log_2(p_k\sigma^{-2})}{\omega\sum_{k=1}^K p_k + \mathscr{P}_s}\\ &= \frac{2BK\log_2 M + B\sum_{k=1}^K log_2(L^2\mathbf{V}_{D,k}^H\mathbf{V}_{D,k}\sigma^{-2})}{\omega M^2\sum_{k=1}^K(L^2\mathbf{V}_{D,k}^H\mathbf{V}_{D,k}) + M\cdot LP_R + (N_0 - M)P_B + KP_U},\end{aligned} \tag{3.62}$$

where N_0 is the total number of BSs and RISs, i.e., $N_0 = N + M$.

Therefore, the derivative of energy efficiency η with respect to the number of RISs M can be given as

$$\frac{\partial\eta}{\partial M} = \frac{c_1c_5 + Mc_1c_4 - M\ln Mc_1c_4 - Mc_2c_4 + M^2c_1c_3 - 2M^2\ln Mc_1c_3 - 2M^2c_2c_3}{Mc_5^2 + (2M^2c_4 + 2M^3c_3)c_5 + M^3c_4^2 + 2M^4c_3c_4 + M^5c_3^2}, \tag{3.63}$$

where $c_1 = \frac{2BK}{\ln 2}$, $c_2 = B\sum_{k=1}^K \log_2(L^2\mathbf{V}_{D,k}^H\mathbf{V}_{D,k}\sigma^{-2})$, $c_3 = \omega L^2\mathrm{Tr}(\mathbf{V}_D^H\mathbf{V}_D)$, $c_4 = LP_R - P_B$, and $c_5 = KP_U + N_0P_B$. The denominator can be rewritten as $(M^{\frac{3}{2}}c_4 + M^{\frac{1}{2}}c_5 + M^{\frac{5}{2}}c_3)^2$ which is positive.

We now prove the numerator of (3.63) is negative. Through simplification, the numerator can be rewritten as (3.64),

$$\underbrace{(c_1 - c_2)(Mc_1c_4 + M^2c_3)}_{①} + \underbrace{c_1c_5 - M^2c_2c_3}_{②} + \underbrace{(-M\ln M(c_1c_4 + 2Mc_1c_3))}_{③}, \tag{3.64}$$

where term ③ is obvious non-positive. The valuence of terms ① and ② are proved as follows.

Proof of ① < 0: The term ① can be negative if $c_1 < c_2$. According to Jensens inequality [35], we have

$$\log_2(L^2\mathrm{Tr}(\mathbf{V}_D^H\mathbf{V}_D)\sigma^{-2}) - \sum_{k=1}^{K}\log_2(L^2\mathbf{V}_{D,k}^H\mathbf{V}_{D,k}\sigma^{-2}) \leq K\log_2 K. \tag{3.65}$$

Consider a high-SNR case, $K\log_2 K$ is small enough to be neglected compared to the terms in the left side which are equivalent to the sum rate in a cell-free system with one RIS. Hence, c_2 can be rewritten as

$$c_2 = B\sum_{k=1}^{K}\log_2(L^2\mathbf{V}_{D,k}^H\mathbf{V}_{D,k}\sigma^{-2}) \approx B\log_2(L^2\mathrm{Tr}(\mathbf{V}_D^H\mathbf{V}_D)\sigma^{-2}). \tag{3.66}$$

Let $c_1 < c_2$, we have

$$\frac{2BK}{\ln 2} < BK\log_2\left[\frac{L^2\mathrm{Tr}(\mathbf{V}_D^H\mathbf{V}_D)}{\sigma^2}\right] \tag{3.67}$$

Through simplification, it can be given as

$$\frac{L^2\mathrm{Tr}(\mathbf{V}_D^H\mathbf{V}_D)}{\sigma^2} > e^2 \tag{3.68}$$

where e is Euler number, $\mathrm{Tr}(\mathbf{V}_D^H\mathbf{V}_D)\sigma^{-2}$ can be considered as a value greater than 1 when the SNR is high. Since each RIS relies on the combination of multiple programmable radiating elements to realize signal reflection, in general, the number of elements of each RIS L is larger than e. Therefore, $c_1 < c_2$ holds, and thus, term ① is negative.

Proof of ② < 0: Based on the assumption in Proposition 3.2, i.e., $M \times L \to \infty$, we have

$$KP_U + NP_B < \omega M^2L^2\mathrm{Tr}(\mathbf{V}_D^H\mathbf{V}_D). \tag{3.69}$$

Specifically, $c_5 < M^2c_3$ holds. Multiply both sides by c_1, we have $c_1c_5 < M^2c_1c_3$. Since $M^2c_1c_3 < M^2c_2c_3$, we have $c_1c_5 < M^2c_2c_3$. Therefore, term ② is negative.

3. Impact of RIS Size: Each RIS relies on the combination of multiple programmable radiating elements to realize a desired transformation on the transmitted, received, or reflected waves [17]. We now reveal the impact of the size of each RIS, i.e., the number of elements of each RIS L, on the energy efficiency performance.

Lemma 3.3 *The energy efficiency of the RIS aided cell-free system tends to zero when size of each RIS tends to infinity, i.e.,* $\lim_{L\to\infty}\eta = 0$.

Proof When the size of each RIS tends to infinity, i.e., $L \to \infty$, based on (3.42) and (3.60), the energy efficiency η at **high SNR** can be rewritten as (3.70).

$$\eta \approx \frac{\sum_{k=1}^{K} \log_2(p_k\sigma^{-2})}{\omega \sum_{k=1}^{K} p_k + \mathscr{P}_s} \approx \frac{2BK \log_2 L + B \sum_{k=1}^{K} log_2(M^2 \mathbf{V}_{D,k}^H \mathbf{V}_{D,k}\sigma^{-2})}{\omega L^2 \sum_{k=1}^{K} (M^2 \mathbf{V}_{D,k}^H \mathbf{V}_{D,k}) + L \cdot M P_R + N P_B + K P_U}. \tag{3.70}$$

It is obvious that the energy efficiency tends to zero when size of each RIS tends to infinity, i.e., $\lim_{L \to \infty} \eta = 0$.

Based on Lemma 3.3, we have the following proposition.

Proposition 3.3 *As the size of each RIS L increases, the energy efficiency of RIS aided cell-free system increases first and then gradually drops to zero if* $\frac{\mathscr{P}_s}{M \cdot P_R} > \ln \frac{N P_T}{k\sigma^2}$.

Proof When the size of each RIS tends to 1, i.e., $L \to 1$, the SNR at user k can be expressed by

$$\begin{aligned} SNR_k &= \frac{|\mathbf{H}_k \mathbf{V}_{D,k}|^2}{\sum_{k=1}^{K} |\mathbf{H}_k \mathbf{V}_{D,k}|^2 + \sigma^2} \\ &\overset{(a)}{=} |\mathbf{H}_k \mathbf{V}_{D,k}|^2/\sigma^2 = \sum_{m=1}^{M} \sum_{l=1}^{L} \mathbf{h}_{m,l,k} \mathbf{h}_{m,l,k}^H \mathbf{V}_{D,k}^H \mathbf{V}_{D,k}/\sigma^2 \\ &= ML\mathbb{E}[\mathbf{h}_{m,l,k} \mathbf{h}_{m,l,k}^H] \mathbf{V}_{D,k}^H \mathbf{V}_{D,k}/\sigma^2 = MLa_k, \end{aligned} \tag{3.71}$$

where $a_k = \mathbb{E}[\mathbf{h}_{m,l,k} \mathbf{h}_{m,l,k}^H] \mathbf{V}_{D,k}^H \mathbf{V}_{D,k}/\sigma^2$ and $\mathbf{h}_{m,l,k}$ is the channel between the BSs to user k via the l-th element of RIS m. (a) is obtained by the properties of ZF beamformer as given in (3.41).

According to the property of the logarithmic function [33], we have $\mathbb{E}[\log_2(1 + SNR_k)] \approx \log_2(1 + \mathbb{E}[SNR_k])$, the energy efficiency can be rewritten as

$$\begin{aligned} \eta = \frac{BR}{\mathscr{P}} &= \frac{B \sum_{k=1}^{K} \log_2(1 + MLa_k)}{\omega \sum_{k=1}^{K} p_k + MLP_R + kP_U + NP_B} \\ &\approx \frac{B \log_2(1 + \frac{\sum_{k=1}^{K} a_k}{K})}{\omega \sum_{k=1}^{K} p_k + MLP_R + kP_U + NP_B}. \end{aligned} \tag{3.72}$$

Therefore, the derivative of energy efficiency η with respect to the size of each RIS L can be given as

$$\frac{\partial \eta}{\partial L}=\frac{\frac{B}{\ln 2}(\omega\sum_{k=1}^{K}p_k+KP_U+NP_B)+[(L-L\ln L)\frac{B}{\ln 2}-LB\log_2(\frac{M}{K}\omega\sum_{k=1}^{K}a_k)]MP_R}{L(\omega\sum_{k=1}^{K}p_k+KP_U+NP_B)^2+2L^2MP_R(\omega\sum_{k=1}^{K}p_k+KP_U+NP_B)+L^3M^2P_R^2}, \tag{3.73}$$

where the denominator is guaranteed to be positive.

We define the numerator as $g(L)$, let $g(L) > 0$, we have

$$\omega\sum_{k=1}^{K}p_k+\mathscr{P}_s > \ln\left(\frac{M}{K}\sum_{k=1}^{K}a_k\right)MP_R \tag{3.74}$$

where the left side is not less than $\mathscr{P}_s$ and the right side is not larger than $\ln\left(\frac{NP_T}{\sigma^2}\right)\cdot MP_R$. Therefore, when the RIS aided cell-free system satisfies that $\frac{\mathscr{P}_s}{M\cdot P_R} > \ln\frac{NP_T}{k\sigma^2}$, we have $g(L) > 0$.

Based on Lemma 3.3, the energy efficiency of RIS aided cell-free system increases when the size of each RIS L is small, and then gradually drops to zero if $\frac{\mathscr{P}_s}{M\cdot P_R} > \ln\frac{NP_T}{k\sigma^2}$.

This completes the proof.

3.2.6 *Simulation Results*

In this section, we evaluate our proposed EEM algorithm for the RIS aided cell-free system in terms of energy efficiency. We show how the energy efficiency is influenced by the transmit power per BS, the number of RISs, the size of each RISs, and the number of quantization bits for discrete phase shifts. For comparison, the following schemes are considered as benchmarks.

1. *Conventional distributed antenna system [36]*: Multiple distributed antennas are deployed remotely rather than centrally at the BSs. For a fair comparison, we assume that the distributed antennas in the DAS and the RIS elements in the proposed scheme have the same number and locations since each RIS element can be viewed as an antenna far away from BSs.
2. *No-RIS case*: Conventional user-centric wireless communication networks without RISs. K single-antenna users are served by N BSs simultaneously.
3. *Conventional cell-free system [12, 13]*: Conventional cell-free communication networks without RISs. Different from the no-RIS case, we replace all RISs by BSs, i.e., the number of BSs is set as $N + M$.
4. *Random Phase Shift Algorithm*: The phase shift of each RIS element is selected randomly from the feasible set $\mathscr{F}$.
5. *Exhaustive Search Algorithm*: We consider the exhaustive search as an upper bound to evaluate the energy efficiency performance of the EEM algorithm where the phase shifts of each RIS element are selected by traversing the feasible set $\mathscr{F}$.

Table 3.2 Simulation parameters

Parameter	Value
Rician factor κ	4
Number of BSs N	4
Number of antennas of each BS N_a	8
Number of users K	8
Noise power σ^2	−90 dBm
Carrier frequency	5.9 GHz
Size of each RIS element	0.02 m
Hardware static power consumption of each BS	10 dBw
Hardware static power consumption of each RIS element with 1-bit phase shifts	5 dBm
Hardware static power consumption of each RIS element with 2-bit phase shifts	10 dBm
Hardware static power consumption of each RIS element with 3-bit phase shifts	15 dBm
Hardware static power consumption of each RIS element with continuous phase shifts	25 dBm
Hardware static power consumption of each distributed antenna in DAS	20 dBm
Hardware static power consumption of each user	10 dBm
Thresholds ϱ and ε	0.001

For the propagations, we use the UMa path loss model in [37] as the distance-dependent channel path loss model. BSs and RISs are uniformly distributed in circles around the users within a radius of 150 m. Users are uniformly distributed in the circle of radius 20 m. The distance between two adjacent antennas at the BS is 1 m. For small-scale fading, we assume the Rician fading channel model for all channels involved. Simulation parameters are set up based on the existing works [20, 38], as given in Table 3.2.

Figure 3.11 shows the energy efficiency of the RIS aided cell-free MIMO system versus the transmit power per BS P_T, obtained by different algorithms with 3 RISs (i.e., $M = 3$). Each RIS consists of 64 elements (i.e., $L = 64$), the number of quantization bits for discrete phase shifts b is set as 3. We observe that the energy efficiency of different systems grows rapidly with a low transmit power per BS P_T and gradually flattens as P_T continues to increase, as proved in Proposition 3.1.

It can be seen that the proposed EEM algorithm achieves a better energy efficiency performance than the random phase shift algorithm and performs very close to that of the exhaustive search algorithm. Compared with the no-RIS case, we observe that these RISs deployed in the cell-free system can effectively improve the energy efficiency of the multi-user communication system, even with random-phase-shift RISs. Compared with the conventional DAS, it can be observed that the energy efficiency performance is improved in the RIS aided cell-free system since signals are transmitted via a larger number of independent paths (i.e., the direct and reflected links), implying that the spatial resources are better utilized. However, with

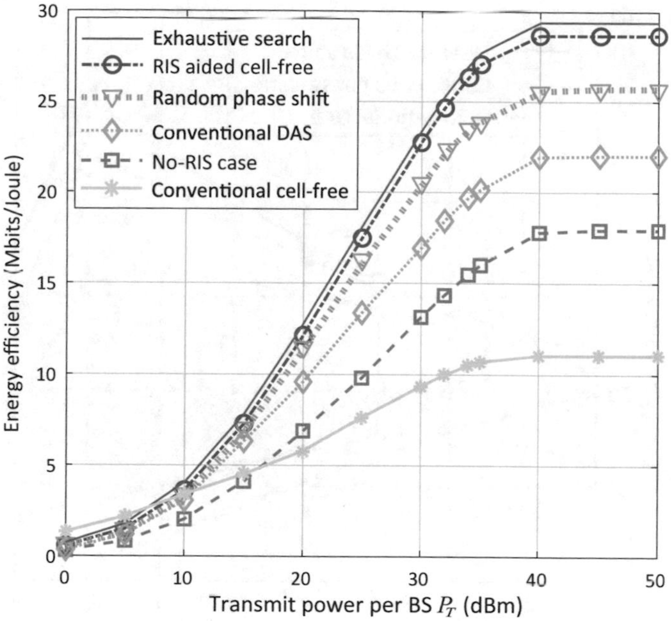

Fig. 3.11 Energy efficiency vs. transmit power per BS of different schemes with $M = 3$, $L = 64$, and $b = 3$

a low transmit power per BS P_T, the energy efficiency of conventional cell-free systems is higher than that of the proposed scheme. This is because conventional cell-free systems provide more transmit power by more BSs, thereby achieving a higher sum rate which is the main impact factor of the energy efficiency when the total transmit power budget available at all BSs is very low. When the transmit power per BS is larger than 10 dBm, the hardware static power consumption becomes one of the dominant items of energy efficiency, and thus, our proposed system can achieve a better energy efficiency performance than the conventional cell-free systems.

To fully explore how RISs as cost-efficient devices influence the energy efficiency of the cell-free system, we evaluate the performance with different number of RISs given a fixed total number of BSs and RISs, i.e., $N_0 = N + M$ is a constant. Specifically, Fig. 3.12 shows the energy efficiency vs. the number of RISs M with $N_0 = 10$, $P_T = 30$ dBm, $L = 64$, and different number of quantization bits b, and the conventional DAS is considered as the benchmark. As the quantization bits of RISs b increases, the hardware static power consumption of each RIS increases, as given in Table 3.2.

It can be seen that the energy efficiency increases with the number RISs M for a small-scale deployment of RISs. This figure also implies that both the slope and value of energy efficiency vs. the number of deployed RISs decrease with the quantization bits of RISs b since the higher resolution of RISs brings a significantly

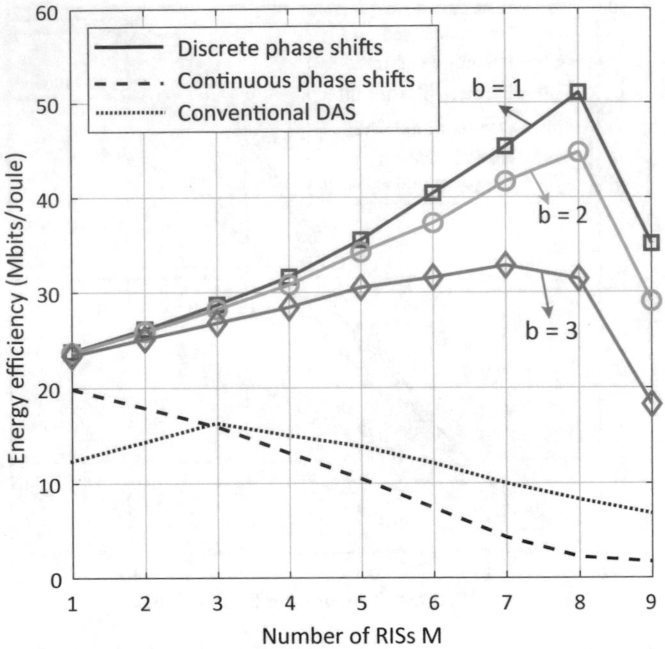

Fig. 3.12 Energy efficiency vs. number of RISs M for different number of quantization bits b ($P_T = 30\,\text{dBm}$)

higher hardware static power consumption. As the number of RISs M continues to increase, i.e., $M \times L \rightarrow \infty$, the energy efficiency decreases as proved in Proposition 3.2. The optimal number of deployed RISs M shows up at a smaller value with a higher level quantization bits of RISs b, especially with the continuous phase shifts, due to the higher energy consumption. For the continuous case, the energy efficiency decreases when more BSs are replaced by RISs due to its high hardware static power consumption and passive character.

Compared with the conventional DAS, it can be observed that the proposed scheme achieves a better energy efficiency performance with different RISs, and the optimal M shows up at a larger value. This is because that signals are transmitted via both the direct and reflected links, implying that the spatial resources are better utilized. Moreover, the lower hardware static power consumption of RISs brings another dimension to enhance the energy efficiency performance.

Figure 3.13 shows the impact of the RIS size L on the energy efficiency performance with $P_T = 30\,\text{dBm}$ and $b = 3$. It can be seen that the RIS aided cell-free system achieves a better energy efficiency performance than the conventional DAS with different size of RISs. We can also observe that as the number of elements of each RIS increases, the energy efficiency first increases and then decreases since the hardware static power consumption of each RIS increases with the number of elements in each RIS L, which verifies our theoretical analysis in the

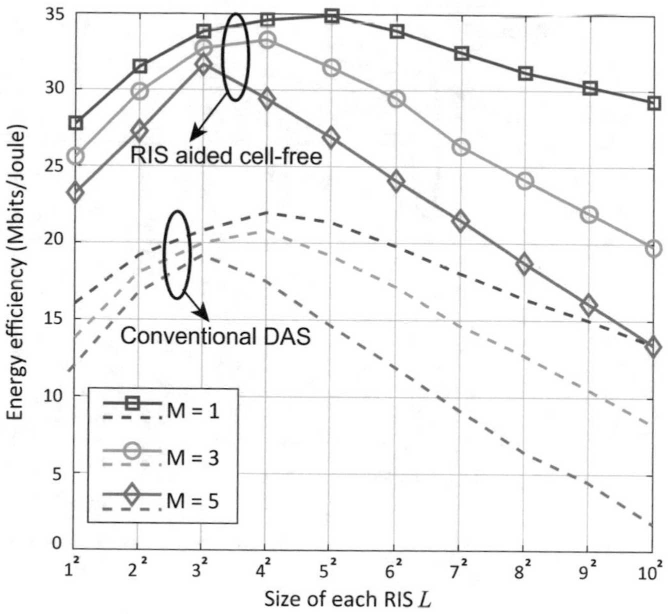

Fig. 3.13 Energy efficiency vs. size of RISs L for different number of RISs M ($P_T = 30\,\text{dBm}$)

Proposition 3.3. The optimal RIS size L shows up at a smaller value when more RISs are deployed in the cell-free system.

Figure 3.14 depicts the sum rate versus the size of each RIS L with different number of RISs in the cell-free system. The transmit power per BS P_T is set as 30 dBm and the number of quantization bits for discrete phase shift b is set as 3. As the number of elements of each RIS increases, the sum rate grows and gradually converges to a stable value. Moreover, when the number of RISs grows, the gap between the curves obtained with different size of each RIS shrinks since a larger scale of RIS deployment usually provides more freedom of generating directional beams.

Figure 3.15 shows the energy efficiency vs. number of iterations for different sizes of each RIS. The transmit power per BS P_T is set as 30 dBm and the number of quantization bits for discrete phase shifts b is set as 3. It can be observed that the convergence speed slows down when the size of each RIS L grows since more variables need to be optimized. We observe that the algorithm can converge within 12 iterations for most cases. Therefore, the convergence analysis provided in Sect. 3.2.5.1 is verified, and the complexity is acceptable.

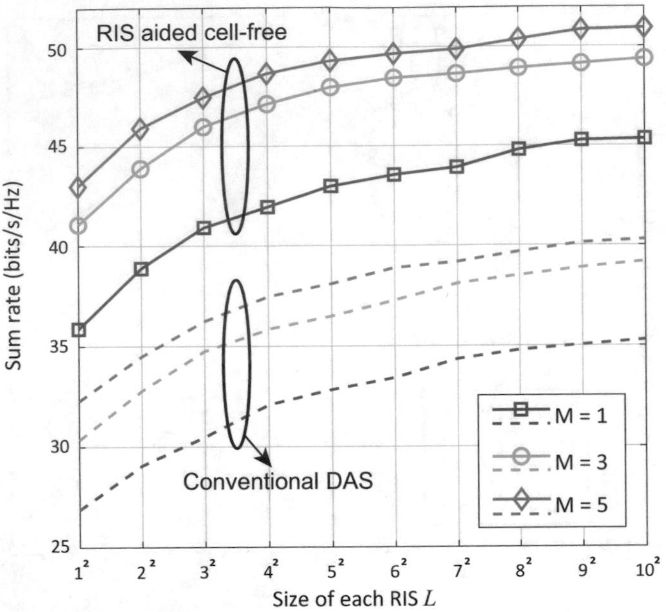

Fig. 3.14 Energy efficiency vs. size of RISs L for different number of RISs M ($P_T = 30\,\text{dBm}$)

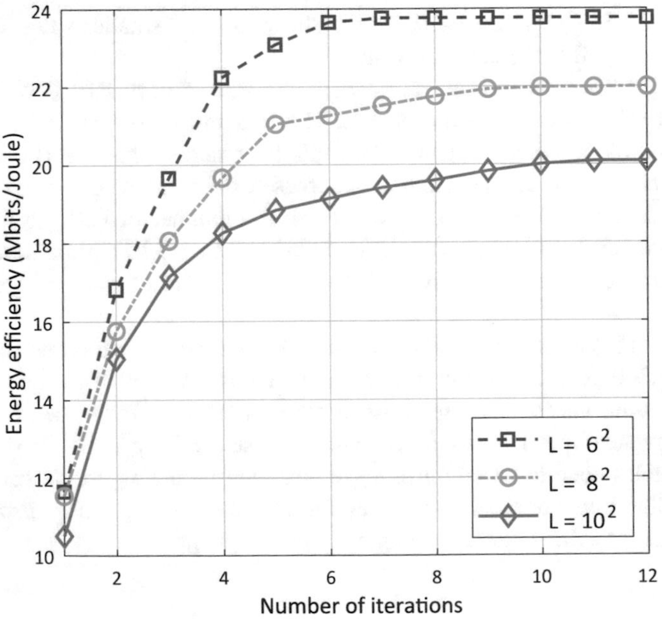

Fig. 3.15 Number of iterations to converge given different RIS sizes ($P_T = 30\,\text{dBm}$)

3.2.7 Summary

In this section, we have considered an RIS aided cell-free MIMO system where several BSs are coordinated to serve various users with the assisting of multiple RISs. To maximize the energy efficiency, we have proposed an HBF scheme where the digital beamforming and the RIS-based analog beamforming are performed at BSs and RISs, respectively. The energy efficiency maximization problem has been formulated and solved by our proposed EEM algorithm in an iterative manner. Simulation results show that, with the same transmit power in the entire system, the RIS aided cell-free system achieves better energy efficiency performance compared with the conventional ones.

Three remarks can be drawn from both *theoretical analysis* and *numerical results*, providing insights for the design of RIS aided cell-free systems.

- The energy efficiency increases rapidly with the transmit power of each BS P_T and gradually flattens when P_T is large enough.
- The energy efficiency increases with the number RISs M for a small-scale deployment of RISs, and then decreases as M continues to increase, indicating that there exists an optimal number of deployed RISs to maximize the energy efficiency.
- With the size of each RIS increases, the energy efficiency first increases and then decreases while the sum rate grows and gradually converges to a stable value, indicating that there exists a trade-off between energy efficiency and sum rate determined by the size of each RIS.

3.2.8 Appendix

Let $\frac{\partial f(\mathbf{Q})}{\partial \theta_{m,l}} = 0$, through simplification, we have

$$b_1 e^{2j\theta_{m,l}} + b_2 e^{j\theta_{m,l}} + b_3 + b_4 e^{-j\theta_{m,l}} + b_5 e^{-2j\theta_{m,l}} = 0, \tag{3.75}$$

where $b_1 = j(a_1a_6 - a_2a_5)$, $b_2 = 2j(a_1a_7 - a_3a_5)$, $b_3 = j(3a_1a_8 + a_2a_7 - a_3a_6 - 3a_4a_5)$, $b_4 = 2j(a_2a_8 - a_4a_6)$, and $b_5 = j(a_3a_8 - a_4a_7)$. According to the Euler's formula and the auxiliary Angle formula, this equation can be rewritten as

$$\begin{aligned} &\sqrt{\mathrm{Re}\{b1 + b5\}^2 + \mathrm{Im}\{b5 - b1\}^2} \sin(2\theta_{m,l} + \gamma_1) \\ &+ \sqrt{\mathrm{Re}\{b2 + b4\}^2 + \mathrm{Im}\{b4 - b2\}^2} \sin(\theta_{m,l} + \gamma_2) = 0. \end{aligned} \tag{3.76}$$

Here, $\tan \gamma_1 = \frac{\mathrm{Im}\{b5-b1\}}{\mathrm{Re}\{b1+b5\}}$ and $\tan \gamma_2 = \frac{\mathrm{Im}\{b4-b2\}}{\mathrm{Re}\{b2+b4\}}$. After expanding the expression to the left of the equals by $\sin \alpha = \frac{2\tan\frac{\alpha}{2}}{1+\tan^2 \alpha}$ and $\sin(\alpha + \beta) = \sin \alpha \cos \beta +$

$\cos\alpha \sin\beta$, (3.76) can be rewritten as a quartic equation of one unknown, i.e., $\tan\frac{\theta_{m,l}}{2}$, which can be easily solved. The solution of the equation is denoted by χ.

3.3 RIS Aided Spatial Equalization

3.3.1 Motivations

An increasing number of mobile devices in the past decade has triggered urgent needs for high-speed data services in future wireless communication systems. Although various technologies have been developed to strengthen target signals such as relay and MIMO systems, network operators have been continuously struggling to build wireless networks that can guarantee to provide high QoS in presence of harsh wireless propagation environments due to uncontrollable interactions of transmitted waves with surrounding objects and their destructive interference at receivers [39]. Fortunately, recent developments of RISs have given a rise to a new opportunity to enable the control of wireless propagation environments [40]. This can be achieved by controlling phase shifts of impinging radios at the RIS such that incident signals can be reflected towards to those intended receivers [16].

In this section, we propose to exploit the potential of the RIS as a spatial equalizer to address the multi-path fading issue. To be specific, we consider a downlink multi-user MISO communication system, where some controllable paths are introduced against the multi-path fading effect through the RIS. Different from traditional communication systems, where equalization can only be done at receivers, the proposed scheme can achieve equalization in the transmission process, and thus multiple users can share the same RIS which is more cost-effective. However, since the objective of the spatial equalizer is to reduce the inter-symbol interference (ISI), the phase shift design of the RIS for beamforming applications cannot be applied directly. To this end, we formulate the ISI minimization problem by optimizing the phase shifts at the RIS and propose an algorithm to solve this problem efficiently. Simulation results verify the effectiveness of the RIS based spatial equalizer, and how the size of the RIS impacts the performance is also discussed.

The rest of this section is organized as follows. In Sect. 3.3.2, we introduce the system model for the RIS-assisted spatial equalization. In Sect. 3.3.3, an ISI minimization problem is formulated and a PSO algorithm is proposed in Sect. 3.3.4. The simulation results are given in Sect. 3.3.5 to evaluate the energy efficiency and sum rate performance to validate our analysis. Finally, we summarize this section in Sect. 3.3.6.

3.3.2 System Model

As shown in Fig. 3.16, we consider a downlink multi-user RIS-assisted MISO communication network consisting of one base station (BS) with M antennas and K single-antenna users [41], denoted by $\mathcal{K} = \{1, \dots, K\}$. To reduce the ISI, an RIS is deployed as a spatial equalizer. The RIS is composed of N electrically controllable elements with the size length being a, denoted by $\mathcal{N} = 1, \dots, N$. Each element can adjust its phase shift by switching Positive-Intrinsic-Negative (PIN) diodes between "ON" and "OFF" states. Due to some physical limitations, the state transition for each PIN diode may cost some time. In this paper, within a considered period, we assume that the phase shift for each element is fixed. Define θ_n as the phase shift for element n, and the reflection factor of element n can be written by $\Gamma_n = \Gamma e^{-j\theta_n}$, where $\Gamma \in [0, 1]$ is a constant.

For each user, it can receive two rays of signals. The first ray is the direct link from the BS, which consists of the scattered signals from the environment. We define $g_k^D(t)$ as the channel impulse response of the direct link from the BS to user k, which models independent fast fading and path loss. To be specific, $g_k^D(t)$ can be written as

$$g_k^D(t) = (\beta_k^D)^{1/2} h_k^D(t), \tag{3.77}$$

where $h_k^D(t)$ is the fast fading coefficient caused by the multi-path effect and β_k^D is the path loss related to distance d_k between the BS and user k, i.e., $\beta_k^D = G d_k^{-\alpha}$. Here, G is a normalized factor for the direct link and α is the path loss exponent.

The second ray is the reflection link through the RIS. Each RIS element will reflect the incident signals from the BS to users to eliminate the multi-path effect.

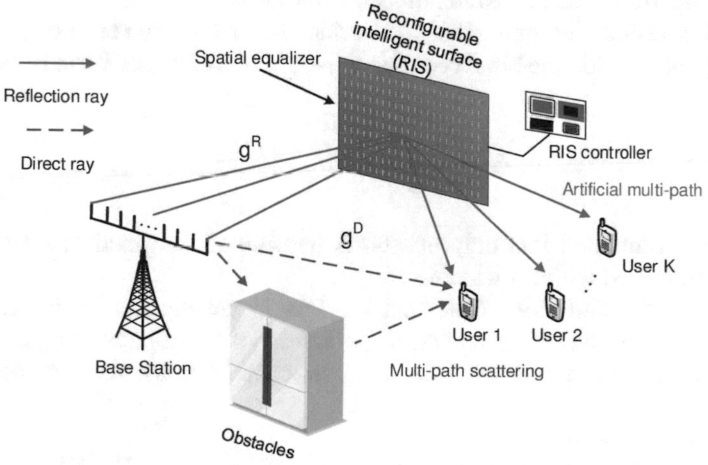

Fig. 3.16 System model for the RIS-assisted spatial equalization

We define $g^R_{n,k}(t)$ as the channel impulse response of the reflection link through RIS element n to user k, which also includes independent fast fading and path loss. Specifically, $g^R_{n,k}(t)$ can be written as

$$g^D_{n,k}(t) = (\beta^R_{n,k})^{1/2}\Gamma_n h^R_{n,k}(t), \tag{3.78}$$

where $h^R_{n,k}(t)$ is the fast fading coefficient and $\beta^R_{n,k}$ is the path loss related to distance l_n between the BS and the n-th RIS element, and distance $l_{n,k}$ between the n-th RIS element and user k. According to the result in [21], we have $\beta^R_{n,k} = G'(l_n l_{n,k})^{-\alpha}$ where G' is a normalized factor for the reflection link. It is worthwhile to point out that we can approximate the distance to different RIS elements as the distance to the center of the RIS when $l_n, l_{n,k} \gg a$ [42]. Therefore, we have $\beta^R_{n,k} \approx \tilde{\beta}^R_k, \forall n \in \mathcal{N}$, where $\tilde{\beta}^R_k$ is the path loss of the link going through the center of the RIS.

Define one-bit signal for user k as $s_k(t)$, and the received signal at user k can be written as[1]

$$y_k(t) = \left(g^D_k(t) + \sum_{n\in\mathcal{N}} g^D_{n,k}(t) \right) * s_k(t), \tag{3.79}$$

where $*$ is the convolution operator.

3.3.3 Problem Formulation

The objective of this paper is to reduce ISI through the RIS-based spatial equalizer. In the following, we will first introduce how to extract ISI using the peak distortion analysis and formulate the ISI minimization problem.

ISI Extraction: Assuming that $y_k(t)$ achieve its maximum at $t = 0$ and T is the sampling interval for one bit. According to [43], the ISI for user k can be written as

$$I_k = \sum_{i=-\infty, i\neq 0}^{\infty} y_k(t - iT)|_{t=0}, \tag{3.80}$$

under the assumption that only one bit is transmitted. In practice, we will only considered the ISI within a window.

Problem Formulation: Note that the RIS is not equipped with any delay components and thus cannot control the spread of multi-paths. In practice, we will select a boundary which includes most significant ISI for the ease of ISI

[1] In this paper, we assume that perfect beamforming is done by the BS, and thus the interference between different users are neglected here. The beamformer design is not studied due to the length limit.

calculation. Therefore, the objective of the spatial equalizer is to reduce the energy of remaining ISI within the considered boundary after equalization. In consideration of the fairness, we will minimize the maximum power of ISI among these users by adjusting phase shifts at the RIS. Mathematically, the optimization problem can be written as

$$(P1): \quad \min_{\{\theta_n\},\eta} \ \eta, \tag{3.81a}$$

$$s.t. \ I_k I_k^* \leq \eta, \forall k \in \mathcal{K}, \tag{3.81b}$$

$$\eta \geq 0, \tag{3.81c}$$

where η is the maximum power of ISI among these users, and I_k^* is the conjugate of I_k.

3.3.4 Algorithm Design

In this section, we will propose a phase shift optimization (PSO) algorithm to solve problem (P1) efficiently. Define $\mathcal{F}(\cdot)$ as the Fourier transformation operator. Let $H_k^D(\omega) = \mathcal{F}(g_k^D(t))$, $H_{n,k}^R(\omega) = \mathcal{F}(g_{n,k}^R(t))$, $S_k(\omega) = \mathcal{F}(s_k(t))$, and $Y_k(\omega) = \mathcal{F}(y_k(t))$. With these notations, we have

$$Y_k(\omega) = \left(H_k^D(\omega) + \sum_{n\in\mathcal{N}} H_{n,k}^R(\omega) \right) S_k(\omega). \tag{3.82}$$

According to the definition of the Fourier transformation, we have

$$Y_k(0) = \int_{-\infty}^{\infty} y_k(t)dt \approx (y_k(0) + I_k)T. \tag{3.83}$$

Therefore, we can have the following equation [44]:

$$I_k = \frac{Y_k(0)}{T} - y_k(0). \tag{3.84}$$

Note that phase shifts of the RIS will not affect $y_k(0)$ since the transmission delay through the RIS is typically longer than the direct one. Motivated by this observation, we optimize $Y_k(0)$ by tuning phase shifts of the RIS. In the following, we will elaborate on how to find the optimal phase shifts.

Given $y_k(0)$, optimization problem (P1) can be solved by the Lagrange-Dual technique. Let μ_k be the Lagrange multiplier corresponding to the ISI constraint for user k, the Lagrangian can be written as

$$L(\theta_n, \eta, \mu_k) = \eta + \sum_{k \in \mathcal{K}} \mu_k \left(\left| \frac{Y_k(0)}{T} - y_k(0) \right|^2 - \eta \right), \tag{3.85}$$

and the dual problem can be written as

$$\max_{\mu_k, \nu_k \geq 0} \min_{\theta_n, \eta} L(\theta_n, \eta, \mu_k). \tag{3.86}$$

The problem can be solved by gradient based method [45]. In the l-th iteration, primal and dual problems are solved in the following way:

Primal Problem: In the primal problem, we solve θ_n and η given the value of μ_k. To be specific, we have

$$\eta^{l+1} = [\eta^l - \delta_\eta \nabla_\eta^l L(\theta_n^l, \eta^l, \mu_k^l)]^+, \tag{3.87}$$

$$\theta_n^{l+1} = \theta_n^l - \delta_\theta \nabla_{\theta_n}^l L(\theta_n^l, \eta^l, \mu_k^l), \tag{3.88}$$

where $[a]^+ = \max\{0, a\}$, δ_η and δ_θ are step sizes of η and θ_n, respectively. Here, the gradients can be calculated by

$$\nabla_\eta^l L(\theta_n^l, \eta^l, \mu_k^l) = 1 - \sum_{k \in \mathcal{K}} \mu_k, \tag{3.89}$$

$$\nabla_{\theta_n}^l L(\theta_n^l, \eta^l, \mu_k^l) = 2 \sum_{k \in \mathcal{K}} \frac{\mu_k}{T^2} \Big(A_{k,n} B_{k,n}^* j e^{j\theta_n^l} - B_{k,n} A_{k,n}^* j e^{-j\theta_n^l} + y_k^*(0) T B_{k,n} j e^{-j\theta_n^l}$$
$$- y_k(0) T B_{k,n}^* j e^{j\theta_n^l} \Big), \tag{3.90}$$

where $Y_k(0) = A_{k,n} + B_{k,n} e^{-j\theta_n}$. The detailed proof of (3.90) is given in the Appendix.

Dual Problem: In the dual problem, we fix the results θ_n and η, and solve the dual variable μ_k. According to [45], μ_k can be updated in the following way:

$$\mu_k^{l+1} = \left[\mu_k^l + \delta_\mu \left(\left| \frac{Y_k^{l+1}(0)}{T} - y_k(0) \right|^2 - \eta^{l+1} \right) \right]^+, k \in \mathcal{K}, \tag{3.91}$$

where $Y_k^{l+1}(0)$ is obtained by θ_n^{l+1} and δ_μ is a step size of μ_k.

The PSO algorithm can be summarized as the flowchart given in Fig. 3.17. In each iteration, we use the primal-dual gradient method to obtain phase shifts θ_n and the maximum power of ISI η for all users. The termination condition is that the difference of the values of the objective for two successive iterations is less than a predefined threshold σ. It is worthwhile to point out that the obtained solution is local-optimal since the original problem is non-convex. The complexity of the

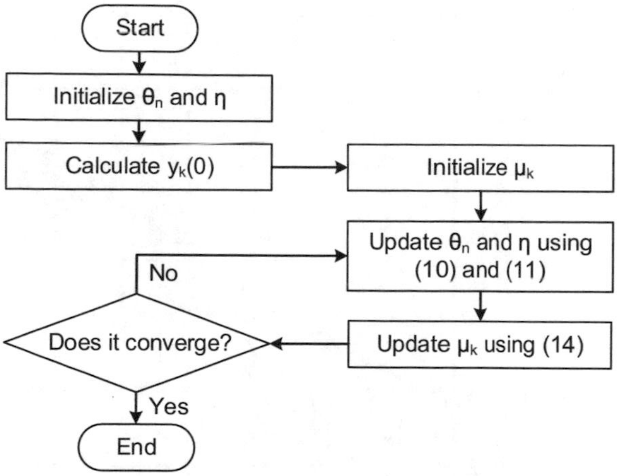

Fig. 3.17 Flowchart of the proposed PSO algorithm

proposed PSO algorithm should be $\mathscr{O}(\sqrt{K}\log(1/\sigma))$ [45]. This implies that we can adjust the complexity according to the requirements of applications by tuning σ.

Remark 3.1 When we neglect the fading and the number of RIS elements is even, the RIS filter can achieve at least the same performance with the one without the RIS in terms of the maximum ISI power. We can achieve this by setting phase shifts of two adjacent elements as 0 and π, respectively.

3.3.5 Simulation Results

In this section, we evaluate the performance of the proposed PSO algorithm. The parameters are selected according to the 3GPP standard [37] and existing work [42]. The height of the BS is 25 m. We set the number of users $K = 4$ and the number of antennas $M = 10$. The users are uniformly located in a square area whose side length is set as 100 m, and distance from the center of this area to the BS is 100 m. The RIS is placed in parallel to the direction from the BS to the center of the user area, and the horizontal distance from the BS to the RIS is 100 m. We assume that the distance between the center of the RIS and the projected point of the BS at the RIS-plane is $D = 50$ m. The center of the RIS is located at the middle between the BS and the square area with the height being 25 m. The carrier frequency is set as 5.9 GHz, the size length of an RIS element is set as $a = 0.02$ m, and the number of RIS elements is set as $N = 100$. We also assume that the RIS is fully reflected, i.e., $\Gamma = 1$. For the channel model, the path loss exponent is set as $\alpha = 2$. The normalized factor $G = G' = -43$ dB. The stochastic model in [46] is used to capture the multi-path effect. For the direct ray, we assume that there exist L paths and each RIS element corresponds to a reflection path. The sampling interval is set

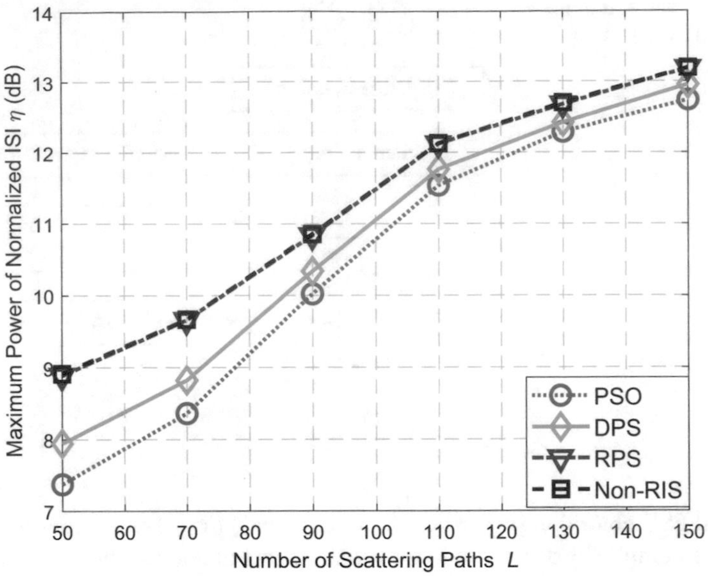

Fig. 3.18 Maximum power of normalized ISI η for different number of scattering paths L

$T = 1$ ms. We set convergence threshold $\delta = 0.01$. All numeral results are obtained by 300 Monte Carlo simulations.

In comparison, we also present the performance of the following schemes: (1) Random phase shift (RPS) scheme: the phase shift for each RIS element is selected randomly; (2) Discrete phase shift (DPS) scheme: the phase shift for each RIS element is discrete, i.e., 2-bit quantified in this simulation. We will select the phase shift value which is closest to the solution obtained by the proposed PSO algorithm. (3) Non-RIS scheme: the spatial equalizer is removed.

In Fig. 3.18, we present the maximum power of normalized ISI η for different number of scattering paths L. Here, we normalize the received power for each user at $t = 0$ as 1. From this figure, we can observe that the proposed PSO algorithm can outperform other benchmark algorithms. We can also learn that even with 2-bit quantization at the RIS, we can reduce 1 dB compared to that without the RIS filter in terms of the maximum power of normalized ISI when $L = 100$. These observations are consistent with Remark 3.1. Moreover, we can observe that random phase shifts at the RIS can achieve almost the same performance as that without the RIS. On the other hand, η will increase as the number of scattering paths L grows, and the benefit brought by the RIS filter will drop due to the limited size of the RIS.

In Fig. 3.19, we plot the maximum power of normalized ISI η for different size of the RIS $\sqrt{N}$. We can observe that η will decrease with a larger size of the RIS since it can provide more diversity to optimize. Moreover, η will be lower with a higher reflection coefficient Γ. Under the assumption that phase shifts at the RIS are

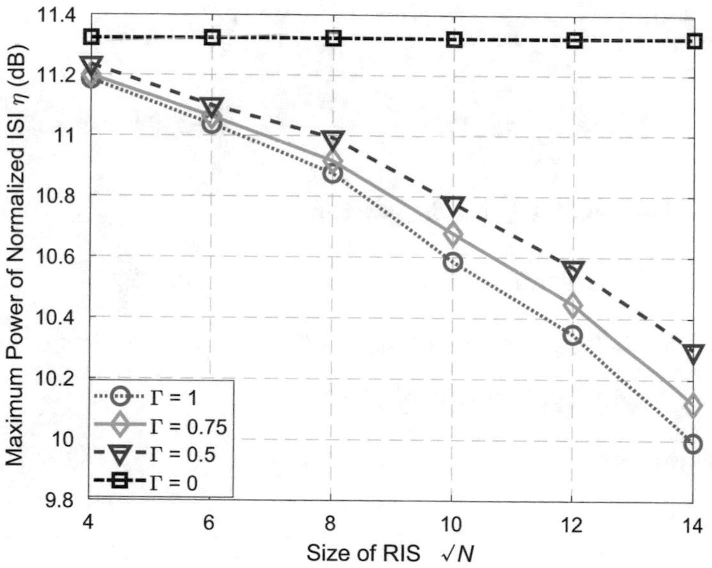

Fig. 3.19 Maximum power of normalized ISI η for different size of the RIS $\sqrt{N}$

continuous, a larger reflection coefficient will provide more options on the amplitude of reflection rays, and thus can achieve a better performance.

3.3.6 *Summary*

In this section, we have proposed the introduction of controllable paths artificially to mitigate multi-path fading through the RIS. As such, equalization can be done before signal reception. To eliminate ISI for multiple users, we have formulated a phase shift optimization problem and proposed an iterative algorithm to solve it. From simulation analysis, we can draw the following conclusions: (1) The proposed RIS-based spatial filter can effectively reduce the ISI. Even with 2-bit quantization, the performance of the proposed scheme is still better than that without the RIS; (2) The ISI will be further reduced with a larger RIS.

3.3.7 *Appendix*

According to the definition in (3.85), we have

$$\begin{aligned}\nabla_{\theta_n} L(\theta_n^l, \eta^l, \mu_k^l) &= \sum_{k\in\mathcal{K}} \mu_k \nabla_{\theta_n} \left(\frac{Y_k(0)}{T} - y_k(0)\right)\left(\frac{Y_k^*(0)}{T} - y_k^*(0)\right) \\ &= \sum_{k\in\mathcal{K}} \mu_k \left(\nabla_{\theta_n}\frac{Y_k(0)Y_k^*(0)}{T^2} - \nabla_{\theta_n}\frac{Y_k(0)}{T} y_k^*(0) - \nabla_{\theta_n} y_k(0)\frac{Y_k^*(0)}{T}\right)\end{aligned} \tag{3.92}$$

With the definitions of $A_{k,n}$ and $B_{k,n}$, we have

$$\begin{aligned}&\nabla_{\theta_n} Y_k(0)Y_k^*(0) = A_{k,n}B_{k,n}^* j e^{j\theta_n^l} - B_{k,n}A_{k,n}^* j e^{-j\theta_n^l}, \\ &\nabla_{\theta_n} Y_k(0)y_k^*(0) = -y_k^*(0)B_{k,n} j e^{-j\theta_n^l}, \\ &\nabla_{\theta_n} y_k(0)Y_k^*(0) = y_k(0)B_{k,n}^* j e^{j\theta_n^l}.\end{aligned} \tag{3.93}$$

This ends the proof.

References

1. G. Zhao, S. Chen, L. Qi, L. Zhao, L. Hanzo, Mobile-traffic-aware offloading for energy- and spectral-efficient large-scale D2D-enabled cellular networks. IEEE Trans. Wirel. Commun. **18**(6), 3251–3264 (2019)
2. L. Song, D. Niyato, Z. Han, E. Hossain, *Wireless Device-to-Device Communications and Networks* (Cambridge University Press, Cambridge, 2015)
3. K.W. Choi, Z. Han, Device-to-device discovery for proximity-based service in LTE-advanced system. IEEE J. Sel. Areas Commun. **33**(1), 55–66 (2015)
4. Y. Gu, Y. Zhang, M. Pan, Z. Han, Matching and cheating in device to device communications underlying cellular networks. IEEE J. Sel. Areas Commun. **33**(10), 2156–2166 (2015)
5. S. Zhang, H. Zhang, B. Di, Y. Tan, Z. Han, L. Song, Beyond intelligent reflecting surfaces: reflective-transmissive metasurface aided communications for full-dimensional coverage extension. IEEE Trans. Veh. Technol. **69**(11), 13905–13909 (2020)
6. E. Basar, M. Renzo, J. Rosny, M. Debbah, M. Alouini, R. Zhang, Wireless communications through reconfigurable intelligent surfaces. IEEE Access **7**, 116753–116773 (2019)
7. H. Zhang, Y. Liao, L. Song, D2D-U: device-to-device communications in unlicensed bands for 5G system. IEEE Trans. Wirel. Commun. **16**(6), 3507–3519 (2017)
8. N. Vucic, S.Y. Shi, M. Schubert, DC programming approach for resource allocation in wireless networks, in *WiOpt '10: Modeling and Optimization in Mobile, Ad Hoc, and Wireless Networks*, Avignon (2010)
9. Y. Chen, B. Ai, H. Zhang, Y. Niu, L. Song, Z. Han, H.V. Poor, Reconfigurable intelligent surface assisted device-to-device communications. IEEE Trans. Wirel. Commun. **20**(5), 2792–2804 (2021)
10. S. Deng, M. Samimi, T. Rappaport, 28 GHz and 73 GHz millimeter-wave indoor propagation measurements and path loss models, in *Proceedings of the IEEE International Conference on Communication Workshop*, London (2015)
11. D. Wang, Y. Zhang, H. Wei, X. You, X. Gao, J. Wang, An overview of transmission theory and techniques of large-scale antenna systems for 5G wireless communications. Sci. China Inf. Sci. **59**(8) (2016)
12. E. Nayebi, A. Ashikhmin, T.L. Marzetta, H. Yang, Cell-free massive MIMO systems, in *Proceedings of the Asilomar Conference on Signals, Systems, and Computers*, Pacific Grove (2015)

13. Z. Shaik, E. Björnson, E. Larsson, Cell-free massive MIMO with radio stripes and sequential uplink processing, in *Proceedings of the IEEE International Conference on Communications Workshops*, Dublin (2020)
14. H.Q. Ngo, A. Ashikhmin, H. Yang, E.G. Larsson, T.L. Marzetta, Cell-Free massive MIMO versus small cells. IEEE Trans. Wirel. Commun. **16**(3), 1834–1850 (2017)
15. M.D. Renzo, M. Debbah, D. Huy, A. Zappone, M. Alouini, C. Yuen, V. Sciancalepore, G. Alexandropoulos, J. Hoydis, H. Gacanin, J. Rosny, A. Bounceur, G. Lerosey, M. Fink, Smart radio environments empowered by reconfigurable AI meta-surfaces: an idea whose time has come. EURASIP J. Wirel. Commun. Netw. **2019**(129), 1–20 (2019)
16. M.A. Elmossallamy, H. Zhang, L. Song, K. Seddik, Z. Han, G.Y. Li, Reconfigurable intelligent surfaces for wireless communications: Principles, challenges, and opportunities. IEEE Trans. Cognitive Commun. Netw. **6**(3), 990–1002 (2020)
17. N. Shlezinger, O. Dicker, Y. Eldar, I. Yoo, M. Imani, D. Smith, Dynamic metasurface antennas for uplink massive MIMO systems. IEEE Trans. Commun. **67**(10), 6829–6843 (2019)
18. F. Sohrabi, W. Yu, Hybrid digital and analog beamforming design for large-scale antenna arrays. IEEE J. Sel. Topics Signal Process. **10**(3), 501–513 (2016)
19. Y. Zhang, B. Di, H. Zhang, J. Lin, Y. Li, L. Song, Beyond cell-free MIMO: energy efficient reconfigurable intelligent surface aided cell-free MIMO communications. IEEE Trans. Cognitive Commun. Netw. arxiv: https://arxiv.org/pdf/2011.08473.pdf
20. B. Di, H. Zhang, L. Song, Y. Li, Z. Han, H.V. Poor, Hybrid beamforming for reconfigurable intelligent surface based multi-user communications: achievable rates with limited discrete phase shifts. IEEE J. Sel. Areas Commun. **38**(8), 1809–1822 (2020)
21. S. Zeng, H. Zhang, B. Di, Z. Han, L. Song, Reconfigurable intelligent surface (RIS) assisted wireless coverage extension: RIS orientation and location optimization. IEEE Commun. Lett. **25**(1), 269–273 (2021)
22. J. Hu, H. Zhang, B. Di, L. Li, L. Song, Y. Li, Z. Han, H.V. Poor, Reconfigurable intelligent surface based RF sensing: design, optimization, and implementation. IEEE J. Sel. Areas Commun. **38**(11), 2700–2716 (2020)
23. T.J. Cui, M.Q. Qi, X. Wan, J. Zhao, Q. Cheng, Coding metamaterials, digital metamaterials and programmable metamaterials. Light Sci. Appl. **3**, e218 (2014)
24. N. Kaina, M. Dupré, G. Lerosey, M. Fink, Shaping complex microwave fields in reverberating media with binary tunable metasurfaces. Sci. Rep. **4**, 6693 (2014)
25. H.Q. Ngo, L. Tran, T.Q. Duong, M. Matthaiou, E.G. Larsson, On the total energy efficiency of cell-free massive MIMO. IEEE Trans. Green Commun. Netw. **2**(1), 25–39 (2018)
26. L.D. Nguyen, T.Q. Duong, H.Q. Ngo, K. Tourki, Energy efficiency in cell-free massive MIMO with zero-forcing precoding design. IEEE Commun. Lett. **21**(8), 1871–1874 (2017)
27. C.B. Peel, B.M. Hochwald, A.L. Swindlehurst, A vector-perturbation technique for near-capacity multiantenna multiuser communication-part I: channel inversion and regularization. IEEE Trans. Commun. **53**(1), 195202 (2005)
28. H.P. Benson, Solving sum of ratios fractional programs via concave minimization. J. Optim. Theory Appl. **135**(1), 117 (2007)
29. K. Shen, W. Yu, Fractional programming for communication systems part I: power control and beamforming. IEEE Trans. Signal Process. **66**(10), 2616–2630 (2018)
30. M.S. Bartlett, An inverse matrix adjustment arising in discriminant analysis. Ann. Math. Stat., 107111 (1951)
31. A. Ben-Tal, A. Nemirovski, *Lectures on Modern Convex Optimization: Analysis, Algorithms, and Engineering Applications* (SIAM, Philadelphia, 2001)
32. K.B. Petersen, M.S. Pedersen, *The Matrix Cookbook*, vol. 7 (Technical University of Denmark, Lyngby, 2008)
33. H. Zhang, B. Di, L. Song, Z. Han, Reconfigurable intelligent surfaces assisted communications with limited phase shifts: how many phase shifts are enough? IEEE Trans. Veh. Technol. **69**(4), 4498–4502 (2020)
34. D. Tse, P. Viswanath, *Fundamentals of Wireless Communications* (Cambridge University Press, Cambridge, 2005)

35. E. Björnson, L. Sanguinetti, M. Kountouris, Deploying dense networks for maximal energy efficiency: small cells meet massive MIMO. IEEE J. Selected Areas Commun. **34**(4), 832–847 (2016)
36. X. Chen, X. Xu, X. Tao, Energy efficient power allocation in generalized distributed antenna system. IEEE Commun. Lett. **16**(7), 1022–1025 (2012)
37. 3GPP TR 38.901, Study on channel model for frequencies from 0.5 to 100 GHz (Release 14) (2018)
38. Y. Zhang, B. Di, H. Zhang, J. Lin, Y. Li, L. Song, Reconfigurable intelligent surface aided cell-free MIMO communications. IEEE Wirel. Commun. Lett. **10**(4), 775–779 (2021)
39. E. Basar, I.F. Akyildiz, Reconfigurable intelligent surfaces for Doppler effect and multipath fading mitigation (2019). arXiv:1912.04080
40. M. Di Renzo, A. Zappone, M. Debbah, M.S. Alouini, C. Yuen, J. De Rosny, S. Tretyakov, Smart radio environments empowered by reconfigurable intelligent surfaces: how it works, state of research, and road ahead. IEEE J. Sel. Areas Commun. **38**(11), 2450–2525 (2020)
41. H. Zhang, L. Song, Z. Han, H.V. Poor, Spatial equalization before reception: reconfigurable intelligent surfaces for multi-path mitigation, in *Proceedings of the IEEE ICASSP*, Toronto (2021)
42. B. Di, H. Zhang, L. Li, L. Song, Y. Li, Z. Han, Practical hybrid beamforming with limited-resolution phase shifters for reconfigurable intelligent surface based multi-user communications. IEEE Trans. Veh. Technol. **69**(4), 4565–4570 (2020)
43. B.K. Casper, M. Haycock, R. Mooney, An accurate and efficient analysis method for multi-Gb/s chip-tochip signaling schemes, in *Proceedings of the International Symposium IEEE VLSI Circuits*, Honolulu (2002)
44. E. Song, J. Kim, J. Kim, A passive equalizer optimization method based on time-domain inter-symbol interference (ISI) cancellation technique. IEEE Trans. Electromagn. Compat. **60**(3), 807810 (2018)
45. S. Boyd, L. Vandenberghe, *Convex Optimization* (Cambridge University Press, Cambridge, 2004)
46. A.A.M. Saleh, R. Valenzuela, A statistical model for indoor multipath propagation. IEEE J. Sel. Areas Commun. **5**(2), 128137 (1987)

Chapter 4
RIS Aided RF Sensing and Localization

In future cellular systems, localization and sensing will be built-in with specific applications, and to support flexible as well as seamless connectivity. Driving by this trend, there exists a need for fine resolution sensing solutions and cm-level localization accuracy [1]. Fortunately, with recent development of new materials, reconfigurable intelligent surfaces (RIS) provide an opportunity to reshape and control the electromagnetic characteristics of the environment, which can be utilized to improve the performance of sensing and localization [2].

In this chapter, we first study RIS aided radio-frequency (RF) sensing for posture recognition where the posture set is known in Sect. 4.1. In Sect. 4.2, we study a more generalized and challenging scenario, i.e., 3D RF sensing, where we recognize arbitrary 3D objects from RF signals through the RIS. Finally, we study how to improve the indoor localization accuracy through the RIS in Sect. 4.3.

4.1 2D Sensing

4.1.1 Motivations

Recently, leveraging widespread RF signals for wireless sensing applications is of special interests. Different from methods based on wearable devices or surveillance cameras, the RF sensing techniques need no contact with sensing targets and will raise no privacy concerns [3]. The basic principle behind RF sensing is that the influence of the sensing objectives on the wireless signal propagation can be potentially recognized by the receivers [4]. In RF sensing, posture recognition has been one of the most commonly studied topics with many applications such as surveillance [5], ambient assisted living [6], and remote health monitoring [7]. It is crucial to design RF sensing systems with high posture recognition accuracy.

H. Zhang et al., *Reconfigurable Intelligent Surface-Empowered 6G*, Wireless Networks, https://doi.org/10.1007/978-3-030-73499-2_4

RF posture recognition aims to automatically recognize different human postures such as standing, walking, sitting, and lying by analyzing the propagation characteristics and impacts of different postures on wireless signal propagation between sensors and receivers [8]. Such posture information can be extracted from the received signals. In the literature, some wireless posture recognition systems with single-antenna transceivers have been proposed: In [9], the authors designed a gesture-recognition system named AllSee, which uses an envelope detector to extract the amplitude information of the received signals for gesture classification. In [10], the authors proposed a posture recognition system based on a pair of radio frequency identification (RFID) transceivers and multiple RFID tags, where the postures are recognized by analyzing the received signal strength indicators. In [11], a human fall is detected by analyzing the phase shift of the received signals via a pair of Wi-Fi devices. However, as multi-dimensional posture information is carried in the channels, the accuracy of posture recognition increases with the number of independent reflected paths between the transmitter and receiver [11]. Therefore, to increase the number of paths, several systems with multi-antenna transceivers have been proposed. In [12], multiple transceivers were used to identify different parts of the human body, which can be used to recognize human postures. In [13], the authors used a pair of commodity Wi-Fi devices with 3×3 MIMO transmission to build a human identification system that can discriminate individuals. Besides, in [14], an MIMO array with 16 patch antennas was employed in an RF sensing system to recognize human postures such as standing, sitting, and lying.

However, due to the complicated and unpredictable wireless environments, the accuracy and flexibility of posture recognition are greatly affected by unwanted multi-path fading [15] and the limited number of independent channels from the transmitters to the receivers in the conventional RF sensing systems. Recently, the RIS technique has been developed as a promising technology to actively cusemize the propagation channels to create a favorable propagation environment [2, 16]. By optimizing and programming the configurations, the RIS is able to customize the wireless channels and generate a favorable massive number of independent paths to enhance the posture recognition accuracy [17].

In this section, we propose an RIS-based RF sensing system for human posture recognition. By periodically programming RIS configurations, the developed system can create multiple independent paths carrying out richer information of the human postures to achieve high accuracy of human posture recognition. There are several challenges in designing an RIS-based posture recognition system. *First*, in order to obtain high posture recognition accuracy, RIS configurations need to be carefully designed to create favorable propagation conditions for posture recognition at the receiver. However, the complexity of finding the optimal configuration is extremely high due to the large number of RIS elements and different states in each RIS element. *Second*, the decision function for posture recognition also greatly affects the recognition accuracy and is coupled with the RIS configuration optimizations. Therefore, it is necessary to jointly optimize the configuration and decision function to maximize recognition accuracy.

To tackle the above challenges, we decompose the problem into two subproblems, i.e., configuration optimization subproblem and decision function optimization subproblem, and then apply an alternating optimization algorithm and a supervised learning algorithm to solve these two subproblems, respectively. More importantly, to demonstrate its benefits in practical systems, we implement our designed system using universal software radio peripheral (USRP) devices and carry out simulations and practical experiments which verify the effectiveness of our system design and proposed algorithms.

The rest of this section is organized as follows. In Sect. 4.1.2, we introduce the design of RIS based human posture recognition system. In Sect. 4.1.3, we formulate a problem to optimize the RIS configurations and the decision function to minimize the average cost of false posture reconfiguration, and develop algorithms to solve the problem in Sect. 4.1.4. Performance analysis including convergence, optimality, and false recognition cost are provided in Sect. 4.1.5. In Sect. 4.1.6, we introduce the system implementation. Numerical and experimental results in Sect. 4.1.7 validate our proposed algorithms and analysis. Finally, conclusions are drawn in Sect. 4.1.8.

4.1.2 System Design

In this section, we propose the RIS-based posture recognition system [18]. As shown in Fig. 4.1, the system is composed of a pair of single-antenna transceivers and an RIS. The RIS can reflect and modify the incident narrow-band signals with a certain frequency. Without loss of generality, we assume that the RIS in the system is specifically designed for f_c signals. To perform RF sensing, the transmitter continuously transmits a single-tone RF signal of frequency f_c, which are reflected by the RIS and the human body and received by receiver. As the transmitter sends single-tone signals, the occupied bandwidth of the system is approximately zero, which indicates the system to be spectrally efficient.

In the following, we first introduce the RIS in Sect. 4.1.2.1 and the channel model in Sect. 4.1.2.2. Then, we propose a periodic configuring protocol to perform posture recognition for the proposed system.

4.1.2.1 RIS Model

The RIS is an artificial thin film of electromagnetic and reconfigurable materials, which is composed of N uniformly distributed and electrically controllable *RIS elements*, denoted by $\mathcal{N}$. As shown in Fig. 4.1, the RIS elements are arranged in a two-dimensional array.

Each RIS element is made of multiple metal patches connected by electrically controllable components, e.g., PIN diodes [2], which are assembled on a dielectric surface. Each PIN diode can be switched to either an *ON* or *OFF* state based on the applied bias voltages. The *state* of an RIS element is determined by the states

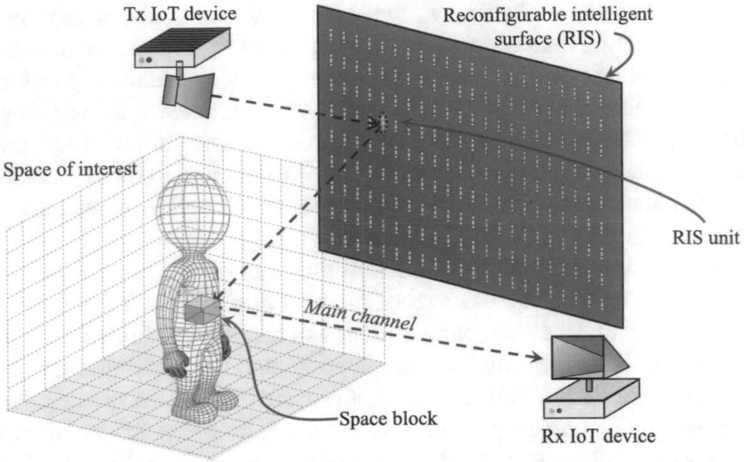

Fig. 4.1 Components of RIS-based posture recognition system

of the PIN diodes on the RIS element. Each state of the RIS element shows its own electrical property, leading to a unique reflection coefficient for the incident RF signals. Suppose that an RIS element contains N_D PIN diodes, and thus the RIS element can be configured into 2^{N_D} possible states. We will describe the detailed implementation of the RIS elements in Sect. 4.1.6.

For simplicity, we refer to the set of all possible states of each RIS element as an *available state set*, denoted by $\mathscr{S}_a$. We denote that the number of available states by N_a, and the i-th ($i \in [1, N_a]$) state in $\mathscr{S}_a$ is denoted by $\hat{s}_i$. In this section, we have the following assumption:

Assumption 1 *Assume that the RIS elements are not correlated with each other, and thus the reflection coefficient of an RIS element is determined by its own state. Then, the reflection coefficient of an RIS element for a certain frequency as $r(\boldsymbol{\theta}_I, \boldsymbol{\theta}_R, s)$, which is a function of the incidence angle $\boldsymbol{\theta}_I = (\theta_{I,1}, \theta_{I,2})$, the reflection angle $\boldsymbol{\theta}_R = (\theta_{R,1}, \theta_{R,2})$, and its state s. The reflection coefficient function of the RIS element is assumed to be known a priori.*

An example of the incidence angle and the reflection angle is depicted in Fig. 4.2. The value of reflection coefficient is a complex number, i.e., $r(\boldsymbol{\theta}^I, \boldsymbol{\theta}^R, s_n) \in \mathbb{C}$, and $|r(\boldsymbol{\theta}^I, \boldsymbol{\theta}^R, s)|$ and $\angle r(\boldsymbol{\theta}^I, \boldsymbol{\theta}^R, s)$ denote the amplitude ratio and the phase shift between the reflection and incidence signals, respectively, which can be obtained through the simulation for the given RIS element. As the model does not rely on the RIS having a specific reflection coefficient for the Tx signals, our model and proposed system can apply to scenarios where an RIS designed for a different frequency is used.

However, since the number of RIS elements N is usually large, it is costly and inefficient to control each RIS element independently. To alleviate the controlling complexity, we divide the RIS elements into *groups*. Each group contains the same

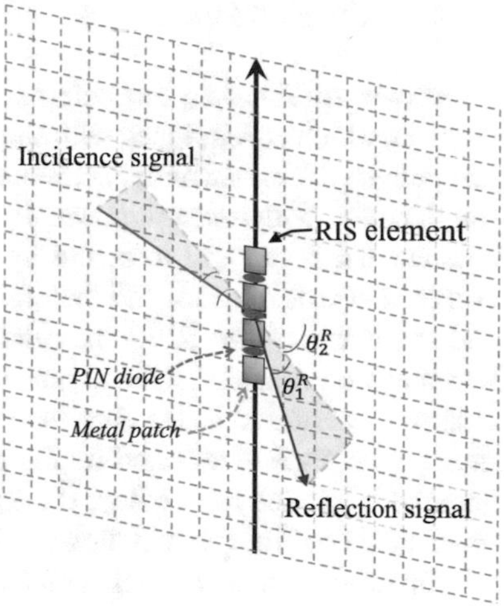

Fig. 4.2 Incidence and reflection angles for signals reflected on an RIS element

number of RIS elements, and the RIS elements in the same group are located within a square region, as shown in Fig. 4.1. Specifically, the RIS elements are equally divided into L groups, and the set of RIS elements in the l-th group is denoted $\mathcal{N}_l$, which satisfies $\mathcal{N}_l \cap \mathcal{N}_{l'} = \emptyset$ $(\forall l, l' \in [1, L], l \neq l')$ and $\bigcup_{l=1}^{L} \mathcal{N}_l = \mathcal{N}$. Moreover, we denote the size of each group by $N_G = N/L$.

We control the RIS elements in the basic unit of the group, that is, the RIS elements in the same group are in the same state, and different groups of RIS elements are controlled independently. The state of the l-th group is denoted by s_l. We refer to the state vector of the L groups, i.e., $\boldsymbol{s} = (s_1, \ldots, s_L)$, as the *configuration* of the RIS. Through changing the configurations, the RIS is able to modify the waveforms of the reflected signals to form beamforming [19]. By using the beamforming capability, the RIS is then able to generate various different waveforms which enhance posture recognition.

Remark 4.1 We assume that the correlation between different RIS elements can be neglected, and thus the reflection coefficient of an RIS element depends only on its own state. As we control the RIS elements in a group to be in the same state, the influence of the RIS elements at different states is reduced, which makes the actual situation more consistent with the assumption. Besides, the experimental results in Sect. 4.1.7.3 verifies this assumption. It is shown that the measured signals are basically the same with the predicted signals under Assumption 1. Moreover, the experimental results also show that adopting this assumption does not affect the system performance in terms of obtaining high posture recognition accuracy.

4.1.2.2 Channel Model

As shown in Fig. 4.1, the pair of transceivers consists of a transmitter and a receiver, which are equipped with single antennas to transmit and receive RF signals, respectively. The transmitter continuously transmits a unit baseband signal at carrier frequency f_c with transmit power P_t. The antenna at the transmitter is a directional antenna and is referred to as the *Tx antenna*. The main lobe of the Tx antenna is pointed towards the RIS, and therefore, most of the transmitted signals are reflected and modified by the RIS. The modified signals enter the region in front of the RIS and are reflected by the human body at different positions. The antenna of the receiver, referred to as the *Rx antenna*, is an omni-directional vertical antenna, and thus all the signals reflected by the human body can be received. Besides, the Rx antenna is located right below the RIS, so the signals reflected by the RIS are not received by the Rx antenna directly.

The transmission channel from the transmitter to the receiver can be modeled as a *Rice channel* [20], which is composed of a *direct line-of-sight (LoS) component*, multiple *reflection dominated components*, and a *multi-path component*. As shown in Fig. 4.1, the direct LoS component accounts for the signal path from the Tx antenna to the Rx antenna without any reflection; the reflection dominated component indicates the signals transmitted from the transmitter to the receiver via the shortest paths reflected from the RIS elements and human body. The multi-path components account for the signals after the complex environment reflection and scattering.

To better describe the received signals at the receiver, we first define the region in front of the RIS as the *space of interest*, where the human postures are positioned in. Specifically, the space of interest is an $l_x \times l_y \times l_z$ cuboid region, as shown in Fig. 4.1. Besides, we discretize the space of interest into M equally-sized *space blocks*, as shown in Fig. 4.1. Based on [21], the human body in the space of interest can be considered as a reflector for the wireless signals. For generality, we denote the reflection coefficients of the M space blocks as $\boldsymbol{\eta} = (\eta_1, \ldots, \eta_M)$, which is referred to as the *space reflection vector*.[1] Here, η_m $(m \in [1, M])$ is the reflection coefficient of the m-th space block for the signals reflected from the RIS to the receiver. Intuitively, the space reflection vector is determined by the human postures in the space of interest. For example, for a given posture, if a space block contains part of the human body, its reflection coefficient will be nonzero. Otherwise, if the m-th space block is empty, $\eta_m = 0$. In other words, the space reflection vector carries the information of human postures.

Given configuration $\boldsymbol{s}$ and space reflection vector $\boldsymbol{\eta}$, the received signal can be expressed as

[1]For simplicity of description, we assume a single human in the space of interest. Nevertheless, it can be observed in Sect. 4.1.4 that the proposed algorithms are independent of the number of humans. Therefore, the system and algorithm proposed in this section can also be easily extended to the case of multi-objective posture recognition.

$$y = h_d \cdot P_t \cdot x + \sum_{m\in[1,M]} \sum_{l\in[1,L]} \sum_{n\in\mathcal{N}_l} h_{n,m}(s_l, \eta_m) \cdot P_t \cdot x + h_{rl} \cdot P_t \cdot x + \sigma, \tag{4.1}$$

where the first term indicates the signals of the direct LoS component, the second term is the signals of the $N \times M$ reflection dominated components, the third term represents the multi-path component, and σ denotes the noise signals. Here, h_d is the channel gain of the direct LoS path and can be calculated by

$$h_d = \frac{\lambda}{4\pi} \cdot \frac{\sqrt{g_{T,\text{los}} g_{R,\text{los}}} \cdot e^{-j2\pi d_{\text{los}}/\lambda}}{d_{\text{los}}}, \tag{4.2}$$

where λ is the wavelength of the carrier signal, $g_{T,\text{los}}$ and $g_{R,\text{los}}$ denote the gains of the transmitter and the receiver for the direct LoS component, respectively, and d_{los} is the distance of the direct LoS path from the Tx antenna to the Rx antenna.

Besides, $h_{n,m}(s_l, \eta_m)$ denotes the channel gain for the signals reflected by the n-th RIS element in the l-th group in state s_l ($s_l \in \mathcal{S}_a$) and by the m-th space block and then reach the receiver directly. Based on [22, 23], channel gain $h_{n,m}(s_l, \eta_m)$ which involves an RIS element can be calculated by

$$h_{n,m}(s_l, \eta_m) = \frac{\lambda \cdot r_{n,m}(\nu_l) \cdot \eta_m \cdot \sqrt{g_{T,n} g_{R,m}} \cdot e^{-j2\pi(d_n + d_{n,m})/\lambda}}{4\pi \cdot d_n \cdot d_{n,m}}, \tag{4.3}$$

where $r_{n,m}(s_l) = r(\boldsymbol{\theta}_n^I, \boldsymbol{\theta}_m^R, s_l)$ denotes the reflection coefficient of the n-th RIS element for the incidence signal towards the m-th space block in state s_l, $g_{T,n}$ is the gain of the transmitter towards the n-th RIS element, $g_{R,m}$ is the gain of the receiver towards the m-th space block, d_n is the distance from the Tx to the n-th RIS element, and $d_{n,m}$ denotes the distance from the n-th RIS element to the Rx antenna via the m-th space block. Moreover, in (4.4), h_{rl} denotes the channel gain of the multi-path component, and σ is the noise signal which follows the complex normal distributions, i.e., $\sigma \sim \mathcal{CN}(0, \epsilon)$ with ϵ being the variances.

Based on (1), as there are $(N_a)^L$ different configurations of the RIS, the number of possible transmission channels is $(N_a)^L$. In comparison, in the traditional non-configurable radio environment, the transmission channel remains largely unchanged if the surrounding environment is steady. Therefore, by changing the RIS's configuration, the RIS-aided sensing system can change the gains of the propagation path and increase the diversity of the transmission channel.

4.1.2.3 Protocol Design

To coordinate the RIS and the transceiver in performing the posture recognition, we propose the *periodic configuring protocol* as follows. The timeline in the protocol is in *frames* with time duration δ. Moreover, instead of changing the RIS configurations in frames, in each frame, each RIS element changes from state $\hat{s}_1$ to $\hat{s}_{N_a}$ sequentially. The duration that each RIS element is in each state is the parameter to be designed in frame configuration. Under the proposed protocol, the limitations of the RIS due to the discreteness of the available state sets of RIS elements can be alleviated. In a certain frame, for the l-th group, the time duration for the RIS elements to be in the N_a available states are denoted by $\tilde{\boldsymbol{t}}_l = (\tilde{t}_{l,1}, \ldots, \tilde{t}_{l,N_a})^T$ with $\sum_{i=1}^{N_a} \tilde{t}_{l,i} = \delta$.

Based on this, we can define the *frame configuration* of the RIS by vector $\boldsymbol{t} = (\tilde{\boldsymbol{t}}_1^T, \ldots, \tilde{\boldsymbol{t}}_L^T)^T$, which has size $L \cdot N_a$ and indicates the time duration for L groups to be at the N_a states. In the RIS-based posture recognition system, the reflected waveforms of the RIS need to be carefully designed through designing frame configurations, so that the human postures can be recognized with high accuracy. To alleviate the complexity of the frame configuration design for the RIS, we propose the periodic frame configurations of the RIS. We define the *recognition period* as the time interval that the sequence of frame configurations of the RIS repeats. As illustrated in Fig. 4.3, the recognition period is composed of K frames. The K frame configurations in the recognition period is referred to as the *configuration matrix*, which can be expressed as $\boldsymbol{T} = (\boldsymbol{t}_1, \ldots \boldsymbol{t}_K)^T$, where $\boldsymbol{t}_k = (\tilde{\boldsymbol{t}}_{k,1}^T, \ldots, \tilde{\boldsymbol{t}}_{k,L}^T)^T$ ($k \in [1, K]$) is the configuration of the RIS in the k-th frame of the recognition period with $\tilde{\boldsymbol{t}}_{k,l}$ ($l \in [1, L]$) being the frame configuration of the l-th group in the k-th frame.

To recognize the human postures in a recognition period, the receiver measure the K mean values of the signals received in the K frames. Based on (4.3) and (4.4), in the k-th frame, the mean value of the received signal can be computed as

$$y_k = h_d \cdot P_t \cdot x + \sum_{m=1}^{M} \sum_{l=1}^{L} \sum_{n \in \mathcal{N}_l} \sum_{i \in \mathcal{S}_a} t_{k,l,i} \cdot h_{n,m}(\hat{s}_i, \eta_m) \cdot P_t \cdot x + \bar{h}_{rl} \cdot P_t \cdot x + \bar{\sigma},$$

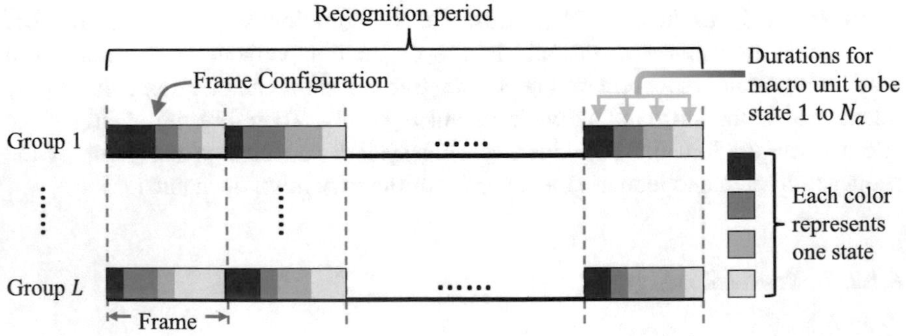

Fig. 4.3 Examples of frame, frame configurations and recognition period with $N_a = 4$

$$= h_d \cdot P_t \cdot x + P_t \cdot x \cdot \boldsymbol{t}_k^T \boldsymbol{A}\boldsymbol{\eta} + h_{rl} \cdot P_t \cdot x + \bar{\sigma}_k. \tag{4.4}$$

Here, $\boldsymbol{A} = (\boldsymbol{\alpha}_1, \ldots, \boldsymbol{\alpha}_M)$ is referred to as the *projection matrix*, where $\boldsymbol{\alpha}_m = (\hat{\boldsymbol{\alpha}}_{m,1}^T, \ldots, \hat{\boldsymbol{\alpha}}_{m,L}^T)^T$ with $\hat{\boldsymbol{\alpha}}_{m,l} = (\hat{\alpha}_{m,l,1}, \ldots, \hat{\alpha}_{m,l,N_a})^T$ and $\hat{\alpha}_{m,l,i} = \sum_{n \in \mathcal{N}_l} \lambda \cdot r_{n,m}(s_l) \cdot \sqrt{g_{T,n} g_{R,m}} \cdot e^{-j2\pi d_{n,m}/\lambda}/(4\pi d_{n,m})$, $(l \in [1, L],\ i \in [1, N_a])$, Besides, $\bar{\sigma}_k$ denotes the average noise signal in the k-th frame, which is a complex Gaussian random variable following $\mathcal{CN}(0, \bar{\epsilon})$. Here, $\bar{\epsilon}$ is determined by ϵ and the number of samples in a frame.

The receiver recognizes the human postures based on the measured K mean values, which constitute a *measurement vector* and can be denoted as $\boldsymbol{y}(\boldsymbol{T}, \boldsymbol{\eta}) = (y_1, \ldots, y_K)^T$ or $\boldsymbol{y}$ for simplicity. Based on (4.4), the measurement vector can be expressed as

$$\boldsymbol{y} = h_d \cdot P_t \cdot \boldsymbol{x} + P_t \cdot \boldsymbol{x} \cdot \boldsymbol{T}\boldsymbol{A}\boldsymbol{\eta} + h_{rl} \cdot P_t \cdot \boldsymbol{x} + \bar{\boldsymbol{\sigma}}. \tag{4.5}$$

Since matrix $\boldsymbol{T}\boldsymbol{A}$ determines how the information of human postures is mapped to the measurement, we denote $\boldsymbol{\Gamma} = \boldsymbol{T}\boldsymbol{A}$ and refer to $\boldsymbol{\Gamma}$ as the *measurement matrix*. More specifically, $\boldsymbol{y}$ is a K-dimensional complex vector, i.e., $\boldsymbol{y} \in \mathbb{C}^K$. Given a measurement vector $\boldsymbol{y}$ and configuration matrix $\boldsymbol{T}$, the receiver adopts a likelihood function to perform the posture recognition, which is referred to as the *decision function* and can be expressed as $\mathcal{L}(\boldsymbol{y})$. Here, $\mathcal{L}(\boldsymbol{y})$ is an N_P-dimensional vector, where $\mathcal{L}_i(\boldsymbol{y})$ $(i \in [1, N_P])$ denotes the probability for the receiver to determine the human posture as the i-th posture.

It can be observed that the performance of the posture recognition depends on the design of the decision function as well as the configuration matrix, which need to be optimized in order to obtain high recognition accuracy. We will formulate the optimizations for the configuration matrix and the decision function in Sect. 4.1.3 and propose algorithms to solve the formulated optimization problems in Sect. 4.1.4.

4.1.3 Problem Formulation of RIS-Based Posture Recognition

In this section, we formulate the problem to optimize the RIS-based posture recognition system by minimizing the average cost of false posture recognition. We then decompose it into two subproblems, i.e., the configuration optimization and the decision function optimization.

4.1.3.1 Problem Formulation

For the i-th posture, denote by $\boldsymbol{\eta}_i$ the corresponding space reflection vector. We assume that the The measurement vector for the i-th posture can be expressed as $\boldsymbol{y}_i = \boldsymbol{y}(\boldsymbol{T}, \boldsymbol{\eta}_i)$. Since the objective of the system is to recognize human postures accurately, we minimize the average cost due to the false recognition of the system. The objective function, which we refer to as the *average false recognition cost*, can be calculated as

$$\Psi_{\mathcal{L}} = \sum_{i,i' \in [1,N_P],\, i \neq i'} p_i \cdot \chi_{i,i'} \cdot \int_{\boldsymbol{y}_i \in \mathbb{C}^K} \Pr(\boldsymbol{y}_i | \boldsymbol{\eta}_i) \cdot \mathcal{L}_{i'}(\boldsymbol{y}_i) \cdot d\boldsymbol{y}_i, \tag{4.6}$$

where $p_i \in [0, 1]$ denotes the probability for the i-th human posture to appear and $\sum_{i \in [1,N_P]} p_i = 1$, $\chi_{i,i'} \in \mathbb{R}^+$ denotes the *cost* for the false recognition of actual i-th posture as the i'-th posture, and $\Pr(\boldsymbol{y}_i | \boldsymbol{\eta}_i)$ denotes the probability for the measurement vector to be $\boldsymbol{y}_i$, given space reflection vector $\boldsymbol{\eta}_i$.

Based on (4.4) and (4.6), the optimization for the RIS-based posture recognition system can be formulated as

$$(\text{P4.1}): \min_{\boldsymbol{T},\mathcal{L}} \Psi_{\mathcal{L}}, \tag{4.7}$$

$$s.t.\ \mathcal{L}_{i'}(\boldsymbol{y}_i) \geq 0,\ \forall \boldsymbol{y}_i \in \mathbb{C}^K,\ i, i' \in [1, N_a], \tag{4.8}$$

$$\sum_{i' \in [1,N_P]} \mathcal{L}_{i'}(\boldsymbol{y}_i) = 1,\ \forall \boldsymbol{y} \in \mathbb{C}^K, \tag{4.9}$$

$$\begin{aligned} \boldsymbol{y}_i &= h_d \cdot P_t \cdot \boldsymbol{x} + P_t \cdot \boldsymbol{x} \cdot \boldsymbol{T}\boldsymbol{A}\boldsymbol{\eta}_i \\ &\quad + \bar{h}_{rl} \cdot P_t \cdot \boldsymbol{x} + \bar{\boldsymbol{\sigma}},\ \forall i \in [1, N_P], \end{aligned} \tag{4.10}$$

$$\boldsymbol{1}^T \tilde{\boldsymbol{t}}_{k,l} = \delta,\ \forall k \in [1, K], l \in [1, L], \tag{4.11}$$

$$\tilde{t}_{k,l,i} \geq 0,\ \forall k \in [1, K], l \in [1, L], i \in [1, N_a]. \tag{4.12}$$

In (P4.1), the optimization variables are the decision function, i.e., $\mathcal{L}$, and the configuration matrix, i.e., $\boldsymbol{T}$. Constraints (4.8) and (4.9) are due to the fact that the decision function returns probabilities. Constraint (4.10) denotes the relationship between the measurement vector, $\boldsymbol{y}_i$, and the space reflection vector of the i-th posture, $\boldsymbol{\eta}_i$. Constraints (4.11) and (4.12) indicate that the time duration for all the states is positive and sum up to δ in each frame. Moreover, when solving the formulated (P4.1), the following assumption is adopted:

Assumption 2 *Assume that the probability of each posture, i.e., p_i ($i \in [1, N_P]$), is known a priori, and that the false recognition has cost* 1 *and the accurate recognition has cost* 0, *i.e., $\chi_{i,i} = 0$ and $\chi_{i,i'} = 1$ ($i \neq i'$).*

4.1.3.2 Problem Decomposition

In (P4.1), the configuration matrix and the decision function are coupled and need to be optimized jointly, which makes (P4.1) hard to solve. To handle this difficulty, we decompose (P4.1) into two sub-problems by separating the configuration matrix optimization and the decision function optimization. The two sub-problems are referred to as the *configuration matrix optimization problem* and the *decision function optimization problem*, which are described as follows.

Configuration Matrix Optimization Given a decision function, the optimization problem for configuration matrix $\boldsymbol{T}$ can be formulated as

$$(\text{P4.2}): \min_{\boldsymbol{T}} \quad \Psi_{\mathscr{L}}, \tag{4.13}$$

$$s.t. \quad \boldsymbol{y}_i = h_d \cdot P_t \cdot \boldsymbol{x} + P_t \cdot \boldsymbol{x} \cdot \boldsymbol{T} \boldsymbol{A} \boldsymbol{\eta}_i + \bar{h}_{rl} \cdot P_t \cdot \boldsymbol{x} + \bar{\boldsymbol{\sigma}}, \ \forall i \in [1, N_P], \tag{4.14}$$

$$\boldsymbol{1}^T \tilde{\boldsymbol{t}}_{k,l} = \delta, \ \forall k \in [1, K], l \in [1, L], \tag{4.15}$$

$$\tilde{t}_{k,l,i} \geq 0, \ \forall k \in [1, K], l \in [1, L], i \in [1, N_a]. \tag{4.16}$$

Decision Function Optimization Given a configuration sequence, the optimization for the decision function $\mathscr{L}$ can then be formulated as

$$(\text{P4.3}): \min_{\mathscr{L}} \quad \Psi_{\mathscr{L}}, \tag{4.17}$$

$$s.t. \quad \mathscr{L}_{i'}(\boldsymbol{y}_i) \geq 0, \ \forall \boldsymbol{y}_i \in \mathbb{C}^K, \ i, i' \in [1, N_a], \tag{4.18}$$

$$\sum_{i' \in [1, N_P]} \mathscr{L}_{i'}(\boldsymbol{y}_i) = 1, \ \forall \boldsymbol{y} \in \mathbb{C}^K. \tag{4.19}$$

In the following Sect. 4.1.4, we design the algorithms to solve the configuration matrix optimization and the decision function optimization problems, respectively.

4.1.4 Algorithms for Configuration Matrix and Decision Function Optimizations

In this section, we first propose the algorithms to solve the configuration optimization and decision function optimization. Then, we analyze the convergence of the proposed algorithms.

4.1.4.1 Configuration Optimization Algorithm

In (P4.2), we optimize the configuration matrix to minimize the average false recognition cost. Solving (P4.2) directly and explicitly requires the space reflection vectors, i.e., $\boldsymbol{\eta}_i$ ($i \in [1.N_P]$) to be known a priori. Besides, the optimized configuration matrix for specific known coefficient vectors may be sensitive to the subtle changes of the postures. Therefore, instead of optimizing RIS configurations given specific space reflection vectors, we will find an optimal configuration matrix for the general posture recognition scenarios.

As the decision function recognizes the human postures based on measurement vector $\boldsymbol{y}$, we consider optimizing the configuration matrix $\boldsymbol{T}$ so that $\boldsymbol{y}$ is able to carry the richest information about the human postures. As indicated in (4.14), the information of human body distribution is contained in space reflection vector $\boldsymbol{\eta}$. Therefore, intuitively, it requires that $\boldsymbol{\eta}$ can be potentially reconstructed from $\boldsymbol{y}$ with the minimum loss.

Since the signals from the multi-path component and the noise are usually much smaller than the reflection channels gains and are random values determined by the environment, we neglect them and consider the relation between $\boldsymbol{\eta}$ and $\boldsymbol{y}$ as $\boldsymbol{y} = P_t \cdot \boldsymbol{x} \cdot \boldsymbol{T}\boldsymbol{A}\boldsymbol{\eta}$. Since the number of space blocks is large, we assume $M \gg K$. In this case, $\boldsymbol{y} = P_t \cdot \boldsymbol{x} \cdot \boldsymbol{T}\boldsymbol{A}\boldsymbol{\eta}$ is an underdetermined equation which has an infinite number of solutions. Therefore, the true $\boldsymbol{\eta}$ cannot be reconstructed from $\boldsymbol{y}$ given $\boldsymbol{T}\boldsymbol{A}$, unless additional constraints on $\boldsymbol{\eta}$ are employed.

One of the usually employed constraints to reconstruct the target signals in an underdetermined equation is that the signal to reconstruct is *sparse*. Intuitively, the target signal, i.e., $\boldsymbol{\eta}$, is sparse when the number of its nonzero entries is sufficiently smaller than the dimension of the signal vector, i.e.,

$$|\mathrm{supp}(\boldsymbol{\eta})| \ll \dim(\boldsymbol{\eta}). \tag{4.20}$$

Here, $\dim(\boldsymbol{\eta})$ denotes the dimension of $\boldsymbol{\eta}$, $\mathrm{supp}(\boldsymbol{\eta}) = \{\eta_i | \eta_i \neq 0, i \in [1, \dim(\boldsymbol{\eta})]\}$ indicates the support set of $\boldsymbol{\eta}$, and $|\cdot|$ provides the cardinality of a set [24]. From Fig. 4.1, it can be observed that in the proposed fall-detection system, most of the space blocks are empty and thus have zero reflection coefficients. Besides, for the space blocks where the human body lies, only those that contain the surfaces of the human body with specific angles can reflect the incidence signals towards the receiver and have non-zero reflection coefficients. Therefore, $\boldsymbol{\eta}$ is a space vector, and condition (4.20) is satisfied. The sparse target signals in an underdetermined equation can be reconstructed efficiently using approaches such as *compressive sensing* [25].

Based on [26], to minimize the loss of reconstruction for sparse target signals, we can minimize the averaged *mutual coherence* of measurement matrix $\boldsymbol{\Gamma} = \boldsymbol{T}\boldsymbol{A}$, which is defined as

$$\mu(\boldsymbol{\Gamma}) = \frac{1}{M(M-1)} \cdot \sum_{m,m' \in [1,M], m \neq m'} \frac{|\boldsymbol{\gamma}_m^T \boldsymbol{\gamma}_{m'}|}{\|\boldsymbol{\gamma}_m\|_2 \cdot \|\boldsymbol{\gamma}_{m'}\|_2}, \tag{4.21}$$

where $\boldsymbol{\gamma}_m = \boldsymbol{T}\boldsymbol{\alpha}_m$ denotes the m-th column of $\boldsymbol{\Gamma}$, and $\|\cdot\|_2$ denotes the l_2-norm.

Therefore, based on (4.21), we can reformulate the configuration sequence optimization as the following mutual coherence minimization problem.

$$\text{(P4.4)} \ \min_{\boldsymbol{T}} \ \mu(\boldsymbol{\Gamma}), \tag{4.22}$$

$$\text{s.t.} \quad \boldsymbol{\gamma}_m = \boldsymbol{T}\boldsymbol{\alpha}_m, \ \forall m \in [1, M], \tag{4.23}$$

$$\mathbf{1}^T \tilde{\boldsymbol{t}}_{k,l} = \delta, \ \forall k \in [1, K], l \in [1, L], \tag{4.24}$$

$$\tilde{t}_{k,l,i} \geq 0, \ \forall k \in [1, K], l \in [1, L], i \in [1, N_a]. \tag{4.25}$$

Due to the non-convex objective function, (P4.4) is a non-convex optimization problem and NP-hard. Besides, it can be observed that the number of variables in (P4.4) is $K \cdot L \cdot N_a$. In practical scenarios, the number of measurements and the number of groups can be large, which makes $K \cdot L \cdot N_a$ a large number. Moreover, the variables in (P4.4) are coupled together, and thus (P4.4) cannot be separated into independent sub-problems, resulting in prohibitive computational complexity.

To solve (P4.4) in an acceptable complexity in practice, we propose a low-complexity algorithm to solve (P4.4) sub-optimally based on the alternating optimization (AO) technique. The proposed algorithm is referred to as the frame configuration alternating optimization (FCAO) algorithm. In the FCAO algorithm, we alternately optimize each of the frame configuration in an iterative manner by fixing the other $K - 1$ frame configurations, until the convergence is achieved.

In each iteration, we need to optimize (P4.4) with respect to $\boldsymbol{t}_k$, with $\boldsymbol{T}_{-k} = (\boldsymbol{t}_1, \ldots, \boldsymbol{t}_{k-1}, \boldsymbol{t}_{k+1}, \ldots, \boldsymbol{t}_K)$ fixed. Besides, we denote the coherence of $\boldsymbol{\gamma}_m$ and $\boldsymbol{\gamma}_{m'}$ by $u_{m,m'}$, i.e.,

$$u_{m,m'} = \frac{\boldsymbol{\gamma}_m^T \boldsymbol{\gamma}_{m'}}{\|\boldsymbol{\gamma}_m\|_2 \cdot \|\boldsymbol{\gamma}_{m'}\|_2}, \tag{4.26}$$

and arrange $u_{m,m'}$ $(m, m' \in [1, M], m \neq m')$ as vector $\boldsymbol{u} = (u_{1,2}, \ldots, u_{1,M}, \ldots, u_{M-1,M})$. The optimization problem in each iteration can then be formulated as

$$\text{(P4.5)} \ \min_{\boldsymbol{t}_k, \boldsymbol{u}} \ \|\boldsymbol{u}\|_1, \tag{4.27}$$

$$\text{s.t.} \quad u_{m,m'} = \frac{\boldsymbol{\gamma}_m^T \boldsymbol{\gamma}_{m'}}{\|\boldsymbol{\gamma}_m\|_2 \cdot \|\boldsymbol{\gamma}_{m'}\|_2}, \ m, m' \in [1, M], m \neq m', \tag{4.28}$$

$$\mathbf{1}^T \tilde{\boldsymbol{t}}_{k,l} = \delta, \ \forall l \in [1, L], \tag{4.29}$$

$$\tilde{\boldsymbol{t}}_{k,l} \succeq 0, \ \forall l \in [1, L], \tag{4.30}$$

where $\| \cdot \|_1$ denotes the l_1-norm of the contained vector.

To solve (P4.5), we can adopt the augmented Lagrangian method [27], where the original constrained optimization problem is handled by solving a sequence of unconstrained minimizations for the augmented Lagrangian function. To express the augmented Lagrangian function, we first define the indicator function for (P4.5) as

$$\mathbb{I}(\boldsymbol{t}_k) = \begin{cases} 1, & \text{if } \tilde{\boldsymbol{t}}_{k,l} \succeq 0 \text{ and } \mathbf{1}^T \tilde{\boldsymbol{t}}_{k,l} = 1, \ \forall l \in [1, L], \\ 0, & \text{otherwise.} \end{cases} \tag{4.31}$$

The augmented Lagrangian function for (P4.5) can be expressed as

$$\begin{aligned}\mathscr{L}_A(\boldsymbol{t}_k, \boldsymbol{u}; \boldsymbol{\beta}, \rho, \boldsymbol{T}_{-k}) &= \|\boldsymbol{u}\|_1 \\ &+ \mathbb{I}(\boldsymbol{t}_k) + \sum_{\substack{m,m'\in[1,M],\\ m\neq m'}} \beta_{m,m'} \left(u_{m,m'} - \frac{\boldsymbol{\gamma}_m^T \boldsymbol{\gamma}_{m'}}{\|\boldsymbol{\gamma}_m\|_2 \cdot \|\boldsymbol{\gamma}_{m'}\|_2}\right) \\ &+ \sum_{\substack{m,m'\in[1,M],\\ m\neq m'}} \frac{\rho}{2} \left| u_{m,m'} - \frac{\boldsymbol{\gamma}_m^T \boldsymbol{\gamma}_{m'}}{\|\boldsymbol{\gamma}_m\|_2 \cdot \|\boldsymbol{\gamma}_{m'}\|_2} \right|^2, \end{aligned} \tag{4.32}$$

where ρ is a positive scaling factor and $\boldsymbol{\beta} = (\beta_{1,2}, \beta_{1,3}, \ldots, \beta_{1,M}, \beta_{2,3}, \ldots, \beta_{M-1,M}) \in \mathbb{C}^{M(M-1)/2}$ is the vector for Lagrange multipliers. The augmented Lagrangian method finds a local optimal solution to (P4.5) by minimizing a sequence of (4.32) where $\boldsymbol{\beta}$ is fixed in each iteration. Specifically, the sequence of unconstrained Lagrangian minimization can be solved using an *alternating minimization procedure*, in which $\boldsymbol{u}$ is updated while $\boldsymbol{t}_k$ is fixed and vice versa. The completed algorithm to solve (P4.5) by the augmented Lagrangian method can be found in Sect. 4.1.9.

However, the augmented Lagrangian method may result in a local optimum far from the global optimum, if the starting point is poorly chosen. Therefore, we adopt an intuitive algorithm named *pattern search* [28], which can obtain a good initial point for the augmented Lagrangian method. The complete FCAO algorithm for (P4.4) can be summarized as Algorithm 11.

Algorithm 11: FCAO algorithm for solving (P4.4)

Input : Initial random feasible configuration matrix $\boldsymbol{T}^{(0)}$.
Output: Optimal averaged mutual coherence μ^* and configuration matrix $\boldsymbol{T}^*$ for (P4.4).

1 Compute initial $\mu^{(0)}$ based on (4.23) given $\boldsymbol{T}^{(0)}$;
2 Set the number of consecutive iterations with no improvements as $N_{\text{non}} = 0$ and current frame index $k = 1$;
3 **for** $i = 1, 2, \ldots$ **do**
4   Invoke pattern search method in [28] for (P4.5) to obtain a initial $\boldsymbol{t}'_k$ which results in low average mutual coherence given fixed $\boldsymbol{T}_{-k}^{(i-1)}$;
5   Using $\boldsymbol{t}_k$ as an initial point, solve (P4.5) by using the augmented Lagrangian method described in Sect. 4.1.9, and denote the resulting minimum mutual coherence as μ' and optimal configuration as $\boldsymbol{t}'_k$;
6   If $\mu' < \mu^{(i)}$, update $\mu^{(i)} = \mu'$, the k-th frame configuration in $\boldsymbol{T}^{(i)}$ as $\boldsymbol{t}_k^{(i)} = \boldsymbol{t}'_k$, $\boldsymbol{T}_{-k}^{(i)} = \boldsymbol{T}_{-k}^{(i-1)}$, and $N_{\text{non}} = 0$; otherwise, set $N_{\text{non}} = N_{\text{non}} + 1$;
7   If $N_{\text{non}} < K$, set $k = \text{mod}(k + 1, K) + 1$; otherwise, return $\mu^* = \mu^{(i)}$ and $\boldsymbol{T}^* = \boldsymbol{T}^{(i)}$;
8 **end**

4.1.4.2 Supervised Learning Algorithm for Solving (P4.3)

To solve (P4.3) efficiently, we parameterize $\mathscr{L}$ by a real-valued parameter vector $\boldsymbol{\theta} \in \mathbb{R}^L$. Let the parameterized function to be denoted as $\mathscr{L}^{\boldsymbol{\theta}}$, and the decision function optimization problem can be formulated as

$$\text{(P4.6)} : \min_{\boldsymbol{\theta}} \quad \Psi_{\boldsymbol{\theta}}(\{s_k\}_{k=1}^K), \tag{4.33}$$

$$s.t. \quad \mathscr{L}_i^{\boldsymbol{\theta}}(\boldsymbol{y}) \geq 0, \ \forall \boldsymbol{y} \in \mathbb{C}^K, \ i \in [1, N_P], \tag{4.34}$$

$$\sum_{i \in [1, N_a]} \mathscr{L}_i^{\boldsymbol{\theta}}(\boldsymbol{y}) = 1, \ \forall \boldsymbol{y} \in \mathbb{C}^K. \tag{4.35}$$

To solve (P4.6), we propose the following algorithm based on the supervised learning approach using neural network (NN) [29]. In the designed human posture recognition system, we adopt a fully connected NN for the decision function $\mathscr{L}^{\boldsymbol{\theta}}$. The fully connected NN consists of the input layer, hidden layers, and the output layer, which are connected successively. The input layer takes the K-dimensional complex-valued measurement vector $\boldsymbol{y}$ in a recognition period and passes it to the first hidden layer. The j-th hidden layer has $n_{\text{hid},j}$ nodes. Each node calculates a biased weighted sum of its input, processes the sum with an *activation function*, and outputs the result to the nodes connected to it in the next layer. In the output layer, the number of nodes is number of posture N_P. The output nodes handle their input with a *softmax* function f_{softmax} [30], which can be expressed as

$$\boldsymbol{y} = f_{\text{softmax}}(\boldsymbol{x}) = \left[\frac{e^{x_1}}{\sum_{n=1}^{N_P} e^{x_n}}, \ldots, \frac{e^{x_{N_P}}}{\sum_{n=1}^{N_P} e^{x_n}} \right]. \tag{4.36}$$

It can be observed that the softmax function converts the input of the output layer $\boldsymbol{x}$ to vector $\boldsymbol{y}$, which satisfies constraints (4.34) and (4.35) and can be considered as the probabilities for the N_P postures. In the NN, the parameters for the decision function, i.e., $\boldsymbol{\theta}$, stands for the weights of the connections and the biases of the nodes.

It can be seen that parameter $\boldsymbol{\theta}$ in NN determines how the input is processed into output and thus determines the performance of the decision function in terms of average false recognition cost (4.33). Therefore, (P4.3) is equivalent to finding the optimal $\boldsymbol{\theta}$ for the NN, which minimize (4.34). To solve (P4.3), we propose the following algorithm based on the back-propagation algorithm [31]. To train the NN requires a training data set of the measurement vectors labeled with the postures, which can be expressed as $\mathscr{D} = \{(\boldsymbol{y}_j, \boldsymbol{L}_j)\}_{j=1}^{N_{data}}$. Here, $\boldsymbol{y}_j$ and $\boldsymbol{L}_j$ denote the j-th measurement vector and its one-hot label, respectively, and N_{data} denotes the size of the data set, which is collected a priori given a certain configuration sequence. The i-th element of $\boldsymbol{L}_j$, i.e., $L_{j,i}$ is equal to 1 if $\boldsymbol{y}_j$ corresponds to the i-th posture, otherwise, $L_{j,i} = 0$. Besides, the number of labeled data for the i-th posture needs to be approximately $p_i \cdot N_{data}$, where p_i is defined in (4.6).

For $\boldsymbol{y}_j$ in a data pair $(\boldsymbol{y}_j, L_j)$, we assume that the output of the NN is denoted by $\tilde{p}_j$, which is a N_P-dimensional vector. The i-th element in $\tilde{p}_j$ is the probability that the posture is detected as the i-th posture. In accordance with (4.6), for $(\boldsymbol{y}_j, \boldsymbol{L}_j)$, the loss function, i.e., the recognition cost incurred by the NN is defined as

$$E_j = \sum_{i=1}^{N_P} \sum_{i'=1}^{N_P} \tilde{p}_{j,i'} \chi_{i,i'} L_{j,i}. \tag{4.37}$$

Based on the back-propagation algorithm, the parameters of the NN need to be updated in the negative gradient direction of the loss function (4.37). The adjustment of $\boldsymbol{\theta}$ can be expressed as

$$\Delta\boldsymbol{\theta} = -\zeta \frac{\partial E_j}{\partial \boldsymbol{\theta}}, \tag{4.38}$$

where $\zeta \in (0, 1)$ denotes the learning rate. The update of parameter $\boldsymbol{\theta}$ proceeds iteratively and repeatedly until the average loss, i.e., the average recognition cost due to the NN over the data set, converges. The algorithm to solve (P4.6) is summarized as Algorithm 12.

Algorithm 12: Supervised learning algorithm for solving (P4.6)

Input : Training data set $\mathscr{D} = \{(\boldsymbol{y}_j, L_j)\}$;
Learning rate ζ.
Output: Trained parameter $\boldsymbol{\theta}^*$, which is the solution for (P4.6).
Obtain initial $\boldsymbol{\theta}^{(0)}$ with random value which follows uniform distribution within (0, 1);
Calculate the current average recognition cost for the training data set $\Psi^{(0)} = \sum_{j=1}^{N_d} E_j$ based on (4.37);
for i = *1,2,...* **do**
  for $(\boldsymbol{y}_j, \boldsymbol{L}_j) \in \mathscr{D}$ **do**
    Input $\boldsymbol{y}_j$ to the NN $\mathscr{L}^{\boldsymbol{\theta}}|_{\boldsymbol{\theta}=\boldsymbol{\theta}^{(i-1)}}$ and obtain output $\tilde{p}_j$;
    Calculate the gradient of the loss with respect to the current parameter $\boldsymbol{\theta}$, i.e. $\partial E_j / \partial \boldsymbol{\theta}$;
    Update parameter by $\boldsymbol{\theta}' = \boldsymbol{\theta}^{(i-1)} + \Delta\boldsymbol{\theta}$ based on (4.38);
  end
  Calculate $\Psi' = \sum_{j=1}^{N_d} E_j$ based on (4.37) using the updated $\mathscr{L}^{\boldsymbol{\theta}}$;
  If $\Psi' \geq \Psi^{(i-1)}$, break and output $\boldsymbol{\theta}^* = \boldsymbol{\theta}^{(i-1)}$; otherwise, set $\Psi^{(i)} = \Psi'$ and $\boldsymbol{\theta}^{(i)} = \boldsymbol{\theta}'$.
end

4.1.5 *Performance Analysis*

In the following, we analyze the convergence of the proposed algorithms in this section as well as the optimality of the decision function. Besides, we provide an upper-bound on the minimal average false recognition cost and analyze the influence of the transmit power on it.

4.1.5.1 Convergence of FCAO Algorithm

Based on Algorithm 11, in the $(i+1)$-th iteration, a better configuration matrix resulting in a lower μ can be obtained given configuration matrix $\boldsymbol{T}^{(i)}$ in the $(i+1)$-th iteration. Therefore, we have $\mu(\boldsymbol{T}^{(i+1)}\boldsymbol{A}) \leq \mu(\boldsymbol{T}^{(i)}\boldsymbol{A})$, which implies that the objective value obtained in Algorithm 11 is non-increasing after each iteration of the FCAO algorithm. Since the average mutual coherence of $\boldsymbol{\Gamma} = \boldsymbol{T}\boldsymbol{A}$ has lower-bound of 0, the proposed FCAO algorithm is guaranteed to converge.

4.1.5.2 Convergence of Supervised Learning Algorithm

Similar to the convergence analysis of the FCAO algorithm, in the $(i+1)$-th iteration, parameter $\boldsymbol{\theta}$ only updates to $\boldsymbol{\theta}'$ when the resulting average cost $\Psi' < \Psi^{(i)}$. Therefore, the objective value obtained in Algorithm 12 is also non-increasing. As the average false recognition cost is lower bounded by zero, the proposed Supervised learning is guaranteed to converge.

4.1.5.3 Optimality of Decision Function

As shown in (4.38), the back-propagation algorithm is based on the gradient descent method which can be trapped in local optima and is not guaranteed to find the global minimum of the loss function. To find the global optimum, the back-propagation algorithm aided by multiple-starting and simulated annealing techniques [32, 33] can be adopted, which is left for future work.

4.1.5.4 Upper-Bound on Minimal Average False Recognition Cost

When the reflection coefficient vectors of the different postures are known a priori, then the optimal decision function can be obtained based on the Bayesian criterion for signal detection [34] as

$$\mathscr{L}_i^*(\boldsymbol{y}) = \begin{cases} 1, & \text{if } I_i(\boldsymbol{y}) \leq I_j(\boldsymbol{y}),\ \forall j \in [1, N_P], \\ 0, & \text{otherwise}, \end{cases} \tag{4.39}$$

where $I_i(\boldsymbol{y}) = \sum_{j=0}^{M} p_j \cdot (\chi_{ij} - \chi_{jj}) \cdot \Pr(\boldsymbol{y}|\boldsymbol{\eta}_j)$. Under the given optimal decision function, the average false recognition cost can be minimized, which is referred as the *minimal average false recognition cost*. Moreover, by proposing the following proposition, the upper-bound on the minimal average false recognition cost can be proven to be a monotonically decreasing function of the transmit power P_t.

Proposition 4.1 *Given that the reflection coefficient vectors are known and* $\chi_{i,i} = 1$, $\chi_{i,j} = 0$ $(i \neq j)$, *an upper-bound on the minimal average false recognition cost can be expressed as*

$$\Psi_{\mathcal{L}^*} \leq \sum_{i=1}^{N_P}\sum_{j=1}^{N_P} p_j \mathscr{P}(K, Z_{ij}(\gamma_{rx})^2/2), \quad i \neq j, \tag{4.40}$$

where γ_{rx} *denotes the SNR of the received signals,* $\mathscr{P}(\cdot)$ *denotes the regularized gamma function [35], and* $Z_{ij}(\gamma_{rx}) = \frac{\sqrt{\gamma_{rx}}(\gamma_{rx}\|\hat{\boldsymbol{y}}_j - \hat{\boldsymbol{y}}_i\|_2^2/\bar{\epsilon} - \ln(p_j/p_i))}{\|\hat{\boldsymbol{y}}_j - \hat{\boldsymbol{y}}_i\|_2}$. *Here,* $\hat{\boldsymbol{y}}_i$ *is the expected received signal of the* i*-th posture, i.e.,* $\boldsymbol{\eta}_i$*, given a unit SNR, i.e.,* $\boldsymbol{\gamma}_{rx} = 1$. *Since* γ_{rx} *is proportional to* P_t*, the upper-bound on the minimal average false recognition cost is a monotonically decreasing function of* P_t.

Proof Based on [34], when the optimal decision function is adopted, the minimal false recognition cost can be calculated as

$$\Psi_{\mathcal{L}^*} = \sum_{i=1}^{N_P}\sum_{j=1}^{N_P} p_j \Pr(\mathscr{H}_i|\boldsymbol{\eta}_j). \quad i \neq j \tag{4.41}$$

Here, $\Pr(\mathscr{H}_i|\boldsymbol{\eta}_j)$ denotes the probability that the decision function determines the i-th posture when the j-th posture is true. Based on (4.5), the received signal vector given the i-th posture can be expressed as $\sqrt{\gamma_{rx}}\hat{\boldsymbol{y}}_i + \boldsymbol{\sigma}$, where $\bar{\boldsymbol{\sigma}}$ indicates the average noise signal vector. Then, an upper bound on the minimal false recognition cost in this case can be calculated as follows. The upper bound of $\Pr(\mathscr{H}_i|\boldsymbol{\eta}_j)$ is

$$\Pr(\mathscr{H}_i|\boldsymbol{\eta}_j) \leq \Pr\{I_j(\sqrt{\gamma_{rx}}\hat{\boldsymbol{y}}_i + \bar{\boldsymbol{\sigma}}) \leq I_i(\sqrt{\gamma_{rx}}\hat{\boldsymbol{y}}_i + \bar{\boldsymbol{\sigma}})\} \tag{4.42}$$

The inequality $I_j(\hat{\boldsymbol{y}}_i + \bar{\boldsymbol{\sigma}}) \leq I_i(\hat{\boldsymbol{y}}_i + \bar{\boldsymbol{\sigma}})$ can be derived as

$$\bar{\boldsymbol{\sigma}}^T(\hat{\boldsymbol{y}}_j - \hat{\boldsymbol{y}}_i) \geq \sqrt{\gamma_{rx}}(\gamma_{rx}\sum_{k=1}^{K}\|\hat{\boldsymbol{y}}_j - \hat{\boldsymbol{y}}_i\|_2^2 - \bar{\epsilon}\ln(p_j/p_i)) \tag{4.43}$$

Since $\bar{\boldsymbol{\sigma}}^T(\hat{\boldsymbol{y}}_j - \hat{\boldsymbol{y}}_i) \leq \|\bar{\boldsymbol{\sigma}}\|_2 \cdot \|\hat{\boldsymbol{y}}_j - \hat{\boldsymbol{y}}_i\|_2$, the upper bound of $\Pr(\mathscr{H}_i|\boldsymbol{\eta}_j)$ is

$$\Pr\left\{\|\bar{\boldsymbol{\sigma}}/\bar{\epsilon}\|_2 \geq \frac{\sqrt{\gamma_{rx}}(\gamma_{rx}\|\hat{\boldsymbol{y}}_j - \hat{\boldsymbol{y}}_i\|_2^2/\bar{\epsilon} - \ln(p_j/p_i))}{\|\hat{\boldsymbol{y}}_j - \hat{\boldsymbol{y}}_i\|_2}\right\}. \tag{4.44}$$

As $\bar{\sigma}/\bar{\epsilon}$ is the complex Gaussian random variable with mean 0 and variance 1, $\|\bar{\sigma}/\bar{\epsilon}\|_2$ follows the *chi distribution*. Therefore, $\Pr(\mathscr{H}_i|\eta_j) \leq P(K, Z(\gamma_{rx})^2/2)$, where $Z(\gamma_{rx}) = \sqrt{\gamma_{rx}}(\gamma_{rx}\|\hat{\mathbf{y}}_j - \hat{\mathbf{y}}_i\|_2^2/\bar{\epsilon} - \ln(p_j/p_i))/\|\hat{\mathbf{y}}_j - \hat{\mathbf{y}}_i\|_2$ and $P(\cdot)$ denotes the regularized gamma function.

Substitute $\Pr(\mathscr{H}_i|\eta_j) \leq P(K, Z(\gamma_{rx})^2/2)$ into (4.41), and (4.40) can be proven. Besides, since the CDF of chi distribution, i.e., the regularized gamma function, is a monotonically increasing function with respect to γ_{rx}, the upper-bound on the minimal false recognition cost is a monotonically decreasing function of the SNR and transmit power. ■

4.1.6 System Implementation

In this section, we elaborate on the implementation of the RIS-based posture recognition system. We first elaborate on the implementation of the RIS, and then describe the implementation of the transceiver module.

4.1.6.1 Implementation of RIS

We adopt the electrically modulated RIS proposed in [36], which is shown in Fig. 4.4. The RIS is with the size of $69 \times 69 \times 0.52$ cm^3 and is composed of a two-dimensional array of electrically controllable RIS elements. Each row/column of the array contains 48 RIS elements, and therefore, the total number of RIS elements is 2304.

Each RIS element has a size of $1.5 \times 1.5 \times 0.52$ cm^3 and is composed of four rectangle copper patches printed on a dielectric substrate (Rogers 3010) with a dielectric constant of 10.2. Any two adjacent copper patches are connected by a PIN diode (BAR 65-02L), and each PIN diode has two operation states, i.e., ON and OFF, which are controlled by applied bias voltages on the via holes. Specifically, when the applied bias voltage is 3.3 V (or 0 V), the PIN diode is at the ON (or OFF) state. Besides, to isolate the DC feeding port and microwave signal, four choke inductors of 30 nH are used in each RIS element. Besides, as shown in Fig. 4.4, an RIS element contains four choke inductors that are used to isolate the DC feeding port and RF signals.

As there are $N_D = 3$ PIN diodes in an RIS element, the total number of possible states of an RIS element is 8. We simulate the S_{21} parameters, i.e., the forward transmission gain, of the RIS element in four selected states for normal-direction incidence RF signals in CST software, Microwave Studio, Transient Simulation Package [37]. Table 4.1 provides the amplitude ratios and phase shifts of the RIS element in four selected states for incidence sinusoidal signals with a frequency of 3.198 GHz. It can be observed that the four selected states have the phase shift values equaling to $\pi/4$, $3\pi/4$, $5\pi/4$, and $7\pi/4$, respectively. We pick these four

Table 4.1 Normal direction forward transmission gain of RIS element in different states

State	Bias voltages on PIN diodes			Normal direction forward transmission gains in CST simulation	
	PIN # 1	PIN #2	PIN #3	Phase shift	Amplitude ratio
$\hat{s}_1$	0V (OFF)	0V (OFF)	0V (OFF)	$\pi/4$	0.97
$\hat{s}_2$	0V (ON)	0V (OFF)	3.3V (ON)	$3\pi/4$	0.97
$\hat{s}_3$	3.3V (ON)	0V (OFF)	3.3V (ON)	$5\pi/4$	0.92
$\hat{s}_4$	3.3V (ON)	3.3V (ON)	0V (OFF)	$7\pi/4$	0.88

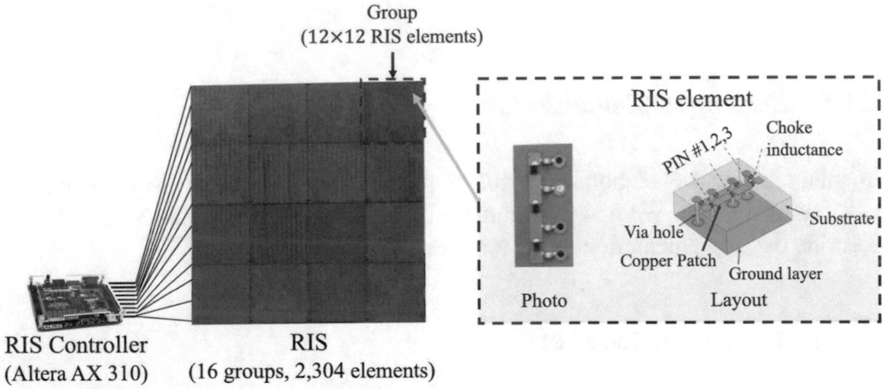

Fig. 4.4 RIS controller, RIS, and RIS element

states with a phase shift interval equaling to $\pi/2$ as the available state set $\mathscr{S}_a$, i.e., $\mathscr{S}_a = \{\hat{s}_1, \hat{s}_2, \hat{s}_3, \hat{s}_4\}$.

The RIS elements are divided into 16 groups, and thus each group contains 12×12 adjacent RIS elements arranged squarely. The RIS elements within the same group are in the same state. As shown in Fig. 4.4, the states of the 16 groups are controlled by a *RIS controller*, which is implemented by a field-programmable gate array (FPGA) (ALTERA AX301). Specifically, we use the expansion ports on the FPGA to control the frame configuration of the RIS. Every three expansion ports control the state of one group by applying bias voltages on the PIN diodes. Besides, the FPGA is pre-loaded with the configuration sequence matrix in Sect. 4.1.4.1. The configurations of the RIS are changed automatically with the control of the FPGA.

4.1.6.2 Implementation of Transceiver Module

As shown in Fig. 4.5, the transceiver module of the designed RIS-based posture recognition system consists of the following components:

- **Tx and Rx USRP devices**: We implement the transmitter and receiver based on two USRPs (LW-N210), i.e., a Tx USRP and an Rx USRP, which are capable of converting baseband signals to RF signals and vise versa. The USRP is composed

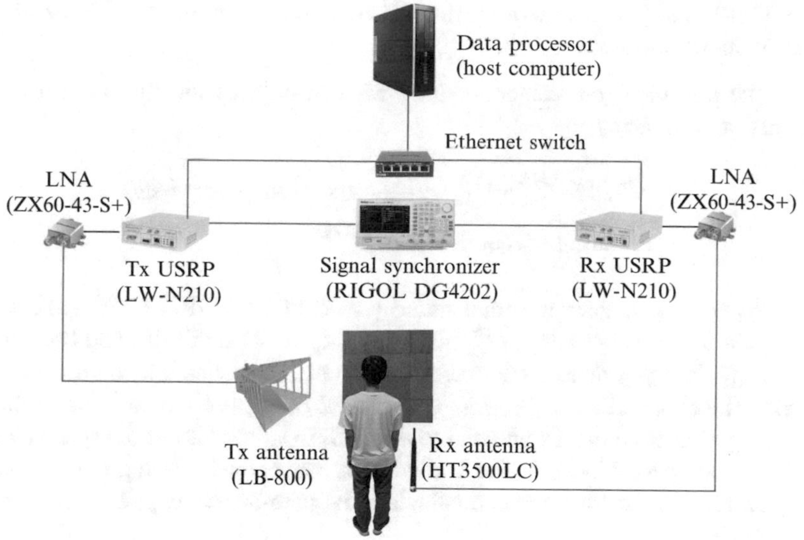

Fig. 4.5 Components of the transceiver module

of the hardwares such as the RF modulation/demodulation circuits and baseband processing unit and can be controlled using software [38].

- **Low-noise amplifiers (LNAs)**: To cope with the large attenuation incurred by the reflections on the RIS and the human body, we connect two LNAs (ZX60-43-S+) to the input and output ports of the Tx and Rx USRPs which amplify the transmitted and received RF signals of the USRPs, respectively.
- **Tx and Rx antennas**: The Tx antenna is a directional double-ridged horn antenna (LB-800), and the Rx antenna an omni-directional vertical antenna (HT3500LC). Both antennas are linearly polarized.
- **Signal synchronizer**: For the Rx USRP to obtain the relative phases and amplitudes of the received signals with respect to the transmitted signals of the Tx USRP, we employ a signal source (RIGOL DG4202) to synchronize the Tx and Rx USRPs. The signal source provides the reference clock signal and the pulses-per-second (PPS) signal to the USRPs, which ensures the modulation and demodulation of the USRPs to be coherent. Besides the phase synchronization, the RF chains of the Tx and Rx USRPs are also connected by a wired link with a fixed gain, which is used to compensate the instrumental error of the USRPs.
- **Ethernet switch**: The Ethernet switch connects the USRPs and a host computer to a common Ethernet, where they exchange the transmission and received signals.
- **Data processor**: The data processor is a host computer which controls the two USRPs by using Python programs based on the GNU packet [39]. Besides, the host computer extracts the measurement vectors from the received signals of the

Rx USRP and handles them by the NN trained by Algorithm 12 to obtain the decisions of postures.

For the designed transceiver module, we can express the link budget of the propagation channel as

$$\begin{aligned} P_{rx} =& P_{tx} + G^{\text{LNA}} + G_{tx}^{A} - L_{tx\to\text{RIS}} - L_{\text{RIS}\to\text{SB}} - L_{\text{SB}\to rx} \\ &- R_{\text{RIS}} - R_{\text{SB}} + G_{tx}^{A} + G^{\text{LNA}}. \end{aligned} \tag{4.45}$$

Here, P_{tx} [dBm] denotes the transmitted power of Tx USRP, G^{LNA} [dB] is the power gain of the Tx/Rx LNA, G_{tx}^{A} [dBi] and G_{rx}^{A} [dBi] are the Tx and Rx antenna gains, R_{RIS} and R_{SB} denote the power loss due to the reflections on the RIS and the space blocks, and $L_{tx\to\text{RIS}}$, $L_{\text{RIS}\to\text{SB}}$, and $L_{\text{SB}\to rx}$ are the path-losses due to the propagation from the Tx antenna to the RIS, from the RIS to the space blocks, and from the space blocks to the Rx antenna, respectively. Using the free-space LoS path-loss model [40], we can calculate the path-losses, e.g., $L_{tx\to\text{RIS}}$ can be calculated by

$$L_{tx\to RIS} = 20\log_{10}(d_{tx\to\text{RIS}}) + 20\log_{10}(f_c) - 144.55,$$

where $d_{tx\to\text{RIS}}$ denotes the distance from the Tx antenna to the RIS.

4.1.7 Simulation and Experimental Results

In this section, we first describe the system setting for the simulation and experiments and then specify the adopted parameters. Simulation results and experimental results are then provided and discussed.

4.1.7.1 System Setting for Simulation and Experiment

In the simulation, the layout of the RIS, the Tx and Rx antennas, and the space of interest is depicted as Fig. 4.6a. Specifically, the origin of the 3D-coordinate is located at the center of the RIS, and the RIS is in the $y-z$ plane. Besides, the z-axis is vertical to the ground and pointing upwards, and the x- and y-axes are parallel to the ground.

The Tx antenna is located more than 10λ (around 0.936 m) away from the RIS, so that the incidence signals on the RIS elements can be approximated as a plane wave. Therefore, the incidence angles of the incidence signals on the RIS elements are approximately the same, which is $(60°, 0°)$. The human body is in the space of interest, which is a cuboid region located at 1 m from the RIS. Since the space of interest is behind the Tx antenna and the Tx antenna is directional horn antenna, no

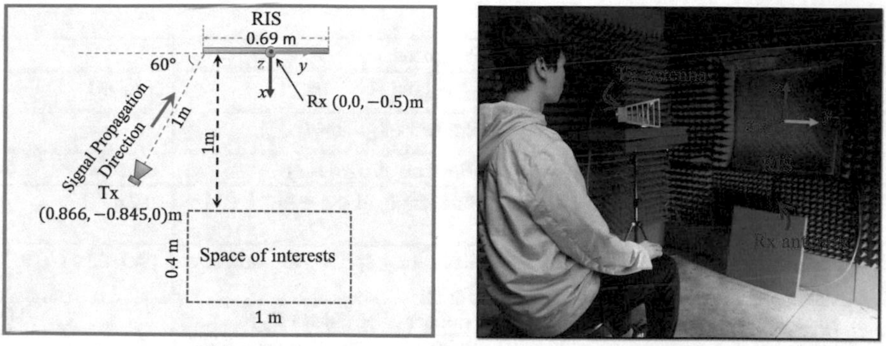

(a) (b)

Fig. 4.6 Layout of the simulation and experiment of the RIS-based posture recognition system. (**a**) is the layout used for simulation, and (**b**) is the layout of the experiment, which is arranged in accordance with the layout in (**a**)

LoS signal path from the Tx antenna to the space of interest exists. The side lengths of the space of interest are $l_x = 0.4\,\text{m}$, $l_y = 1.0\,\text{m}$, and $l_z = 1.6\,\text{m}$. Besides, the space of interest is further divided into $M = 80$ cubics with a side length of 0.2 m.

Besides, we obtain the reflection coefficient function of the RIS elements by using the CST. We model an RIS element according to Fig. 4.4 and simulate the far-field radiation pattern under the stimulation of a z-axis polarized plane wave with the incidence angle $(60°, 0°)$. The simulation is carried out in the frequency domain with the unit-cell boundary. Moreover, we project the results onto the z-axis to obtain the reflection coefficients since the Tx and Rx antenna are both linearly-polarized along the z-axis.

Figure 4.6b shows the experiment environment and layout of the implemented system. We conduct practical experiments on the implemented RIS based posture recognition system in a low-reflection environment, where the walls are covered with the wave-absorbing materials. This environment set-up imitates a vast and empty room, where the RIS is fixed on a wall with a low reflection rate for the 3.198 GHz wireless signals. The Tx and Rx antennas are in the room, while the USRPs, host computer, and the connecting circuits are located behind the RIS.

In Table 4.2, we summarize the adopted parameters in the simulation. The values of the parameters are taken according to the ones that we input to the algorithms and the ones that we obtained from the specifications of the employed devices, i.e., the Tx/Rx antennas and the LNAs.

Based on (4.46) and the experimental setting, we can calculate the link budget of the RIS-based RF sensing system. Assume that the reflections on the RIS and the space blocks incur no energy loss, i.e., $R_{\text{RIS}} = R_{\text{SB}} = 0$, so that the received signal power $P_{rx} = -70.75$ dBm. Based on the calculated link budget, the necessity of adopting LNAs can be further justified: If no LNAs are adopted, the link budget will be $-70.75 - 15.8 \times 2 = -102.35$ dBm and lower than the noise level (about

Table 4.2 Simulation parameters

Parameter	Value	Parameter	Value
Tx antenna gain (G^A_{Tx})	12.6 dBi	LNA gain (G^{LNA})	15.8 dB
Rx antenna gain (G^A_{Rx})	6 dBi	Number of groups (L)	16
Tx Power (P_t)	100 mW	Duration of frame (δ)	0.3 s
Number of RIS elements (N)	2304	Side length of space block (Δ)	0.2 m
Signal frequency (f_c)	3.198 GHz	Size of space of interest (l_x, l_y, l_z)	(0.5, 0.9, 1.7) m
Signal wavelength (λ)	0.0936 m	Number of postures (N_p)	4
Number of available states (N_a)	4	Number of iterations for augmented Lagrangian (L_{AL})	50
Number of iterations for alternating minimization ($N_{A\|M}$)	50	Learning rate (ζ)	0.01

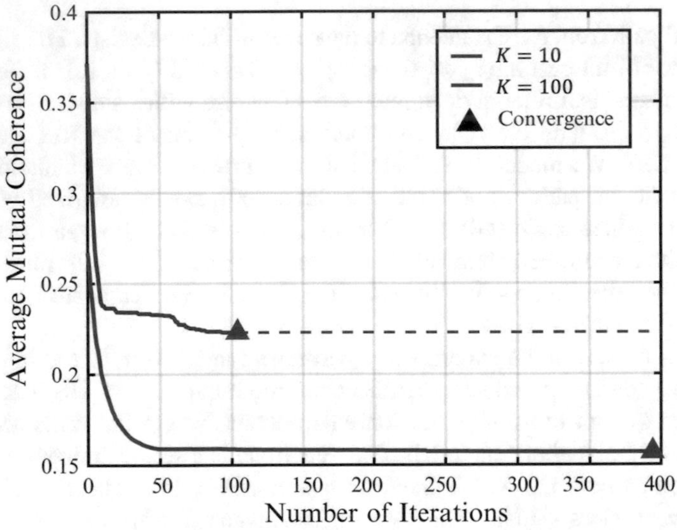

Fig. 4.7 Average mutual coherence of measurement matrix vs. the number of iterations in FCAO algorithm, under different numbers of frames

−100 dBm), which makes the received signals hard to be distinguished from the noise and thus degrades the recognition accuracy.

4.1.7.2 Simulation Results

Figure 4.7 shows the average mutual coherence of $\boldsymbol{\Gamma}$ vs. the number of iterations in Algorithm 11, under different numbers of frames K. It can be observed that average mutual coherence decreases with the number of iterations, which verifies the effectiveness of the FCAO algorithm. Besides, it can also be seen that the converged optimal average mutual coherence of $\boldsymbol{\Gamma}$ decreases with the number of frames, i.e., K. This can be explained as follows. Since $\boldsymbol{\gamma}_m$, i.e., the m-th column of $\boldsymbol{\Gamma}$ indicates the measurement of the m-th space block. Since $\boldsymbol{\gamma}_m = \boldsymbol{T}\boldsymbol{\alpha}_m$ ($\forall m \in [1, M]$), to reduce the mutual coherence of $\boldsymbol{\Gamma}$, it requires that the configuration matrix $\boldsymbol{T}$ maps $\boldsymbol{\alpha}_m$ to $\boldsymbol{\gamma}_m$ where different $\boldsymbol{\gamma}_m$ have large elements at different dimensions. When K is large, the number of dimensions of $\boldsymbol{\gamma}_m$ ($\forall m \in [1, M]$) is large, and therefore, the probability of finding $\boldsymbol{T}$ to distribute the large components of $\boldsymbol{\gamma}_m$ at different dimensions is larger.

Besides, we illustrate the configuration matrix and the corresponding coherence vectors before and after the configuration matrix optimization with $K = 10$. The configuration sequence shown in Fig. 4.8a is a random configuration matrix where the duration of all the states in each configuration follows the same uniform distribution with a fixed sum equaling to δ. Figure 4.8c shows the coherence of column vectors of $\boldsymbol{\Gamma}$, i.e., $u_{m,m'}$ ($\forall m, m' \in [1, M],\ m \neq m'$). The configuration matrix in Fig. 4.8c is the optimized configuration matrix obtained by using Algorithm 11, and Fig. 4.8d shows the mutual coherence of the measurement matrix corresponding to it. Comparing Fig. 4.8b and d, it can be seen that most of the coherence values corresponding to the optimized configuration matrix is lower than that corresponding to the random configuration matrix. Besides, it can also be observed that the average mutual coherence corresponding to the optimized configuration matrix is significantly reduced. Therefore, the effectiveness of the FCAO algorithm is verified.

Figure 4.9a and b shows the optimal average mutual coherence obtained by Algorithm 11 under different sizes of the RIS and different sizes of group, respectively. In Fig. 4.9a, the size of the RIS is determined by the number of groups, L, and each group contains $N_G = 12 \times 12$ RIS elements. It can be seen that the optimal average mutual coherence decreases with the number of groups that the RIS contains, i.e., the size of the RIS. In Fig. 4.9b, the size of the RIS is fixed, and the RIS contains $N = 48 \times 48$ elements. The value of N_G determines the number of groups of the RIS, which can be controlled independently. It can be seen that the optimal average mutual coherence increases with the size of groups.

Based on Sect. 4.1.4.1, the average mutual coherence is negatively related to the recognition accuracy, which is proved in the experimental results in the following subsection. Therefore, the results shown in Fig. 4.9a and b also indicate that the posture recognition accuracy of the system increase with the size of the RIS and the number of independently controllable groups.

4.1.7.3 Experimental Results

We use the implemented posture recognition system described in Sect. 4.1.6 to perform the experiments. We first carry out an experiment which proves the capability of the RIS to increase the number of transmission channels and supports

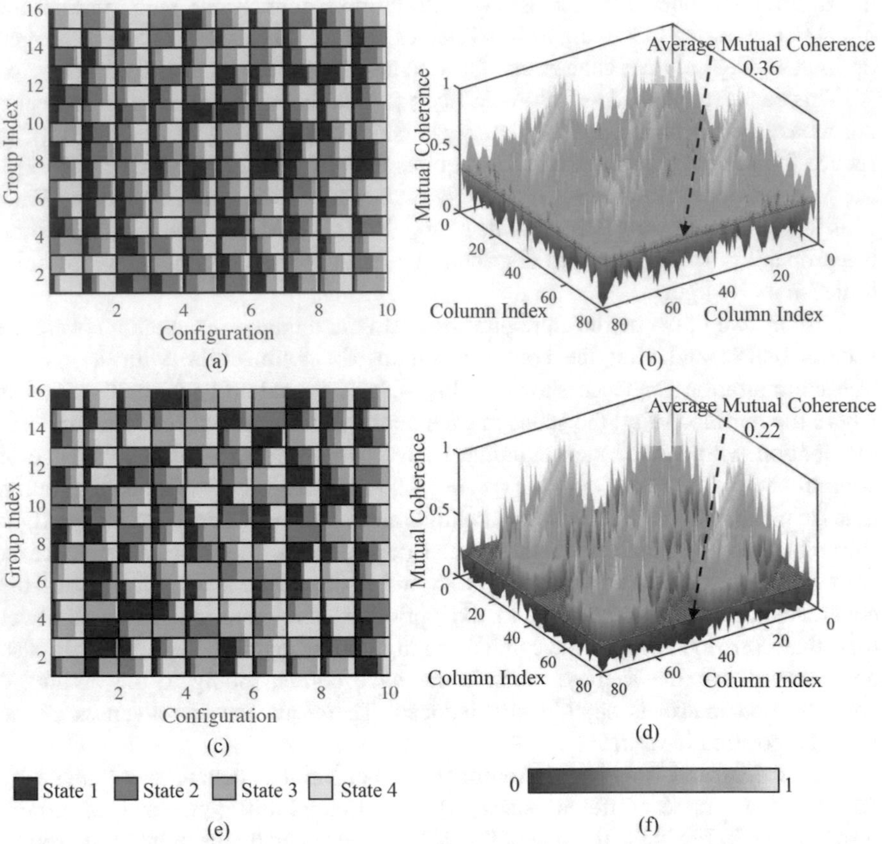

Fig. 4.8 (**a**) Illustration of a random configuration; (**b**) Mutual coherence of the measurement matrix given the random configuration; (**c**) Illustration of the optimized configuration obtained by Algorithm 11; (**d**) Mutual coherence of the measurement matrix given the optimized configuration; (**e**) Colors that represents the duration of different states; (**f**) Color scaling representing the coherence to be from 0 to 1

the assumption that the correlation between RIS elements can be neglected. To be specific, we place a $10 \times 10\,\text{cm}^2$ copper patch at the center of the space of interest, and measure the received signals for 40 random configurations. In Fig. 4.10, we compare the measured signals with the expected signals based on (4.5). To facilitate observation, the amplitudes of the signals are normalized so that the average values are equal to 1. Figure 4.10 shows that by altering the configurations of the RIS, the received signals at the Rx can be changed, which proves the capability of the RIS to increase the diversity of the transmission channel over time. Besides, it can also be observed that the measured signals and the expected signals by (4.5) under the assumption that the correlation can be neglected are approximately the same. Therefore, by controlling the RIS elements in the unit of groups, the influence of

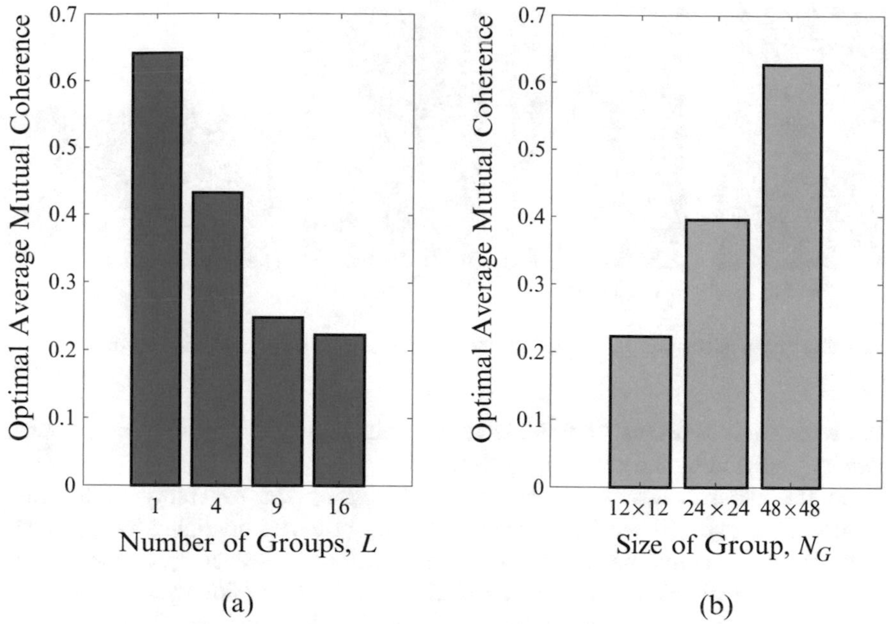

Fig. 4.9 Optimal average mutual coherence values vs. (**a**) number of groups, L, given $N_G = 12 \times 12$; and (**b**) size of group, N_G, given number of elements $N = 48 \times 48$

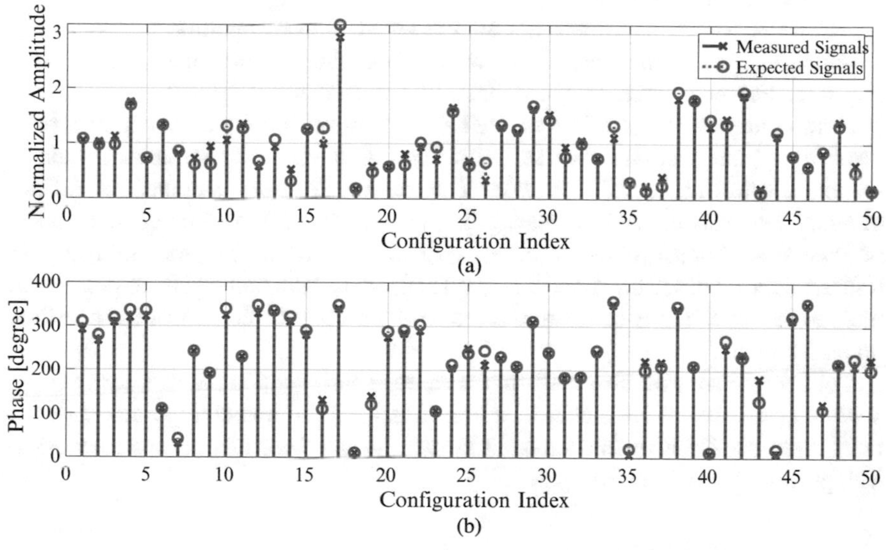

Fig. 4.10 (**a**) Normalized amplitudes and (**b**) phases of the measured signals and the expected signals by (4.5) with the assumption that the RIS elements are uncorrelated

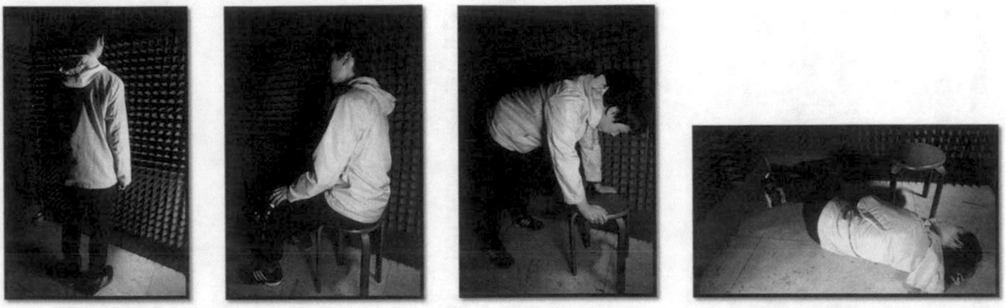

Standing Sitting Bending Lying down

Fig. 4.11 Human postures for recognition: standing, sitting, bending, and lying down

the correlation between the RIS elements is alleviated, and the measured signals can be predicted by the proposed system model.

In the experiments, we focus on demonstrating the capability of the proposed RIS-based RF sensing system with the optimized configuration to obtain high-accuracy human posture recognition. Specifically, we compare the posture recognition accuracies of the proposed RIS-based RF sensing system under the optimized and random configurations, and that of an RF sensing system in the non-configurable environment. Besides, the variable-controlling approach is adopted, i.e., the experimental environments, including the human and his clothing in different cases, are the same.[2]

We set $K = 10$, i.e., there are 10 frames in each recognition period. As for the human postures for recognition, we consider the $N_P = 4$ postures shown in Fig. 4.11. For each posture, we collect 150 labeled measurement vectors in the random configuration matrix case and the optimized configuration matrix case, respectively, and form the data sets. Then, we divide the data set into the training set and the testing set in each case. The training data set contains 120 labeled measurement vectors, and the testing data set contains 30 measurement vectors to be processed. In each case, we train the decision function, i.e., the NN using the training data set based on Algorithm 12. Then, we use the trained NN to process the measurement vectors in the testing set and record the probabilities of the decisions on each posture.

For comparison, we also perform the same experiments in the *non-configurable environment* case, which serves as a benchmark. In the non-configurable environment case, the RIS elements are fixed to state $\hat{s}_1$, and therefore, the system work as a single-antenna RF sensing system.

[2] As the person and clothes impact the space reflection vector and the measurement vector, they influence the training of the decision function, which indicates that the experimental results are specific to the person and clothes. Nevertheless, the designed system can be applied when multiple people with different clothes appear in the environment. In that case, the decision function needs to be trained by Algorithm 12 with a data set containing the measurement vectors corresponding to each person and clothes.

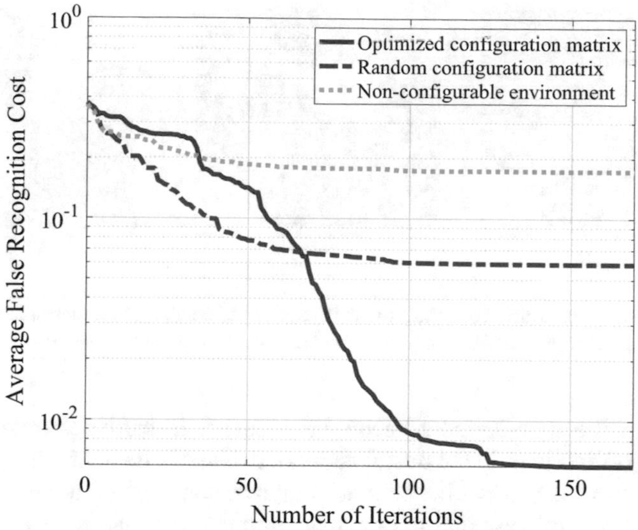

Fig. 4.12 Average false recognition cost vs. the number of iterations of Algorithm 12 given optimized configuration matrix and random configuration matrix. $K = 10$

Figure 4.12 shows the average false recognition cost vs. the number of iterations of Algorithm 12, where the costs of true and false recognition are set to 0 and 1, respectively, i.e., $\chi_{i,i'} = 0$ if $i = i'$; otherwise, $\chi_{i,i'} = 1$. It can be observed that the average false recognition cost decreases with the number of iterations. Besides, the converged value of average false recognition cost using the optimized configuration matrix is about 10 times smaller than that using a random configuration matrix. This verifies that the optimized configuration matrix, which has a measurement matrix with low average mutual coherence, can results in lower false recognition cost compared to the random configuration matrix. Moreover, by comparing with the benchmark case, we can observe that the capability of the RIS to customize the environment helps the RF sensing system to reduce the average false recognition cost.

Besides, it can be seen that when the number of training epochs is small, the training results of the optimized configuration matrix are worse than those of the random configuration matrix and the non-configurable radio environment, which can be explained as follows. In the non-configurable environment, since the configuration of the RIS is fixed, the measurement matrix maps the reflection coefficient vector to a one-dimensional complex number. Besides, compared with that of the optimized configuration matrix, the measurement matrix corresponding to the random configuration matrix has a higher average mutual coherence. Based on [26], this indicates that the received signals of the optimized configuration matrix contain more information about the reflection coefficient vector in space compared to the random one and the one-dimensional complex number. As the information

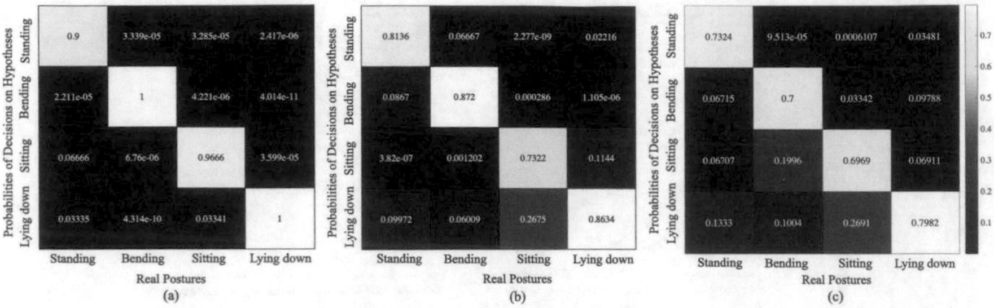

Fig. 4.13 Posture recognition accuracy in the (**a**) optimized configuration matrix case; (**b**) random configuration matrix case; and (**c**) non-configurable environment case. $K = 10$

contained in the measurement vector increases, it is harder for the supervised learning algorithm to train the neural network to classify them [29].

Figure 4.13a and b shows the accuracy of the posture recognition in the optimized configuration matrix case and the random configuration matrix case, respectively. It can be seen that in the optimized configuration matrix case, the recognition accuracy is much higher than that in the random configuration matrix case. This verifies that the optimized configuration matrix, which has a measurement matrix with low average mutual coherence, can result in higher recognition accuracy in the practical posture recognition system. Besides, it can be observed that the system with optimized configuration can achieve 14.6% higher recognition accuracy compared with that with random configuration. Moreover, by comparing with the benchmark case, we can observe that the capability of the RIS to customize the environment increases the posture recognition accuracy of RF sensing systems with 23.5%.

4.1.8 Summary

In this section, we have designed an RIS-based posture recognition system. To facilitate the configuration design, we have proposed a frame-based periodic configuring protocol. Based on the protocol, we have formulated the optimization problem for false recognition cost minimization. To solve the problem, we have decomposed it into the configuration matrix and the decision function optimization problems, and proposed the FCAO algorithm and the supervised learning algorithm to solve them, respectively. Besides, based on USRPs, we have implemented the designed system and executed posture recognition experiments in practical environments. Simulations have verified that the FCAO algorithm can obtain the optimal configuration matrix, which leads to a measurement matrix with low average mutual coherence. The experimental results prove that the configuration matrix with lower average mutual coherence has higher recognition accuracy and a lower false recognition cost. Besides, combining the simulation and experimental results, we have shown that the posture recognition accuracy increase with the size of the RIS and the number of independently controllable groups. Moreover, compared with the random configuration and the non-configurable environment cases, the optimized configuration can achieve 14.6% and 23.5% higher recognition accuracy, respectively.

4.1.9 Appendix

The augmented Lagrangian method finds a local optimal solution to (P4.5) by iteratively minimizing a sequence of (4.32) where $\boldsymbol{\beta}$ and ρ are fixed in each iteration. Specifically, the sequence of augmented Lagrangian minimization can be solved using an *alternating minimization procedure*, in which $\boldsymbol{u}$ is updated while $\boldsymbol{t}_k$ is held fixed, and vice versa. The alternating minimization procedure can be described as follows.

1. **Update Step for $\boldsymbol{u}$**: We consider the $\boldsymbol{u}$ update step in the alternating minimization procedure for (4.32) minimization given $\boldsymbol{\beta}$, ρ and fixed $\boldsymbol{T}_{-k}$, and $\boldsymbol{t}_k$. As the first step, we introduce the auxiliary variable $z_{m,m'} = \frac{\boldsymbol{\gamma}_m^T \boldsymbol{\gamma}_{m'}}{\|\boldsymbol{\gamma}_m\|_2 \cdot \|\boldsymbol{\gamma}_{m'}\|_2} - \beta_{i,j}/\rho$ and arrange the auxiliary variables into a vector $z = (z_{1,2}, z_{1,3}, \ldots, z_{1,M}, z_{2,3}, \ldots, z_{M-1,M})$. Then, update $\boldsymbol{u}$ to solve the minimization for (equ: convenient aug. Lagrangian) can be handled by the proximal gradient method [41], which is equivalent to solving
$$\mathrm{P}_{\boldsymbol{u}}: \min_{\boldsymbol{u}} \|\boldsymbol{u}\|_1 + \frac{\rho}{2}\|\boldsymbol{u} - z\|_2^2. \tag{4.46}$$
Since the sum of norm functions are convex, ($\mathrm{P}_{\boldsymbol{u}}$) is an unconstrained convex optimization problem, which can be solved efficiently by using existing convex optimization algorithms [42].
2. **Update Step for $\boldsymbol{t}_k$**: We then consider the $\boldsymbol{t}_k$ update step. We first introduce the auxiliary variable $\kappa_{m,m'} = u_{m,m'} + \beta_{m,m'}/\rho$. Then, the minimization for the augmented Lagrangian given fixed $\boldsymbol{u}$ can be reduced to the following non-convex optimization problem:

$$\mathrm{P}_{\boldsymbol{t}_k} \min_{\boldsymbol{d}_k} \sum_{1 \le m < m' \le M} \frac{\rho}{2} \left| \frac{\boldsymbol{\gamma}_m^T \boldsymbol{\gamma}_{m'}}{\|\boldsymbol{\gamma}_m\|_2 \cdot \|\boldsymbol{\gamma}_{m'}\|_2} - \kappa_{m,m'} \right|^2 \tag{4.47}$$

$$s.t. \quad \boldsymbol{1}^T \tilde{\boldsymbol{t}}_{k,l} = 1, \ \forall l \in [1, L], \tag{4.48}$$

$$\tilde{\boldsymbol{t}}_{k,l} \succeq 0, \ \forall l \in [1, L]. \tag{4.49}$$

Due to the complicated objective function, the optimum to (P_{t_k}) is hard to solve. Nevertheless, an approximate update for $\boldsymbol{t}_k$ can be found using the proximal gradient method proposed in [43].

After the alternating minimization procedure, dual variable $\beta_{m,m'}$ is updated by

$$\beta_{m,m'} = \beta_{m,m'} + \rho \left(u_{m,m'} - \frac{\boldsymbol{\gamma}_m^T \boldsymbol{\gamma}_{m'}}{\|\boldsymbol{\gamma}_m\|_2 \cdot \|\boldsymbol{\gamma}_{m'}\|_2} \right). \tag{4.50}$$

Besides, ρ is updated by using the method described in [27]. In summary, the algorithm to solve the augmented Lagrangian function minimization is proposed as Algorithm 13.

Algorithm 13: Alternating minimization algorithm for solving (P4.5)

Input : Number of iterations for solving augmented Lagrangian function (N_{AL});
Number of iterations for the alternating minimization procedure (N_{AM});
Maximum iteration number $N_{\max}$;
$\boldsymbol{t}_k^{(0)}$, $\boldsymbol{u}^{(0)}$.
Output: Optimized $\boldsymbol{d}_k^*$ and $\boldsymbol{u}^*$ which are a sub-optimal solution for (P4.5).

Calculate $u_{m,m'}^{(0)} = \frac{(\gamma_m^{(0)})^T \gamma_{m'}^{(0)}}{\|\boldsymbol{\gamma}_m^{(0)}\|_2 \cdot \|\boldsymbol{\gamma}_{m'}^{(0)}\|_2}$ based on $\boldsymbol{t}_k^{(0)}$, and set random $\beta_{m,m'}^{(1)} \in (0, 1)$, $\forall m, m' \in [1, M]$, $m < m'$;
for $a = 1, \ldots, N_{\mathrm{AL}}$ **do**
  Set $\boldsymbol{u}^{(a,0)} = \boldsymbol{u}^{(a-1)}$, $\boldsymbol{t}_k^{(a,0)} = \boldsymbol{t}_k^{(a-1)}$;
  for $b = 1, \ldots, N_{\mathrm{AD}}$ **do**
    Update $\boldsymbol{u}$: Obtain $\boldsymbol{u}^{(a,b+1)}$ by solving $P_{\boldsymbol{u}}$ using convex optimization algorithm in [42], given $\boldsymbol{t}_k^{(a,b)}$, $\boldsymbol{\beta}^{(a)}$, and $\rho^{(a)}$;
    Update $\boldsymbol{t}_k$: Obtain $\boldsymbol{t}_k^{(a,b+1)}$ by solving $P_{\boldsymbol{t}_k}$ using proximal gradient method in [43], given $\boldsymbol{u}^{(a,b+1)}$, $\boldsymbol{\beta}^{(a)}$, and $\rho^{(a)}$;
  end
  Compute the dual variables $\boldsymbol{\beta}^{(a+1)}$ by (4.50) and compute $\rho^{(a+1)}$ by using the method in [27].
end

4.2 3D Sensing

4.2.1 Motivations

As we have introduced in the previous section, leveraging widespread RF signals for wireless sensing applications has attracted growing research interest. Many RF-based sensing methods based on WiFi signals or millimeter wave signals have been proposed for sensing and recognizing human being and objects. However, using RF signals for sensing usually encompasses a signal collection and analysis process which passively accept the radio channel environment. The radio environment is unpredictable and usually unfavorable, and thus the sensing accuracy of conventional RF sensing methods is usually affected by unwanted multi-path fading [15, 44], and/or unfavorable propagation channels from the RF transmitters to the receivers.

RISs have been proposed as a promising solution for turning unwanted propagation channels into favorable ones [2, 45]. By programming the reconfigurable elements, an RIS deployed in the environment can change the RF propagation channel and create favorable signal beams for sensing [46, 47]. Instead of employing complex and sophisticated RF transmitters and receivers [48, 49], RIS assisted RF sensing paves a new way of developing RF sensing methods, which have the capabilities of controlling, programming, and hence customizing the wireless channel. In literature, the authors of [36] explored the use of RISs to assist RF sensing and obtain 2D images for human beings. Nevertheless, no research works have tackled the analysis and design of RIS assisted 3D RF sensing, which is more challenging to analyze and optimize than 2D RF sensing introduced in the previous section.

In this section, we consider an RIS assisted RF 3D sensing scenario, which can sense the existence and locations of 3D objects in a target space. Specifically, by programming the beamformer patterns, the RIS performs beamforming and provides desirable RF propagation properties for sensing. However, there are two major challenges in obtaining high sensing accuracy in RIS assisted RF sensing scenarios.

- *First*, the beamformer patterns of the RIS need to be carefully designed to create favorable propagation channels for sensing.
- *Second*, the *mapping of the received signals*, i.e., the mapping from the signals received at the RF receiver to the sensing results of the existence and locations of the objects, needs to be optimized as well.

Nevertheless, the complexity of finding the optimal beamformer patterns is extremely high because the associate optimization problem is a discrete nonlinear programming with a large number optimization variables. Besides, the optimization of the beamformer patterns and the mapping of the received signals are closely coupled together, which makes optimizing the sensing accuracy in RIS assisted RF sensing scenarios even harder.

To tackle these challenges, we formulate an optimization problem for sensing accuracy maximization by minimizing the cross-entropy loss of the sensing results with respect to the beamformer patterns and the mapping of the received signals. In order to solve the problem efficiently, we formulate a Markov decision process (MDP) for the optimization problem and propose a deep reinforcement learning algorithm. The proposed deep reinforcement learning algorithm is based on the policy gradient algorithm [50] and is referred to as the progressing reward policy gradient (PRPG) algorithm, since the reward function of the MDP is consistently being improved during the learning process. The computational complexity and the convergence of the proposed algorithm are analyzed. Moreover, we derive a non-trivial lower-bound for the sensing accuracy for a given set of beamformer patterns of the RIS. Simulation results verify the effectiveness of the proposed algorithm and showcase interesting performance trends about the sensing accuracy with respect to the sizes of the RIS and the target space.

The rest of this section is organized as follows. In Sect. 4.2.2, we introduce the model of the RIS assisted RF sensing scenario. In Sect. 4.2.3, we formulate the optimization problem to optimize the sensing accuracy by minimizing the cross-entropy loss of the sensing results. In Sect. 4.2.4, we formulate an MDP for the optimization problem and then proposed the PRPG algorithm to solve it. In Sect. 4.2.5, the complexity and convergence of the PRPG algorithm are analyzed, and a lower-bound for the sensing accuracy is derived. Simulation results are provided in Sect. 4.2.6 and we summarize this section in Sect. 4.2.7.

4.2.2 System Model

In this section, we introduce the RIS assisted 3D RF sensing scenario [51], which is illustrated in Fig. 4.14. In this scenario, there exist a pair of single-antenna RF transceivers, an RIS, and a target space where the objects are located. The RIS reflects and modifies the incident narrow-band signals at a certain frequency f_c. The Tx unit and Rx unit of the transceiver keep transmitting and receiving at f_c. The target space is a cubical region that is discretized into M equally-sized *space grids*. Each space grid is of size $\Delta l_x \times \Delta l_y \times \Delta l_z$.

The sensing process adopted in the considered scenario can be briefly described as follow. The signals transmitted by the Tx unit are reflected and modified by the RIS before entering into the target space. The modified signals are further reflected

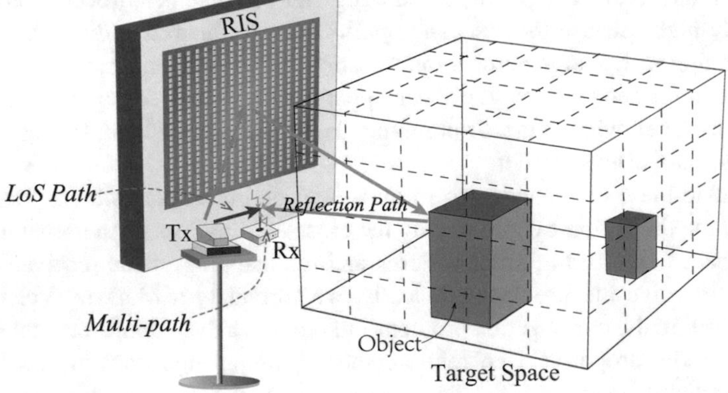

Fig. 4.14 Illustration of the RIS assisted RF sensing scenario

by the objects in the target space and received by the Rx unit. Then, the Rx unit maps the received signals to the sensing result, which indicates whether an object exists in each space grid.

4.2.2.1 RIS Model

As shown in Fig. 4.14, the reconfigurable element of the RIS are arranged in a two-dimensional array. By controlling the PIN diodes coupled with each reconfigurable element, the reconfigurable element can adjust its electromagnetic response to the incident RF signals. For each reconfigurable element, we refer to the different responses to incident RF signals as the reconfigurable element's *configuration* as in [52]. By changing the configuration of each reconfigurable element, the RIS is able to modify the reflected signals and perform beamforming [23].

We assume that each reconfigurable element has N_S configurations, and each configuration of an element has a unique reflection coefficient for the incident RF signals. To be specific, we assume that each row and column of the RIS contain the same number of reconfigurable elements, and the total number of reconfigurable elements is denoted by N. We denote the reflection coefficient of the n-th reconfigurable element corresponding to the incident signal from the TX unit and the reflected signal towards the m-th space grid by $r_{n,m}(c_n)$. Here, $c_n \in [1, N_S]$ denotes the configuration of the n-th reconfigurable element and $c_n \in \mathbb{Z}$, where $\mathbb{Z}$ denotes the set of integers.

4.2.2.2 Channel Model

In the RIS assisted RF sensing scenario, the Tx unit and Rx unit adopt single antennas to transmit and receive RF signals. The Tx antenna is a directional antenna, which points towards the RIS so that most of the transmitted signals are reflected by the RIS and propagate into the target space. The signals reflected by the RIS are reflected by the objects in the target space and then reach the Rx antenna. The Rx antenna is assumed to be omni-directional and located right below the RIS, as shown in Fig. 4.14. This setting ensures that the signals reflected by the RIS are not directly received by the Rx antenna, and thus most of the received signals contain the information of the objects in the target space.

As shown in Fig. 4.14, the transmission channel from the Tx antenna to the Rx antenna is composed of three types of paths, i.e., the LoS path, the reflection paths, and the environmental scattering paths. The LoS path indicates the direct signal path from the Tx antenna to the Rx antenna. The reflection paths are the paths from the Tx antenna to the Rx antenna via the reflections from the RIS and the objects in the target space. The environmental scattering paths account for the signals paths between the Tx antenna and the Rx antenna which involve complex reflection and scattering in the surrounding environment. Then, the equivalent baseband representation of the received signal containing the signals from all these three types of paths is denoted by y and can be expressed as

$$y = h_{\text{los}} \cdot \sqrt{P} \cdot x + \sum_{m=1}^{M} \sum_{n=1}^{N} h_{n,m}(c_n, \nu_m) \cdot \sqrt{P} \cdot x + h_{rl} \cdot \sqrt{P} \cdot x + \sigma, \tag{4.51}$$

where P is the transmit power, and x denotes the transmitted symbol.

The component terms of (4.51) can be explained in detail as follows. The first term, i.e., $h_{\text{los}} \cdot P \cdot x$, corresponds to the signal received in the LoS path, where h_{los} denotes the gain. Based on [40], h_{los} can be expressed as

$$h_{los} = \frac{\lambda}{4\pi} \cdot \frac{\sqrt{g_T g_R} \cdot e^{-j2\pi d_{\text{los}}/\lambda}}{d_{\text{los}}}, \tag{4.52}$$

where λ is the wavelength of the signal, g_T and g_R denote the gains of the Tx and Rx antennas, respectively, and d_{los} is the distance from the Tx antenna to the Rx antenna.

The second term in (4.51) corresponds to the signals that reach the Rx antenna via $N \cdot M$ reflection paths. In the second term, $h_{n,m}(c_n, \nu_m)$ denotes the gain of the reflection path via the n-th reconfigurable element in configuration c_n and the m-th space grid with reflection coefficient ν_m. Based on [22], $h_{n,m}(c_n, \nu_m)$ can be formulated as follows

$$h_{n,m}(c_n, \nu_m) = \frac{\lambda^2 \cdot r_{n,m}(c_n) \cdot \nu_m \cdot \sqrt{g_T g_R} \cdot e^{-j2\pi(d_n + d_{n,m})/\lambda}}{(4\pi)^2 \cdot d_n \cdot d_{n,m}}, \tag{4.53}$$

where d_n denotes the distance from the Tx antenna to the n-th reconfigurable element and $d_{n,m}$ denotes the distance from the n-th reconfigurable element to the Rx antenna via the center of the m-th space grid.

Finally, the third and forth terms in (4.51) correspond to the signals from the environmental scattering paths and the additive noise at the Rx antenna, respectively. The symbol $h_{rl} \in \mathbb{C}$ denotes the equivalent gain of all the environmental scattering paths, and σ is a random signal that follows the complex normal distribution, $\sigma \sim \mathcal{CN}(0, \epsilon)$ with ϵ being the power of the noise.

Moreover, we refer to the vector of configurations selected for the N reconfigurable elements as a *beamformer pattern* of the RIS, which can be represented by a $N \times N_S$-dimensional binary row vector $\boldsymbol{c} = (\hat{\boldsymbol{o}}(c_1), \ldots, \hat{\boldsymbol{o}}(c_N))$. Specifically, $\hat{\boldsymbol{o}}(i)$ ($\forall i \in [1, N_S]$) denotes the N_S-dimensional row vector whose i-th element is 1 and the other elements are 0. Based on the definition of the beamformer pattern, the received signal in (4.4) can be reformulated as

$$y = h_{\text{los}} \cdot \sqrt{P} \cdot x + \boldsymbol{c}\boldsymbol{A}\boldsymbol{v} \cdot \sqrt{P} \cdot x + h_{rl} \cdot \sqrt{P} \cdot x + \sigma, \tag{4.54}$$

where $\boldsymbol{\nu} = (\nu_1, \ldots, \nu_M)$ denotes the vector of reflection coefficients of the M space grids, $\boldsymbol{A} = (\boldsymbol{\alpha}_1, \ldots, \boldsymbol{\alpha}_M)$ is referred to as the *projection matrix*, and $\boldsymbol{\alpha}_m = (\hat{\boldsymbol{\alpha}}_{m,1}, \ldots, \hat{\boldsymbol{\alpha}}_{m,N})^T$ with $\hat{\boldsymbol{\alpha}}_{m,n} = (\hat{\alpha}_{m,n,1}, \ldots, \hat{\alpha}_{m,n,N_S})$. Here, for all $m \in [1, M]$, $n \in [1, N]$, and $i \in [1, N_S]$, $\hat{\alpha}_{m,n,i}$ denotes the channel gain of the reflection path via the n-th reconfigurable element in configuration i and the m-th space grid with a unit reflection coefficient, which can be expressed as follows based on (4.53).

$$\hat{\alpha}_{m,n,i} = \frac{\lambda^2 \cdot r_{n,m}(i) \cdot \sqrt{g_T g_R}}{(4\pi)^2 d_n d_{n,m}} \cdot e^{-j2\pi(d_n + d_{n,m})/\lambda}. \tag{4.55}$$

4.2.2.3 RF Sensing Protocol

To describe the RF sensing process in the RIS assisted scenario clearly, we formulate the following *RF sensing protocol*. In the protocol, the timeline is slotted and divided into *cycles*, and the Tx unit, the Rx unit, and the RIS operate in a synchronized and periodic manner. As shown in Fig. 4.15, each cycle consists of four phases: a *synchronization phase*, a *calibration phase*, a *data collection phase*, and a *data processing phase*. During the synchronization phase, the Tx unit transmits a synchronization signal to the RIS and to the Rx unit, which identifies the start time of a cycle.

Then, in the calibration phase, the Tx unit transmits a narrow band constant signal, i.e., symbol x, at frequency f_c. The RIS sets the beamformer pattern to be $\boldsymbol{c}_0 = (\hat{\boldsymbol{o}}(1), \ldots, \hat{\boldsymbol{o}}(1))$, i.e., the N reconfigurable elements are in their first/default configuration. Besides, the received signal of the Rx unit is recorded as y_0.

The data collection phase is divided into K *frames* that are evenly spaced in time. During this phase, the Tx unit continuously transmits the narrow band RF signal, while the RIS changes its beamformer pattern at the end of each frame. As shown in Fig. 4.15, we denote the beamformer patterns of the RIS corresponding to the K

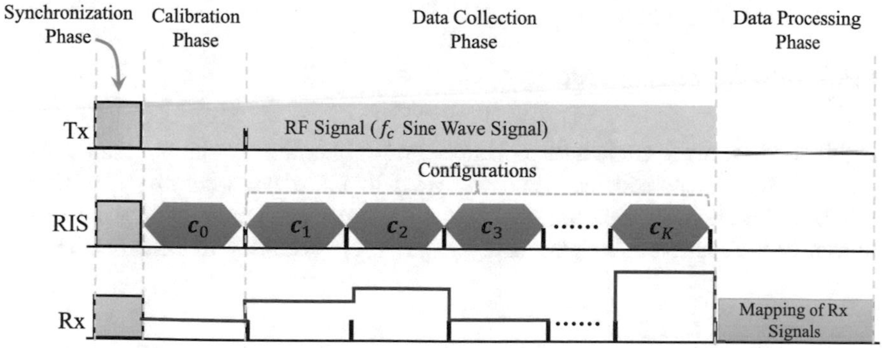

Fig. 4.15 A cycle of the RF sensing protocol

frames by binary row vectors $\boldsymbol{c}_1, \dots \boldsymbol{c}_K$. Specifically, the K beamformer patterns of the RIS during the data collection phase constitutes the *beamformer patternmat*, which is denoted by $\boldsymbol{C} = (\boldsymbol{c}_1^T, \dots, \boldsymbol{c}_K^T)^T$. Besides, as $\boldsymbol{c}_k$ is a binary row vector, beamformer patternmat is a binary matrix.

To remove the signal form the LoS path which contains no information of the target space, the received signals in the K frames are subtracted by y_0. The K differences constitute the *measurement vector*, which is a noisy linear transformation of $\boldsymbol{v}$ by the matrix $\boldsymbol{\Gamma}$, i.e.,

$$\tilde{\boldsymbol{y}} = \boldsymbol{y} - \boldsymbol{y}_0 = \boldsymbol{\Gamma}\boldsymbol{v} + \tilde{\boldsymbol{\sigma}}, \tag{4.56}$$

where $\boldsymbol{\Gamma} = \sqrt{P} \cdot x \cdot (\boldsymbol{C} - \boldsymbol{C}_0)\boldsymbol{A}$ with $\boldsymbol{C}_0 = (\boldsymbol{c}_0^T, \dots, \boldsymbol{c}_0^T)^T$, $\boldsymbol{y}$ is a K-dimensional vector consisting of the sampled received signals during the K frames that can be calculated by (4.54), $\boldsymbol{y}_0$ is a K-dimensional vector with all the elements being y_0, and $\tilde{\boldsymbol{\sigma}}$ is the difference between the noise signals and environmental scattering signals of $\boldsymbol{y}$ and $\boldsymbol{y}_0$. In this article, we assume that the environment in the considered scenario is static or changing slowly. In this case, the signals from the environmental scattering paths, i.e., $h_{\mathrm{rl}} \cdot \sqrt{P} \cdot x$ is subtracted in (4.56), and $\tilde{\boldsymbol{\sigma}}$ contains the difference between the Gaussian noise signals of $\boldsymbol{y}$ and $\boldsymbol{y}_0$.[3] Specifically, the k-th element of $\tilde{\boldsymbol{\sigma}}$ is $\tilde{\sigma}_k \sim \mathcal{CN}(0, 2\epsilon)$. We refer to $\tilde{\boldsymbol{y}}$ as the *measurement vector*. Since $\boldsymbol{\Gamma}$ determines how the reflection characteristics of the objects are mapped to the measurement vector, we refer to $\boldsymbol{\Gamma}$ as the *measurement matrix*.

Finally, during the data processing phase, the receiver maps the measurement vector obtained in the data collection phase to the sensing results, which is a vector indicating the probabilities that objects exist in the M space grids. Given beamformer patternmat $\boldsymbol{C}$, the mapping is modeled through a parameterized function, i.e., $\hat{\boldsymbol{p}} = \boldsymbol{f}^{\boldsymbol{w}}(\tilde{\boldsymbol{y}})$ with $\boldsymbol{w}$ being the parameter vector that is referred to as the *mapping of the received signals*. Moreover, the result of the mapping, i.e., $\hat{\boldsymbol{p}}$, is an M-dimensional real-valued vector. Specifically, its m-th element, i.e., $\hat{p}_m \in [0, 1]$, indicates the probability that an object exists at the m-th space grid; therefore $(1 - \hat{p}_m)$ indicates the probability that the m-th space grid is empty.

4.2.3 Problem Formulation

In this section, we formulate the optimization problem for maximizing the sensing accuracy for the considered scenario. We adopt the *cross-entropy loss* as the objective function to measure the sensing accuracy, as minimizing the cross-entropy loss function can significantly improve the accuracy of classification and

[3] If the environment is changing rapidly, $h_{\mathrm{rl}} \cdot \sqrt{P} \cdot x$ can be considered as an additional complex Gaussian noise [20], and $\tilde{\boldsymbol{\sigma}}$ in (4.56) is composed of the difference of the noise signals at the Rx and that of the environmental scattering signals, and thus its variance is $2\epsilon + 2\epsilon_{\mathrm{hl}}$.

prediction [29]. In other words, the sensing accuracy is inversely proportional to the cross-entropy loss.

We define the cross-entropy loss in the considered scenario as

$$L_{\mathrm{CE}} = -\mathbb{E}_{\boldsymbol{v}\in\mathcal{V}}\Big[\sum_{m=1}^{M} p_m(\boldsymbol{v})\ln(\hat{p}_m) + (1 - p_m(\boldsymbol{v}))\ln(1 - \hat{p}_m)\Big], \tag{4.57}$$

where $\mathcal{V}$ denotes the set of all possible reflection coefficient vectors corresponding to the existence of objects in the target space, and $p_m(\boldsymbol{v})$ is a binary variable indicating the object existence in the m-th space grid. Specifically, $p_m(\boldsymbol{v})$ can be expressed as

$$p_m(\boldsymbol{v}) = \begin{cases} 0, & \text{if } |v_m| = 0, \\ 1, & \text{otherwise.} \end{cases} \tag{4.58}$$

In (4.57), $\hat{\boldsymbol{p}}$ is determined by $\boldsymbol{f}^{\boldsymbol{w}}(\tilde{\boldsymbol{y}})$. Generally, parameterized function $\boldsymbol{f}^{\boldsymbol{w}}(\tilde{\boldsymbol{y}})$ can take any form. For example, it can be a linear function, i.e., $\boldsymbol{f}^{\boldsymbol{w}}(\tilde{\boldsymbol{y}}) = \boldsymbol{W}\tilde{\boldsymbol{y}} + \boldsymbol{w}'$, where $\boldsymbol{W}$ and $\boldsymbol{w}'$ are determined by $\boldsymbol{w}$ and obtained by minimizing the mean squared error of the sensing results [42]. Besides, $\boldsymbol{f}^{\boldsymbol{w}}(\tilde{\boldsymbol{y}})$ can also be a nonlinear decision function, which determines the sensing results of $\tilde{\boldsymbol{y}}$ by using conditional probabilities [53]. In this section, we consider that $\boldsymbol{f}^{\boldsymbol{w}}(\tilde{\boldsymbol{y}})$ is nonlinear and modeled as a neural network, where the elements of $\boldsymbol{w}$ stand for the weights of the connections and the biases of the nodes. We refer to the neural network for $\boldsymbol{f}^{\boldsymbol{w}}(\tilde{\boldsymbol{y}})$ as the *sensing network*.

The optimization problem for the RIS assisted scenario that maximizes the sensing accuracy can be formulated as the following cross-entropy minimization problem, where the beamformer patternmat and the mapping of the received signals parameter are the optimization variables, i.e.,

$$(\text{P4.7}): \min_{\boldsymbol{C},\boldsymbol{w}} L_{\mathrm{CE}}(\boldsymbol{C}, \boldsymbol{w}), \tag{4.59}$$

$$s.t.\ (\hat{p}_1, \ldots, \hat{p}_M) = \boldsymbol{f}^{\boldsymbol{w}}(\tilde{\boldsymbol{y}}), \tag{4.60}$$

$$\tilde{\boldsymbol{y}} = \sqrt{P} \cdot x \cdot (\boldsymbol{C} - \boldsymbol{C}_0)\boldsymbol{A} + \tilde{\boldsymbol{\sigma}}, \tag{4.61}$$

$$\boldsymbol{C} = (\boldsymbol{c}_1^T, \ldots, \boldsymbol{c}_K^T)^T, \tag{4.62}$$

$$\boldsymbol{c}_k = (\hat{\boldsymbol{o}}(c_{k,1}), \ldots, \hat{\boldsymbol{o}}(c_{k,N})),\ \forall k \in [1, K], \tag{4.63}$$

$$c_{k,n} \in [1, N_S],\ \forall k \in [1, K], n \in [1, N]. \tag{4.64}$$

In (P4.7), (4.59) indicates that the objective is to minimize the cross-entropy loss by optimizing $\boldsymbol{C}$ and $\boldsymbol{w}$. As $\hat{\boldsymbol{p}}$ is determined by $\boldsymbol{f}^{\boldsymbol{w}}(\tilde{\boldsymbol{y}})$ and $\tilde{\boldsymbol{y}}$ is determined by control matrix $\boldsymbol{C}$, L_{CE} defined in (4.57) can be expressed as a function of $\boldsymbol{C}$ and $\boldsymbol{w}$. Constraint (4.60) indicates that the probabilities for the M space grids to contain objects are calculated by the mapping of the received signals, i.e., $\boldsymbol{f}^{\boldsymbol{w}}(\tilde{\boldsymbol{y}})$. Constraint (4.61) indicates that the measurement vector is determined by beamformer patternmat $\boldsymbol{C}$ as in (4.56). Besides, constraints (4.62)∼(4.64) are due to the definition of the beamformer patternmat in Sect. 4.2.2.3. Since the beamformer patternmat is a binary matrix and $\boldsymbol{w}$ is a real-valued vector, (P4.7) is a mixed-integer optimization problem and is NP-hard.

To tackle it efficiently, we decompose (P4.7) into two sub-problems, i.e., (P4.8), and (P4.9), as follows:

$$\text{(P4.8)} : \min_{\boldsymbol{w}} \; L_{\text{CE}}(\boldsymbol{C}, \boldsymbol{w}), \quad s.t.\ (4.60). \tag{4.65}$$

$$\text{(P4.9)} : \min_{\boldsymbol{C}} \; L_{\text{CE}}(\boldsymbol{C}, \boldsymbol{w}), \quad s.t.\ (4.61) \text{ to } (4.64). \tag{4.66}$$

In (P4.8), we minimize the cross-entropy loss by optimizing $\boldsymbol{w}$ given $\boldsymbol{C}$, and in (P4.9), we minimize the cross-entropy loss by optimizing $\boldsymbol{C}$ given $\boldsymbol{w}$. Based on the alternating optimization technique [54], a locally optimal solution of (P4.7) can be solved by iteratively solving (P4.8) and (P4.9). Nevertheless, given $\boldsymbol{w}$, (P4.9) is still hard to solve due to the large number of integer variables in the beamformer patternmat. Moreover, the number of iterations for solving (P4.8) and (P4.9) can be large before converging to the local optimum of (P4.7). If traditional methods, such as exhaustive search and branch-and-bound algorithms, are applied, they will result in a high computational complexity. To solve (P4.8) and (P4.9) efficiently, we develop an MDP framework and solve it by proposing an PRPG algorithm, which are discussed in the next section. Furthermore, the convergence of the proposed algorithm to solve (P4.7) is analyzed in Sect. 4.2.5.

4.2.4 Algorithm Design

In this section, we formulate an MDP framework for (P4.8) and (P4.9) in Sect. 4.2.4.1 and propose a deep reinforcement learning algorithm named PRPG to solve it in Sect. 4.2.4.2.

4.2.4.1 MDP Formulation

In (P4.9), the optimization variable $\boldsymbol{C}$ is composed of a large number of binary variables satisfying constraints (4.62)∼(4.64), which makes (P4.9) an integer optimization problem which is NP-hard and difficult to solve. Nevertheless, the RIS can be considered as an intelligent agent who determines the configuration of each reconfigurable element for each beamformer pattern sequentially, and is rewarded by the negative cross-entropy loss. In this regard, the integer optimization problem (P4.9) can be considered as a decision optimization problem for the RIS, which can be solved efficiently by the deep reinforcement learning technique, since it is efficient to solve highly-complexed decision optimization problems for intelligent agents [55, 56]. As the deep reinforcement learning algorithm requires the target problem to be formulated as an MDP, we formulate (P4.8) and (P4.9) as an MDP, so that we can solve them by proposing an efficient deep learning algorithm.

An MDP encompasses an environment and an agent, and consists of four components: the set of states $\mathscr{S}$, the set of available actions $\mathscr{A}$, the state transition function $\mathscr{T}$, and the reward function $\mathscr{R}$ [50]. The states in $\mathscr{S}$ obey the Markov property, i.e., each state only depends on the previous state and the adopted action. Suppose the agent takes action a in state s, and the consequent state s' is given by the transition function $\mathscr{T}$, i.e., $\boldsymbol{s}' = T(\boldsymbol{s}, a)$. After the state transition, the agent receives a reward that is determined by reward function $\mathscr{R}$, i.e., $\mathscr{R}(\boldsymbol{s}', \boldsymbol{s}, a)$.

To formulate the MDP framework for (P4.8) and (P4.9), we view the RIS as the agent, and the RF sensing scenario including the surroundings, the RF transceiver, and the objects in the target space are regarded, altogether, as the environment.

We consider the state of the RIS the current beamformer patternmat, i.e., $\boldsymbol{C}$ and the action of the RIS as selecting the configuration of a reconfigurable element for a beamformer pattern. Thus, actions of the RIS determine the elements in beamformer patternmat $\boldsymbol{C}$. Therefore, the next state of the MDP is determined by the current state and the action, and the Markov property is satisfied. In the following, we describe the components of the MDP framework in detail.

State In the MDP of the RIS assisted RF sensing scenarios, the state of the environment is defined as

$$\boldsymbol{s} = (k, n, \boldsymbol{C}), \tag{4.67}$$

where $k \in [1, K]$ and $n \in [1, N]$ are the row and column indexes for beamformer patternmat indicating the configuration that the RIS aims to select. Besides, $\boldsymbol{C}$ in (4.67) denotes the current beamformer patternmat of the RIS in state $\boldsymbol{s}$. The initial state of the MDP framework is denoted by $s_0 = (1, 1, \boldsymbol{C}_0)$, where $\boldsymbol{C}_0$ is the beamformer patternmat of the RIS whose reconfigurable elements are in the first/default configuration. We refer to the states with indices $(k, n) = (K+1, 1)$ as the *terminal states*. When the terminal states are reached, all the configurations in the beamformer patternmat have been selected.

Action In each state $\boldsymbol{s} = (k, n, \boldsymbol{C})$, the RIS selects the state of the n-th reconfigurable element in the k-th frame. The action set of the RIS in each state can be expressed as $\mathscr{A} = \{1, \ldots, N_S\}$, where the j-th action ($i \in [1, N_S]$) indicates that the RIS selects the target configuration to be the j-th configuration. In other words, the RIS sets $(\boldsymbol{C})_{k,n} = \hat{\boldsymbol{o}}(a), a \in \mathscr{A}$.

State Transition Function After the RIS selects the action, the MDP framework transits into the next state, $\boldsymbol{s}' = (k', n', \boldsymbol{C}')$, if $(k, n) \neq (K, N)$. Or, if $(k, n) = (K, N)$, the agent enters the terminal state of the MDP. For the non-terminal states, the elements of state $\boldsymbol{s}'$ given $\boldsymbol{s}$ and a can be expressed as follows

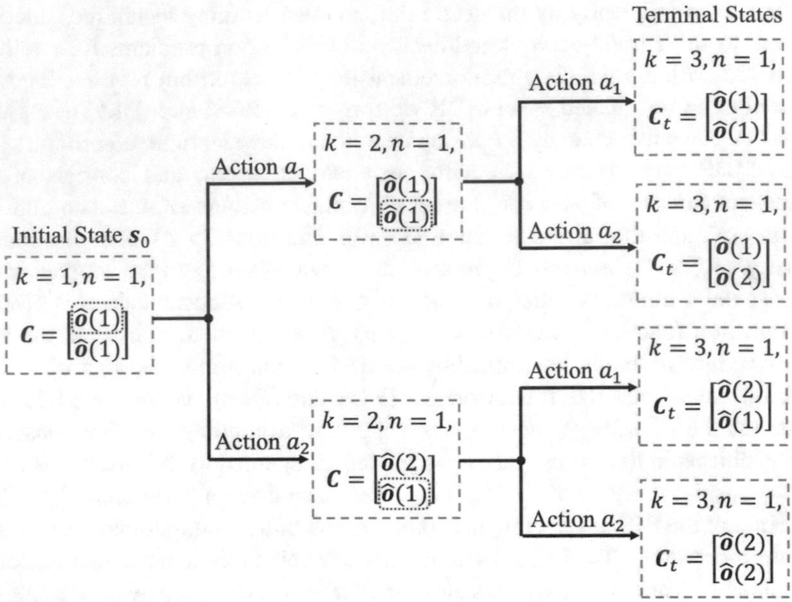

Fig. 4.16 Example of the state transition in the formulated MDP, with $K = 2$, $N = 1$, and $N_S = 2$

$$k' = k + 1, \quad n' = \text{mod}(n + 1, N) + 1, \tag{4.68}$$

$$(\boldsymbol{C}')_{k'',n''} = \begin{cases} (\boldsymbol{C})_{k'',n''}, \text{ if } (k'', n'') \neq (k, n), \\ \hat{o}(a) \text{ if } (k'', n'') = (k, n), \forall k'' \in [1, K], \ n'' \in [1, N]. \end{cases} \tag{4.69}$$

An example of the state transition is illustrated in Fig. 4.16, where $N_S = 2$, $K = 2$, and $N = 1$. In Fig. 4.16, the red dotted box indicates the element of $\boldsymbol{C}$ that is determined by the action in the current state. If $(k, n) = (3, 1)$, it can be observed that all the configurations of the beamformer patternmat have been determined, and the MDP transits into the terminal states, where beamformer patternmat is denoted by $\boldsymbol{C}_t$.

Reward Function In general MDP frameworks, the reward is a value obtained by the agent from the environment and quantifies the degree to which the agent's objective has been achieved [50]. The reward for the agent is defined as the negative cross-entropy loss of the mapping of the received signals given the beamformer patternmat determined in the terminal states. If the terminal state has not been reached, the reward for the state transition is set to be zero. Specifically, given parameter $\boldsymbol{w}$, the reward in state $\boldsymbol{s}$ is defined as

$$\mathscr{R}(s|\boldsymbol{w}) = \begin{cases} -L_{\text{CE}}(\boldsymbol{C}_t, \boldsymbol{w}), & \text{if } \boldsymbol{s} \text{ is a terminal state,} \\ 0, & \text{otherwise.} \end{cases} \tag{4.70}$$

In the formulated MDP, the RIS aims for obtaining an optimal policy to obtain the maximum reward in the terminal states. To be specific, the policy of the agent is a mapping from the state set to the available action set, i.e., $\boldsymbol{\pi} : \mathscr{S} \rightarrow \mathscr{A}$. To define the optimal policy $\boldsymbol{\pi}^*$, we first define the *state-value function* given policy $\boldsymbol{\pi}$ and parameter vector $\boldsymbol{w}$, which indicates the accumulated reward of the agent via a certain state. Based on (4.70), the state-value function can be expressed as

$$V(\boldsymbol{s}|\boldsymbol{\pi}, \boldsymbol{w}) = \begin{cases} -L_{\text{CE}}(\boldsymbol{C}, \boldsymbol{w}), \ \text{if } \boldsymbol{s} \text{ is a terminal state,} \\ V(\boldsymbol{s}'|\boldsymbol{\pi}, \boldsymbol{w})|_{\boldsymbol{s}'=\mathscr{T}(\boldsymbol{s},\boldsymbol{\pi}(\boldsymbol{s}))}, \ \text{otherwise,} \end{cases} \tag{4.71}$$

The state-value function for $\boldsymbol{\pi}$ in state $\boldsymbol{s}$ indicates the accumulated rewards of the agent after state $\boldsymbol{s}$. Based on (4.71), the state-value function for the initial state can be expressed as

$$V(\boldsymbol{s}_0|\boldsymbol{\pi}, \boldsymbol{w}) = -L_{\text{CE}}(\boldsymbol{C}_t^{\pi}, \boldsymbol{w}), \tag{4.72}$$

where $\boldsymbol{C}_t^{\pi}$ denotes the terminal state of the RIS adopting policy $\boldsymbol{\pi}$.

Therefore, given parameter vector $\boldsymbol{w}$, the optimal policy of the agent in the MDP framework is given by

$$\boldsymbol{\pi}^*(\boldsymbol{w}) = \arg\max_{\pi} V(\boldsymbol{s}_0|\boldsymbol{\pi}, \boldsymbol{w}) \iff \arg\min_{\boldsymbol{C}} L_{\text{CE}}(\boldsymbol{C}, \boldsymbol{w}). \tag{4.73}$$

In (4.73), it can be observed that finding the optimal policy of the agent in the formulated MDP framework is equivalent to solving the optimal beamformer patternmat for (P4.9). Besides, solving (P4.8) is equivalent to solving the optimal $\boldsymbol{w}$ given the policy $\boldsymbol{\pi}$.

4.2.4.2 Progressing Reward Policy Gradient Algorithm

To jointly solve (P4.8) and (P4.9) under the formulated MDP framework, we propose a novel PRPG algorithm. The proposed algorithm can be divided into two phase, i.e., the *action selection phase* and the *training phase*, which proceed iteratively.

Action Selection Process In the proposed algorithm, the agent, i.e., the RIS, starts from the initial state $\boldsymbol{s}_0$ and adopts the policy for selecting action in each state until reaching the terminal state. To select the current action in each state, the RIS use policy $\boldsymbol{\pi}$ that maps the current state to a probability vector. To be specific, for a given state $\boldsymbol{s}$, the policy results in an N_S-dimensional probability vector denoted by $\boldsymbol{\pi}(\boldsymbol{s}|\boldsymbol{w})$, which we refer to as the *policy function*. The i-th element of $\boldsymbol{\pi}(\boldsymbol{s}|\boldsymbol{w})$ ($i \in$

$[1, N_S]$), i.e., $\pi_i(\boldsymbol{s}|\boldsymbol{w})$, is in range [0, 1] and denotes the probability of selecting the action a_i in state $\boldsymbol{s}$. Besides, $\boldsymbol{\pi}(\boldsymbol{s}|\boldsymbol{w})$ $(i \in [1, N_S])$ satisfies $\sum_{i=1}^{N_S} \pi_i(\boldsymbol{s}|\boldsymbol{w}) = 1$.

However, since the state contains the current beamformer patternmat that contains $K \cdot N \cdot N_S$ binary variables, the agent faces a large state space, and the policy function is hard to be modeled by using simple functions. To handle this issue, we adopt a neural network to model the policy function as neural networks are a powerful tool to handle large state space [57]. The adopted neural network is referred to as the policy network, and we train the policy network by using the policy gradient algorithm [55]. Specifically, the policy network is denoted by $\boldsymbol{\pi}^{\boldsymbol{\theta}}(\boldsymbol{s}|\boldsymbol{w})$, where $\boldsymbol{\theta}$ denotes the parameters of the policy network and comprises the connection weights and the biases of the activation functions in the neural network.

The structure of the policy network is shown in Fig. 4.17. In state $\boldsymbol{s}$, k and n are embedded as a K-dimensional and an N-dimensional vectors, respectively, where the k-th and n-th elements in the vectors are ones and the other elements are zeros. Specifically, we refer to the resulted vectors as the *one-hot* vectors. As for $\boldsymbol{C}$, since the RF sensing for the target space is determined by $\boldsymbol{C}\boldsymbol{A}$ as shown by (4.54), we first divide $\boldsymbol{C}$ to its real and imaginary parts and right-multiply them by the real and imaginary parts of $\boldsymbol{A}$, respectively. Then, driven by the concept of model-based learning, we process the result, i.e., $\boldsymbol{C}\boldsymbol{A}$, by multi-layer perceptrons (MLPs). Besides, since the K beamformer patterns are symmetric in their physical meaning and changing their order does not impact the sensing performance, the MLPs that extract feature vectors from $\boldsymbol{c}_1$ to $\boldsymbol{c}_K$ need to be symmetric. This can be achieved by utilizing two *symmetric MLP groups*, each containing K MLPs with shared parameters. This significantly reduces the number of parameters and thus facilitates

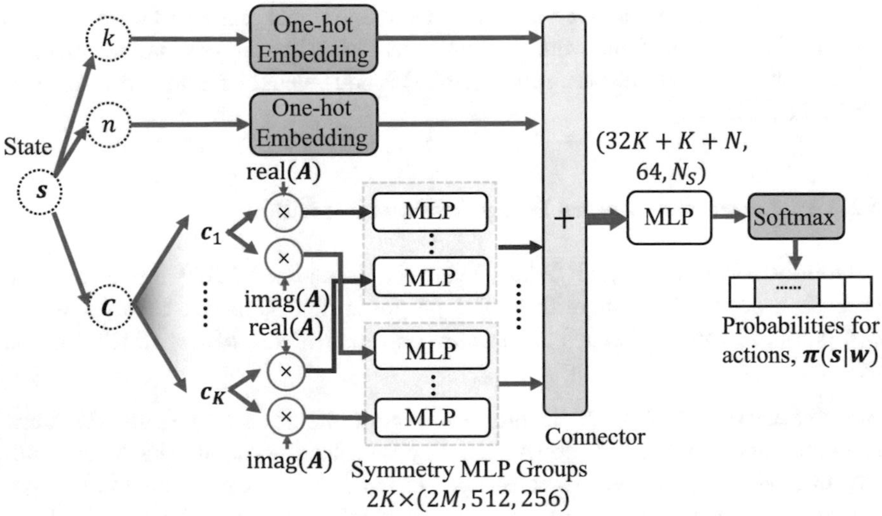

Fig. 4.17 Network structure of the policy network used in the proposed algorithm

the training of the policy network. The sizes of the MLPs are labeled in Fig. 4.17. For example, $(2M, 512, 256)$ indicates that each MLP in a symmetric group that has three layers whose sizes are $2M$, 512, and 256, respectively. Then, the one-hot vectors and the $2K$ extracted feature vectors are connected and input to the final MLP. The result of the final MLP is fed into the *softmax* layer which produces an N_S-dimensional vector indicating the probability of selecting the N_S actions.

Training Process The purpose of the training process is twofold: (a) To make the policy network improves the current policy in action selection based on (4.73). (b) To make the mapping of the received signals incur lower cross-entropy loss. Accordingly, the training process consists of two parts, i.e., training of the policy network and training of the sensing network. In the training of the policy network, we adopt the policy gradient method [50]. Besides, the training of the sensing network results in that the rewards for the terminal states progress during the training of the policy. Due to these characteristics, the proposed algorithm is named as *progressing reward policy gradient* algorithm.

Training of the Policy Network To collect the training data for the policy network, a *replay buffer* is adopted in order to store the experiences of the agent during state transitions. The replay buffer of the agent is denoted by $\mathscr{B} = \{\boldsymbol{e}\}$. The stored experience in the replay buffer is given by $\boldsymbol{e} = (\boldsymbol{s}, a)$. It is worth noting that, differently from the replay buffer in traditional deep reinforcement learning algorithms [55], the experience in the replay buffer does not record the reward obtained during the state transitions. This is because the rewards are determined by the current mapping of the received signals, which changes as $\boldsymbol{w}$ being updated. Thus, we propose that the rewards are calculated when the training process is invoked, instead of being recorded in the replay buffer.

We define a *training epoch* (or epoch in short) as the state transition process from the initial state to a terminal state. The experience of the agent within an epoch is stored into the replay buffer and used for training, which is discarded after being used. Based on the policy gradient theorem[50], in the training process, the gradient of $V(s_0|\boldsymbol{\pi}, \boldsymbol{w})$ with respect to $\boldsymbol{\theta}$ satisfies

$$\nabla_{\boldsymbol{\theta}} V(s_0|\boldsymbol{\pi}, \boldsymbol{w}) \propto \mathbb{E}_{\mathscr{B},\boldsymbol{\pi}^{\boldsymbol{\theta}}}\left[V(\mathscr{T}(S_t, A_t)|\boldsymbol{\theta}, \boldsymbol{w}) \frac{\nabla_{\boldsymbol{\theta}} \pi^{\boldsymbol{\theta}}_{A_t}(S_t|\boldsymbol{w})}{\pi^{\boldsymbol{\theta}}_{A_t}(S_t|\boldsymbol{w})} \right], \tag{4.74}$$

where $(S_t, A_t) \in \mathscr{B}$ are the samples of the state and action in the replay buffer of an agent following policy $\boldsymbol{\pi}^{\boldsymbol{\theta}}$, and $Q(S_t, A_t|\boldsymbol{\theta}, \boldsymbol{w})$ denotes the reward for the agent after selecting the action A_t in S_t and then following $\boldsymbol{\pi}^{\boldsymbol{\theta}}$.

To calculate the gradient in (4.74), the rewards for the agent in (4.70) need to be calculated. If $\boldsymbol{s}$ is a terminal state, the reward $R(\boldsymbol{s}|\boldsymbol{w})$ is calculated by using the Monte Carlo methods [58], i.e.,

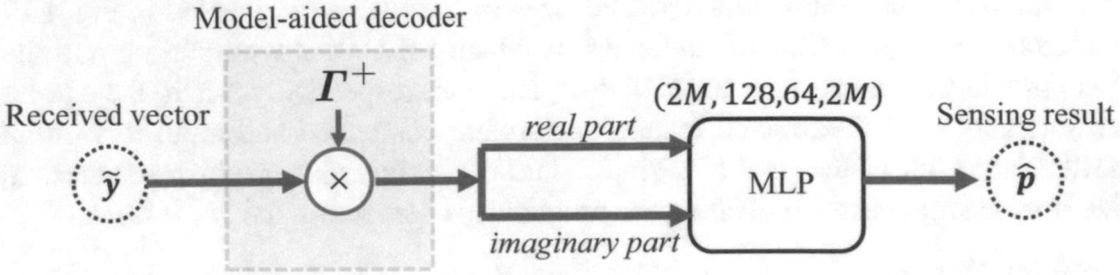

Fig. 4.18 Sensing network of the RIS

$$R(s|\boldsymbol{w}) = -\sum_{\boldsymbol{v}\in\mathcal{V}}\sum_{i=1}^{N_{\mathrm{mc}}}\Big(\sum_{m=1}^{M} p_m(\boldsymbol{v})\ln(\hat{p}_m) + (1-p_m(\boldsymbol{v}))\ln(1-\hat{p}_m)\Big)\Big|_{\hat{\boldsymbol{p}}=f^{\boldsymbol{w}}(\boldsymbol{\Gamma}\boldsymbol{v}+\tilde{\boldsymbol{\sigma}}_i)}. \tag{4.75}$$

Otherwise, $R(s|\boldsymbol{w}) = 0$. In (4.75), N_{mc} indicates the number of sampled noise vectors, and $\tilde{\boldsymbol{\sigma}}_i$ is the i-th sampled noise vector. As the rewards in the non-terminal states are zero, $V(\mathcal{T}(S_t, A_t)|\boldsymbol{\theta}, \boldsymbol{w})$ is equal to the reward at the final state for S_t, A_t, and policy $\boldsymbol{\pi}^{\boldsymbol{\theta}}$.

Specifically, in (4.75), $\hat{\boldsymbol{p}}$ is generated by the sensing network, which is shown in Fig. 4.18. The sensing network consists of two parts, i.e., the *model-aided decoder* and an MLP. Firstly, the received vector is left-multiplied by the *pseudo inverse* of $\boldsymbol{\Gamma}$, which is denoted by $\boldsymbol{\Gamma}^{+}$ and can be calculated based on [59]. According to the least-square method [42], the model-aided decoder, i.e., $\hat{\boldsymbol{v}} = \boldsymbol{\Gamma}^{+}\boldsymbol{y}$, is the optimal linear decoder that results in the minimum mean square error (MSE) for the actual reflection vector $\boldsymbol{v}$, and thus can potentially increase the sensing accuracy of the sensing network. Then, $\hat{\boldsymbol{v}}$ is fed into a fully-connected MLP, which reconstructs the probability vector $\hat{\boldsymbol{p}}$.

In each process, $\boldsymbol{\theta}$ is updated as follows

$$\boldsymbol{\theta} = \boldsymbol{\theta} + \alpha \cdot \mathbb{E}_{e\in\mathcal{B}}\left[V(\mathcal{T}(S_t, A_t)|\boldsymbol{\theta}, \boldsymbol{w})\frac{\nabla_{\boldsymbol{\theta}}\pi^{\boldsymbol{\theta}}_{A_t}(S_t|\boldsymbol{w})}{\pi^{\boldsymbol{\theta}}_{A_t}(S_t|\boldsymbol{w})}\right], \tag{4.76}$$

where the gradient $\nabla_{\boldsymbol{\theta}}\pi^{\boldsymbol{\theta}}_{A_t}(S_t|\boldsymbol{w})$ is calculated by using the back-propagation algorithm [31], and α denotes the training rate.

Training of the Sensing Network After updating $\boldsymbol{\theta}$, the training of the sensing network is executed. The calculated rewards from (4.75) are used to train the sensing network which reduces the cross-entropy loss. To be specific, the loss function used to train the sensing network can be expressed as follows, which is in accordance with the objective function in the optimization problem (P4.8), i.e.,

$$\mathcal{L}_I(\boldsymbol{w}) = \mathbb{E}_{(s,a)\in\mathcal{B}}[R(s|\boldsymbol{w})]. \tag{4.77}$$

In each training process, $\boldsymbol{w}$ is updated by

Algorithm 14: Proposed PRPG algorithm

Input : Random initial network parameter vectors $\boldsymbol{\theta}$ and $\boldsymbol{w}$;
Empty replay buffer $\mathscr{B} = \emptyset$;
Maximum number of training epochs N_{ep};
Set of reflection coefficient vectors $\mathscr{V}$;
Number of Monte Carlo samples for noise N_{mc};
Initial learning rate α_0;
Maximum number of training epochs N_{ep}.
Output: Optimized sensing network parameter vector $\boldsymbol{w}^*$ and the optimized policy network parameter $\boldsymbol{\theta}^*$.

for $n_{\mathrm{ep}} = 1$ *to* N_{ep} **do**
  Set the current state to be the initial state, i.e., $\boldsymbol{s} = \boldsymbol{s}_0$;
  # Action selection phase
  while $\boldsymbol{s}$ *is not a terminal state* **do**
    Select the configuration of the n-th reconfigurable element in the k-th frame following the probability distribution given by $\boldsymbol{\pi}^{\boldsymbol{\theta}}(\boldsymbol{s}|\boldsymbol{w})$.
    Set action a as the selected configuration, and enter into the transited state $\boldsymbol{s}' = \mathscr{T}(\boldsymbol{s}, a)$;
    Store experience $\boldsymbol{e} = (\boldsymbol{s}, a)$ into replay buffer $\mathscr{B}$;
  end
  # Training phase
  Collect all the experiences from $\mathscr{B}$, and calculate the reward for each sampled experience by using (4.75);
  Update parameter $\boldsymbol{\theta}$ and $\boldsymbol{w}$ by (4.76) and (4.78), respectively, where the learning rate $\alpha = \frac{\alpha_0}{1 + n_{\mathrm{ep}} \cdot 10^{-3}}$;
end

$$\boldsymbol{w} = \boldsymbol{w} + \alpha \nabla_{\boldsymbol{w}} \mathscr{L}_I(\boldsymbol{w}), \tag{4.78}$$

where the gradient $\nabla_{\boldsymbol{w}} \mathscr{L}_I(\boldsymbol{w})$ is calculated by using the back-propagation algorithm.

In summary, the proposed PRPG algorithm is summarized in Algorithm 14.

4.2.5 *Algorithm Analysis*

In this section, we analyze the computational complexity and the convergence of the proposed algorithm in Sects. 4.2.5.1 and 4.2.5.2, respectively. In addition, in Sect. 4.2.5.3, we derive a non-trivial lower-bound for the sensing accuracy based on an upper-bound for the cross-entropy loss given a beamformer patternmat.

4.2.5.1 Computational Complexity

Since the PRPG algorithm consists of two main phases, i.e., the action selection phase and the training phase, we analyze their respective computational complexities. The computational complexities are analyzed with regard to the number of beamformer patterns, K, the number of reconfigurable elements, N, the number of available configuration, N_S, and the number of space grids, M.

Complexity of the Action Selection Phase In the proposed algorithm, the computationally most expensive part is the estimation of the action probabilities of the policy network. For each action selection phase, the computational complexity is given in Theorem 4.1.

Theorem 4.1 (Computational Complexity of the Action Selection Phase) *In the PRPG algorithm, for the agent in each state, the complexity to calculate the action probabilities and determine the action is* $\mathcal{O}(KNN_SM)$.

Proof The complexities of embedding K and N as one-hot vectors is $\mathcal{O}(K)$ and $\mathcal{O}(N)$, respectively. Based on [60], the complexity of multiplex the K beamformer patterns that are $N \cdot N_S$-dimensional by $\boldsymbol{A} \in \mathbb{C}^{NN_S \times M}$ is $\mathcal{O}(K \cdot N \cdot N_S \cdot M)$. For a fully connected neural network with a fixed number of hidden layers and neurons, the computational complexity of the back-propagation algorithm is proportional to the product of the input size and the output size [61]. Therefore, the computational complexities of the symmetric MLP group and the Q-value MLP are $\mathcal{O}(K \cdot M)$ and $\mathcal{O}(K \cdot N_S + N \cdot N_S)$, respectively. As the connecting operation has complexity $\mathcal{O}(1)$ and finding the maximum Q-value is $\mathcal{O}(N_S)$, the total computational complexity has complexity $\mathcal{O}(KNN_SM)$ and is therefore dominated by the matrix multiplication.

Complexity of the Training Process The computational complexity of (4.75) is provided in Lemma 4.1.

Lemma 4.1 *The computational complexity of the reward calculation in (4.75) is* $\mathcal{O}\big(KNN_SM + K^2M + M^2\big)$.

Proof We consider the worst case scenario for the computation, i.e., the former states in all the samples are terminal states. In this case, the rewards are calculated from (4.75). The term inside the second summation consists of two part, i.e., the cross-entropy calculation which has computational complexity $\mathcal{O}(M)$, and the calculation of $\hat{\boldsymbol{p}}$ by using the sensing network. The computational complexity of calculating $\boldsymbol{CA}$ is $\mathcal{O}(KNN_SM)$.

Based on [62], calculating the pseudo-inverse matrix $\boldsymbol{\Gamma}^+$, where $\boldsymbol{\Gamma}$ is a $K \times M$ matrix, has complexity $\mathcal{O}(K^2M)$. Similar to the analysis of the computational complexity of MLPs in the proof of Theorem 4.1, the computational complexity of the MLP is $\mathcal{O}(M^2)$. Therefore, the computational complexity of calculating $\hat{\boldsymbol{p}}$ is $\mathcal{O}(KNN_SM + K^2M + M^2)$, which proves Lemma 4.1.

The computational complexities of training the policy network and the sensing network are given in Lemma 4.2.

Lemma 4.2 *After calculating the rewards, the complexity of the training the sensing network and the policy network are* $\mathcal{O}(M^2)$ *and* $\mathcal{O}(N_S(K + N + M))$, *respectively. If a single MLP is used to substitute the symmetric MLP group, the computational complexity of training the policy network is* $\mathcal{O}(KMN_S + NN_S)$.

Proof For a fully connected neural network with a fixed number of hidden layers and neurons, the computational complexity of the back-propagation algorithm is proportional to the product of the input size and the output size [61]. Therefore, the computational complexity of using the back-propagation algorithm for updating the parameter vector of the sensing network is $\mathcal{O}(M^2)$.

The policy network can be considered as two connected MLPs: the first one takes the one-hot embedding vectors of k and n as the input, and the second one takes the K measuring vectors with $2M$ dimensions as the input. Moreover, as a symmetric MLP group is considered, the actual size of the input vector for the second MLP is $2M$ instead of $2KM$. Therefore, the computational complexity of training the first and second MLP of the policy network are $\mathcal{O}(N_S \cdot (K + N))$ and $\mathcal{O}(N_S M)$, respectively, and the total computational complexity is thus $\mathcal{O}(N_S \cdot (K + N + M))$.

Furthermore, if a single large MLP with layer sizes $(2KM, 64, 32K)$ is used to substitute the symmetric MLP group, the computational complexities of training the second MLP is $\mathcal{O}(KN_S M)$, and the total computational complexity of training the policy network is $\mathcal{O}(KMN_S + NN_S)$. Therefore, Lemma 4.2 is proved.

It can be observed from Lemma 4.2 that using a symmetric MLP group instead of a single large MLP in the policy network can reduce the complexity of the training process.

Based on Lemmas 4.1 and 4.2, the total computational complexity of each training process is provided in Theorem 4.2.

Theorem 4.2 *The computational complexity of each training phase of the PRPG algorithm is* $\mathcal{O}\big(KNN_S M + K^2 M + M^2\big)$.

Proof Based on (4.76) and (4.77), the complexity of calculating the loss functions are determined by the computation of the reward and action probabilities. From Theorem 4.1 and Lemma 4.2, it follows that the complexity of calculating the reward is of higher order than that of calculating the action probabilities. Therefore, the computational complexity of the training phase is dominated by the calculation of the N_b rewards, which is $\mathcal{O}\big(KNN_S M + K^2 M + M^2\big)$. Theorem 4.2 is thus proved.

4.2.5.2 Convergence Analysis

The detailed convergence analysis of the PRPG algorithm is based on the convergence analysis of the block stochastic gradient (BSG) algorithm. We denote $\boldsymbol{w}$ by $\boldsymbol{x}_1$ and denote $\boldsymbol{\theta}$ by $\boldsymbol{x}_2$, and thus the objective function in (P4.7) can be denoted by $F(\boldsymbol{x}_1, \boldsymbol{x}_2) = L_{\mathrm{CE}}(\boldsymbol{C}_t^{\pi^{\boldsymbol{\theta}}}, \boldsymbol{w})$, where $\boldsymbol{C}_t^{\pi^{\boldsymbol{\theta}}}$ indicates the beamformer patternmat in the terminal state for the RIS with policy $\boldsymbol{\pi}^{\boldsymbol{\theta}}$. Based on [63], a BSG algorithm for solving (P4.7) is formulated as Algorithm 15, whose convergence analysis can be given by Lemma 4.3.

Lemma 4.3 *Algorithm 15 converges to a locally optimal* $\boldsymbol{x}_1^*$ *and* $\boldsymbol{x}_2^*$ *as the number of iterations* $N_{\mathrm{itr}} \to \infty$*, given that the following conditions are satisfied:*

1. *There exist a constant c and a constant* ε *such that, for each iteration indexed by* j*, the inequalities* $\|\mathbb{E}[\tilde{\boldsymbol{g}}_i^j - \nabla_{\boldsymbol{x}_i} F(\boldsymbol{x}_1, \boldsymbol{x}_2)]\|_2 \leq c \cdot \max_i(\alpha_i^j)$ *and* $\mathbb{E}[\|\tilde{\boldsymbol{g}}_i^j - \nabla_{\boldsymbol{x}_i} F(\boldsymbol{x}_1, \boldsymbol{x}_2)\|^2] \leq \varepsilon^2$, $i = 1, 2$ *are fulfilled.*
2. *There exists a uniform Lipschitz constant* $\varrho > 0$ *such that*

$$\sum_{i=1,2} \|\nabla_{\boldsymbol{x}_i} F(\boldsymbol{x}_1, \boldsymbol{x}_2) - \nabla_{\boldsymbol{x}_i} F(\boldsymbol{x}_1', \boldsymbol{x}_2')\|_2^2 \leq \varrho^2 \sum_{i=1,2} \|\boldsymbol{x}_i - \boldsymbol{x}_i'\|_2^2.$$

3. *There exists a constant* ψ *such that* $\mathbb{E}[\|\boldsymbol{x}_1^j\|_2^2 + \|\boldsymbol{x}_2^j\|_2^2] \leq \psi^2, \forall j$.

Proof Please refer to Corollary 2.12 in [63], where the assumptions required in Corollary 2.12 in [63] are equivalent to the three conditions in Lemma 4.3.

Comparing Algorithms 14 and 15, we can observe that the only difference between the two algorithms is in the functions for updating parameters. Nevertheless, solving the minimization problem (4.80), we can derive that (4.80) is equivalent to that

$$\boldsymbol{x}_i^j = \boldsymbol{x}_i^{j-1} - \alpha_i^j \tilde{\boldsymbol{g}}_i^j. \tag{4.79}$$

As the learning rate sequence $\{\alpha_i^j\}_j$ in Algorithm 15 can be arbitrarily selected, the parameter update of Algorithms 14 and 15 are essentially equivalent. In this regard, the proposed PRPG algorithm can be categorized as an BSG algorithm, whose convergence analysis follows Lemma 4.3.

However, since neural networks are encompassed in the mapping of the received signals and the policy function, the conditions in Lemma 4.3 are hard to be proven theoretically. Therefore, in additional to the theoretical analyses provided above, we also analyze the convergence through practical simulations in Sect. 4.2.6.

Moreover, the obtained solution by the proposed deep learning algorithm is a locally optimal solution of (P4.7). As shown in Algorithm 14, we iteratively solve (P4.8) and (P4.9) by updating $\boldsymbol{\theta}$ using (4.76) and updating $\boldsymbol{w}$ using (4.78), respectively. Based on the Q-learning algorithm [50], updating $\boldsymbol{\theta}$ with the aim to maximize the total reward is equivalent to finding $\boldsymbol{C}$ minimizing L_{CE} given $\boldsymbol{w}$. Besides, it can be observed that updating $\boldsymbol{w}$ directly minimizes L_{CE} given $\boldsymbol{C}$. When the iteration terminates, updating the variables of $\boldsymbol{C}$ or $\boldsymbol{w}$ will not lead to a lower objective function value, i.e., the cross-entropy loss. Therefore, the solution obtained by the proposed Algorithm 14 is a locally optimal solution of the original problem (P4.7).

4.2.5.3 Lower Bound for Sensing Accuracy

In this section, we compute a lower-bound for the sensing accuracy in (P4.8) given beamformer patternmat $\boldsymbol{C}$. To derive a lower bound, we assume that the mapping of the received signals maps the received RF signals to the sensing results by using

Algorithm 15: BSG algorithm for solving (P4.7)

Input : Starting point $\boldsymbol{x}_i^0, i = 1, 2$;
Learning rate sequence $\{\alpha_i^j; i = 1, 2\}_{j=1,2,\ldots}$;
Maximum number of iterations N_{itr};
Monte Carlo sampling size of the random noise N_{mc}.
Output: Optimized $\boldsymbol{x}_1^*$ and $\boldsymbol{x}_2^*$ for (P4.7).

for $j = 1, 2, \ldots, N_{\text{itr}}$ **do**
  for $i = 1, 2$ **do**
    Compute sample gradient for the $\boldsymbol{w}$ in the j-th iteration by $\tilde{\boldsymbol{g}}_i^j = \nabla_{\boldsymbol{x}_i} F(\boldsymbol{x}_{<i}^j, \boldsymbol{x}_{\geq i}^{(j-1)})$
    Update parameter $\boldsymbol{x}_i$ by

$$\boldsymbol{x}_i^j = \arg\min_{\boldsymbol{x}_i} (\tilde{\boldsymbol{g}}_i^j)^T (\boldsymbol{x}_i - \boldsymbol{x}_i^{j-1}) + \frac{1}{2\alpha_i^j} \|\boldsymbol{x}_i - \boldsymbol{x}_i^{j-1}\|_2^2. \tag{4.80}$$

  end
end
Output $(\boldsymbol{x}_1^{N_{\text{itr}}}, \boldsymbol{x}_2^{N_{\text{itr}}})$ as $(\boldsymbol{x}_1^*, \boldsymbol{x}_2^*)$;

an optimal linear decoder and a threshold judging process. In the following, we first provide the detection criterion for sensing, and then derive a lower-bound for sensing accuracy by leveraging an upper-bound for the cross-entropy loss.

Detection Criterion for Sensing The reconstructed reflection coefficient vector from the linear decoder can be expressed as

$$\hat{\boldsymbol{v}} = \boldsymbol{\Gamma}^+ \tilde{\boldsymbol{y}} = \boldsymbol{\Gamma}^+ \boldsymbol{\Gamma} \boldsymbol{v} + \boldsymbol{\Gamma}^+ \tilde{\boldsymbol{\sigma}}. \tag{4.81}$$

Based (4.81), we analyze the probability distribution of the random variable $\hat{v}_m$, i.e., the m-th element of $\hat{\boldsymbol{v}}$. We denote the m-th row vectors of $\boldsymbol{\Gamma}^+$ and $\boldsymbol{\Gamma}^+\boldsymbol{\Gamma}$ as $\boldsymbol{\gamma}_m$ and $\boldsymbol{\xi}_m$, respectively. Then, $\hat{v}_m = \boldsymbol{\xi}_m \boldsymbol{v} + \boldsymbol{\gamma}_m \tilde{\boldsymbol{\sigma}}$. The emptiness of the space grids other than the m-th space grid is modeled by the vector $\boldsymbol{q}_{-m}$, where $q_{-m,m'} = 0$ and 1 indicate that the m' space grid is empty and nonempty, respectively, ($m' \in [1, M]$, $m' \neq m$). When the m-th space grid is empty (or nonempty), we denote the probability density functions (PDFs) of the real and imaginary parts of $\hat{v}_m$, i.e., $\hat{v}_{R,m}$ and $\hat{v}_{I,m}$, by $\mathscr{P}_{R,i}^0(x)$ and $\mathscr{P}_{I,i}^0(x)$ (or $\mathscr{P}_{R,i}^1(x)$ and $\mathscr{P}_{I,i}^1(x)$), respectively.

We judge the emptiness of the m-th space grid according to the sum of $\hat{v}_{R,m}$ and $\hat{v}_{I,m}$, i.e., $\mu_m = \hat{v}_{R,m} + \hat{v}_{I,m}$. When the m-th space grid is empty, given $\boldsymbol{q}_{-m}$, the sum of $\hat{v}_{R,m}$ and $\hat{v}_{I,m}$, i.e., μ_m, follows a normal distribution, i.e., $\mu_m \sim \mathcal{N}(0, \epsilon_m^0(\boldsymbol{q}_{-m}))$, where

$$\epsilon_m^0(\boldsymbol{q}_{-m}) = \sum_{\substack{m' \neq m, \\ m' \in \mathcal{M}}} q_{-m,m'} \cdot \epsilon_{\text{ref},m'} \cdot (\|\boldsymbol{\xi}_{R,m'}\|^2 + \|\boldsymbol{\xi}_{I,m'}\|^2)$$

$$+ \sum_{m' \in \mathcal{M}} \epsilon \cdot (\|\boldsymbol{\gamma}_{R,m'}\|^2 + \|\boldsymbol{\gamma}_{I,m'}\|^2). \tag{4.82}$$

Here, $\mathcal{M}$ is the set of indexes of M space grids, and subscripts R and I indicate the real and imaginary parts of a vector, respectively. The first summation term in (4.82) corresponds to the variance due to the reflection coefficients at the space grids other than the m-th space grid, and the second summation term in (4.82) corresponds to the variance due to the noise at the Rx unit.

On the other hand, when the q-th space grid is nonempty, the variance due to reflection coefficient of the m-th space grid needs to be added. Denote the variance of the reflection coefficient of the m-th space grid by $\epsilon_{\text{ref},m}$, and the variance of μ_m can be expressed as

$$\epsilon_m^1(\boldsymbol{q}_{-m}) = \epsilon_m^0(\boldsymbol{q}_{-m}) + \epsilon_{\text{ref},m} \cdot (\|\boldsymbol{\xi}_{R,m}\|^2 + \|\boldsymbol{\xi}_{I,m}\|^2). \tag{4.83}$$

Given the emptiness of the m-th space grid, the PDF of μ_m can be written as follows

$$\mathscr{P}_m^i(x) = \sum_{\boldsymbol{q}_{-m} \in \mathcal{Q}_{-m}} P_m(\boldsymbol{q}_{-m}) \mathscr{P}_{norm}(x; 0, \epsilon_m^i(\boldsymbol{q}_{-m})), \ i = 0, 1 \tag{4.84}$$

where $\mathcal{Q}_{-m}$ indicates the set of all possible $\boldsymbol{q}_{-m}$, $\mathscr{P}_{norm}(x; 0, \epsilon_m^i(\boldsymbol{q}_{-m}))$ $(i = 0, 1)$ denotes the PDF of a normal distribution with zero mean and variance $\epsilon_m^i(\boldsymbol{q}_{-m})$, and $P_m(\boldsymbol{q}_{-m})$ denotes the probability for the existence indicated by $\boldsymbol{q}_{-m}$ to be true, i.e.,

$$P_m(\boldsymbol{q}_{-m}) = \prod_{m' \neq m, m' \in \mathcal{M}} Pr_{m'}(q_{-m,m'}). \tag{4.85}$$

Here, $Pr_{m'}(x)$ with x being 0 and 1 indicates the probabilities that the m'-th space grid are empty and nonempty, respectively.

We use the difference between $\mathscr{P}_m^1(\boldsymbol{q}_{-m})$ and $\mathscr{P}_m^0(\boldsymbol{q}_{-m})$ as the *judgement variable* to determine whether the m-th space grid is empty or not. To facilitate the analysis, we adopt the *log-sum* as a substitute for the sum in (4.84). Therefore, the judgement variable can be calculated as

$$\tau_m = \sum_{\boldsymbol{q}_{-m} \in \mathcal{Q}_{-m}} \ln\left(p_m(\boldsymbol{q}_{-m}) \mathscr{P}_{norm}(x; 0, \epsilon_m^1(\boldsymbol{q}_{-m}))\right) \tag{4.86}$$
$$- \sum_{\boldsymbol{q}_{-m} \in \mathcal{Q}_{-m}} \ln\left(p_m(\boldsymbol{q}_{-m}) \mathscr{P}_{norm}(x; 0, \epsilon_m^0(\boldsymbol{q}_{-m}))\right).$$

It can be observed from (4.86) that τ_m increases as $\mathscr{P}_m^1(\mu_m)$ increases, and that it decreases as $\mathscr{P}_m^0(\mu_m)$ increases. Therefore, we can judge the emptiness of the

m-th space grid through the value of τ_m. Specifically, the sensing result of the m-th space grid is determined by comparing the judging variable τ_m with the *judging threshold*, which is denoted by ρ_m. If $\tau_m \leq \rho_m$, the sensing result of the m-th space grid is "empty", which is denoted by the hypothesis $\mathscr{H}_0$. Otherwise, if $\tau_m > \rho_m$, the sensing result is "non-empty", which is denoted by the hypothesis $\mathscr{H}_1$. After simplifying (4.86), the *detection criterion* for $\mathscr{H}_0$ and $\mathscr{H}_1$ can be expressed as

$$\tau_m = \mu_m^2 \sum_{\boldsymbol{q}_{-m} \in Q_{-m}} \frac{\epsilon_m^1(\boldsymbol{q}_{-m}) - \epsilon_m^0(\boldsymbol{q}_{-m})}{2\epsilon_m^1(\boldsymbol{q}_{-m})\epsilon_m^0(\boldsymbol{q}_{-m})} - \frac{1}{2} \sum_{\boldsymbol{q}_{-m} \in \mathscr{Q}_{-m}} \ln\left(\frac{\epsilon_m^1(\boldsymbol{q}_{-m})}{\epsilon_m^0(\boldsymbol{q}_{-m})}\right) \underset{\mathscr{H}_0}{\overset{\mathscr{H}_1}{\gtrless}} \rho_m. \tag{4.87}$$

Since $\mu_m^2 > 0$, the range of ρ_m can be expressed $[-\frac{1}{2}\sum_{\boldsymbol{q}_{-m} \in \mathscr{Q}_{-m}} \ln(\frac{\epsilon_m^1(\boldsymbol{q}_{-m})}{\epsilon_m^0(\boldsymbol{q}_{-m})}), \infty]$.

Upper Bound of Cross Entropy Loss We analyze the cross-entropy loss incurred by the detection criterion in (4.87), which can be considered as a non-trivial upper-bound for the cross-entropy loss defined in (4.57). As the sensing result given by (4.87) is either 0 or 1, if the sensing result is accurate, the incurred cross-entropy loss will be $-\ln(1) = 0$; otherwise, the incurred cross-entropy loss will be $-\ln(0) \to \infty$. In practice, the cross-entropy loss due to an inaccurate sensing result is bounded by a large number C_{In0}. Given $\mathscr{H}_0$ (or $\mathscr{H}_1$) being true, the probability for the sensing result to be inaccurate is the probability of $\tau_m > \rho_m$, i.e., $\Pr\{\tau_m > \rho_m|\mathscr{H}_0\}$ (or $\Pr\{\tau_m \leq \rho_m|\mathscr{H}_1\}$). Denote the probability for an object to be at the m-th space grid by $\tilde{p}_m$, and the cross-entropy loss of the m-th space grid can be calculated as

$$L_m = C_{\text{In0}} \cdot (1 - \tilde{p}_m) \cdot \Pr\{\tau_m > \rho_m|\mathscr{H}_0\} + C_{\text{In0}} \cdot \tilde{p}_m \cdot \Pr\{\tau_m \leq \rho_m|\mathscr{H}_1\}, \tag{4.88}$$

where $\Pr\{\tau_m > \rho_m|\mathscr{H}_0\}$ and $\Pr\{\tau_m \leq \rho_m|\mathscr{H}_1\}$ can be calculated by using Proposition 4.2.

Proposition 4.2 *The conditional probability for sensing the m-th space grid inaccurately can be calculated as follows*

$$\Pr\{\tau_m > \rho_m|\mathscr{H}_0\} = \Pr\{\mu_m^2 > \hat{\rho}_m|\mathscr{H}_0\} = 1 - \sum_{\boldsymbol{q}_{-m} \in \mathscr{Q}_{-m}} P_m(\boldsymbol{q}_{-m}) \cdot \text{erf}\left(\sqrt{\frac{\hat{\rho}_m}{2\epsilon_m^0(\boldsymbol{q}_{-m})}}\right), \tag{4.89}$$

$$\Pr\{\tau_m \leq \rho_m|\mathscr{H}_1\} = \Pr\{\mu_m^2 \leq \hat{\rho}_m|\mathscr{H}_1\} \tag{4.90}$$

$$= \sum_{\boldsymbol{q}_{-m} \in \mathcal{Q}_{-m}} P_m(\boldsymbol{q}_{-m}) \cdot \mathrm{erf}\left(\sqrt{\frac{\hat{\rho}_m}{2\epsilon_m^1(\boldsymbol{q}_{-m})}}\right), \tag{4.91}$$

where erf(·) *denotes the* error function *[53], and*

$$\hat{\rho}_m = \frac{\frac{1}{2}\sum_{\boldsymbol{q}_{-m} \in \mathcal{Q}_{-m}} \ln(\epsilon_m^1(\boldsymbol{q}_{-m})/\epsilon_m^0(\boldsymbol{q}_{-m})) + \rho_m}{\sum_{\boldsymbol{q}_{-m} \in \mathcal{Q}_{-m}} \frac{\epsilon_m^1(\boldsymbol{q}_{-m}) - \epsilon_m^0(\boldsymbol{q}_{-m})}{\epsilon_m^1(\boldsymbol{q}_{-m}) \cdot \epsilon_m^0(\boldsymbol{q}_{-m})}}. \tag{4.92}$$

Proof Based on (4.87), the judging condition $\tau_m \underset{\mathcal{H}_0}{\overset{\mathcal{H}_1}{\gtrless}} \rho_m$ is equivalent to $\mu_m^2 \underset{\mathcal{H}_0}{\overset{\mathcal{H}_1}{\gtrless}} \hat{\rho}_m$. Therefore, $\Pr\{\mu_m^2 > \hat{\rho}_m|\mathcal{H}_0\} = \Pr\{\tau_m > \rho_m|\mathcal{H}_0\}$ and $\Pr\{\mu_m^2 \le \hat{\rho}_m|\mathcal{H}_1\} = \Pr\{\tau_m \le \rho_m|\mathcal{H}_1\}$. Also, given $\boldsymbol{q}_{-m}$, μ_m^2 follows a chi-squared distribution with one degree of freedom. Therefore, the cumulative distribution function of μ_m^2 is a weighted sum of error functions, and thus the conditional probabilities can be calculated by using (4.89) and (4.90).

Besides, we can observe in (4.88) that L_m is determined by the judgment threshold ρ_m. Then, based on (4.88) to (4.92), $\partial L_m/\partial\rho_m$ can be calculated as

$$\partial L_m/\partial\rho_m = -\frac{2C_{\mathrm{In}0}}{\sqrt{\pi}} \cdot \frac{\partial\hat{\rho}_m}{\partial\rho_m} \cdot \sum_{\boldsymbol{q}_{-m} \in \mathcal{Q}_{-m}} P_m(\boldsymbol{q}_{-m}) \cdot \phi_m(\boldsymbol{q}_{-m}), \tag{4.93}$$

$$\phi_m(\boldsymbol{q}_{-m}) = \frac{(1-\tilde{p}_m) \cdot e^{-\hat{\rho}_m/2\epsilon_m^0(\boldsymbol{q}_{-m})}}{\sqrt{8\epsilon_m^0(\boldsymbol{q}_{-m})\hat{\rho}_m}} - \frac{\tilde{p}_m \cdot e^{-\hat{\rho}_m/2\epsilon_m^1(\boldsymbol{q}_{-m})}}{\sqrt{8\epsilon_m^1(\boldsymbol{q}_{-m})\hat{\rho}_m}}. \tag{4.94}$$

Then, the optimal ρ_m^* can be obtained by solving $\partial L_m/\partial\rho_m = 0$. Denoting the minimal L_m corresponding to ρ_m^* as L_m^*, the upper bound for the cross-entropy loss in (4.57) can be calculated as

$$L_{\mathrm{ub}} = \sum_{m \in \mathcal{M}} L_m^*. \tag{4.95}$$

When the emptiness of the space grids other than the m-th is given, the upper bound of the cross-entropy loss can be calculated from Proposition 4.3. Since the sensing accuracy is inversely proportional to the cross-entropy loss, a lower-bound for the sensing accuracy is derived.

Proposition 4.3 *When the emptiness of the space grids other than the m-th is given, i.e.,* $\mathcal{Q}_{-m} = \{\boldsymbol{q}_{-m}\}$*, the optimal judging threshold for the m-th space grid is*

$$\rho_m^*(\boldsymbol{q}_{-m}) = \begin{cases} \frac{1}{2}\ln(\frac{\epsilon_m^0(\boldsymbol{q}_{-m})}{\epsilon_m^1(\boldsymbol{q}_{-m})}), \ \textit{if}\ \tilde{p}_m > \sqrt{\frac{\epsilon_m^1(\boldsymbol{q}_{-m})}{\epsilon_m^0(\boldsymbol{q}_{-m})+\epsilon_m^1(\boldsymbol{q}_{-m})}}, \\ 2\ln(\frac{1-\tilde{p}_m}{\tilde{p}_m}) - \frac{1}{2}\ln(\frac{\epsilon_m^0(\boldsymbol{q}_{-m})}{\epsilon_m^1(\boldsymbol{q}_{-m})}), \ \textit{otherwise}. \end{cases} \tag{4.96}$$

Proof The sign of $\partial L_m/\partial p_m$ is determined by $\phi_m(\boldsymbol{q}_{-m})$. We calculate the ratio between the two terms of $\phi_m(\boldsymbol{q}_{-m})$, which can be expressed as

$$\iota_m(\boldsymbol{q}_{-m}) = \frac{1-\tilde{p}_m}{\tilde{p}_m} \cdot \sqrt{\frac{\epsilon_m^1(\boldsymbol{q}_{-m})}{\epsilon_m^0(\boldsymbol{q}_{-m})}} \cdot e^{-\hat{\rho}_m \cdot \frac{\epsilon_m^1(\boldsymbol{q}_{-m})-\epsilon_m^0(\boldsymbol{q}_{-m})}{2\epsilon_m^0(\boldsymbol{q}_{-m})\cdot\epsilon_m^1(\boldsymbol{q}_{-m})}}. \tag{4.97}$$

Since $\epsilon_m^1(\boldsymbol{q}_{-m}) > \epsilon_m^0(\boldsymbol{q}_{-m})$ and $\hat{\rho}_m \propto \rho_m$, $\iota_m(\boldsymbol{q}_{-m})$ is a monotonic decreasing function with respect to ρ_m and $\iota_m(\boldsymbol{q}_{-m}) \geq 0$. Also, $\phi_m(\boldsymbol{q}_{-m}) \geq 0 \iff \iota_m(\boldsymbol{q}_{-m}) \geq 1$, and thus, $\partial L_m/\partial p_m \geq 0$ if and only if $\iota_m(\boldsymbol{q}_{-m}) \geq 1$. Therefore, the minimal H_m is obtained when ρ_m satisfies the condition $\iota_m(\boldsymbol{q}_{-m}) = 1$. Then, we can prove Proposition 4.3 by solving $\iota_m(\boldsymbol{q}_{-m}) = 1$ and considering that $\rho_m \geq -\frac{1}{2}\ln(\frac{\epsilon_m^1(\boldsymbol{q}_{-m})}{\epsilon_m^0(\boldsymbol{q}_{-m})})$.

However, since the number of possible $\boldsymbol{q}_{-m}$ can be large, (typically, $|\mathscr{Q}_{-m}| = 2^{M-1}$), calculating the exact $\partial L_m/\partial \rho_m$ in (4.93) is time-consuming, which makes it hard to find the exact ρ_m^* and L_m^*. Therefore, in practice, we approximate H_{ub} by using a random sampled subset of $\mathscr{Q}_{-m}^{\text{sam}}$, which is denoted by $\mathscr{Q}_{-m}^{\text{sam}} \subset \mathscr{Q}_{-m}$.

Moreover, since the sign of $\partial L_m/\partial \rho_m$ is determined by the sum of $\phi_m(\boldsymbol{q}_{-m})$, and $\phi_m(\boldsymbol{q}_{-m})$ has a zero point, which can be calculated by (4.96). If ρ_m is less than the zero point of $\phi_m(\boldsymbol{q}_{-m})$, $\phi_m(\boldsymbol{q}_{-m}) \geq 0$; and otherwise $\phi_m(\boldsymbol{q}_{-m}) < 0$. Therefore, we use the mean of the optimal $\rho_m^*(\boldsymbol{q}_{-m})$ for each $\boldsymbol{q}_{-m} \in \mathscr{Q}_{-m}^{\text{sam}}$ to estimate ρ_m^*, and approximate the upper-bound accordingly. The estimated ρ_m^* is denoted by $\tilde{\rho}_m^*$, which can be formulated as follows

$$\tilde{\rho}_m^* = \frac{1}{|\mathscr{Q}_{-m}^{\text{sam}}|} \sum_{\boldsymbol{q}_{-m}\in\mathscr{Q}_{-m}^{\text{sam}}} \rho_m^*(\boldsymbol{q}_{-m}), \tag{4.98}$$

where $\rho_m^*(\boldsymbol{q}_{-m})$ can be obtained by Proposition 4.3. When $|\mathscr{Q}_{-m}^{\text{sam}}|$ is large enough, $\tilde{\rho}_m^*$ in (4.98) can approximate ρ_m^*.

Finally, given the approximated upper bound of the cross-entropy loss as $\tilde{L}_{\text{ub}}$, then it can be observed from (4.88) that the upper bound of average probability of sensing error for a space grid is $P_{\text{err, ub}} = \tilde{L}_{\text{ub}}/C_{\text{In0}}$. Therefore, the lower bound of the average sensing accuracy for a space grid is $P_{\text{acc, lb}} = 1 - P_{\text{err, ub}}$.

4.2.6 *Simulation and Evaluation*

In this section, we first describe the setting of the simulation scenario and summarize the simulation parameters. Then, we provide simulation results to verify the effectiveness of the proposed PRPG algorithm. Finally, using the proposed algorithm, we evaluate the cross-entropy loss of the RIS assisted RF sensing scenario with respect to different numbers of sizes of the RIS, and numbers of space grids. Besides, we also compare the proposed method with the benchmark, i.e., the MIMO RF sensing systems.

4.2.6.1 Simulation Settings

The layout of the considered scenario is provided in Fig. 4.19. The RIS adopted here is the same as the one used in the previous section, and the reflection coefficients of the reconfigurable element in different configurations are simulated in CST software, Microwave Studio, Transient Simulation Package [37], by assuming 60° incident RF signals with vertical polarization. Besides, to increase the reflected signal power in the simulation, we combine G reconfigurable elements as an independently controllable group. The reconfigurable elements of an independently controllable group are in the same configuration, and thus they can be considered as a single one. Therefore, the proposed algorithm is suitable for this case. The number of independently controllable group is denoted by N_G.

The origin of the coordinate is at the center of the RIS, and the RIS is in the y-z plane. In addition, the z-axis is vertical to the ground and pointing upwards, and the x- and y-axes are parallel to the ground. The Tx and Rx antennas are located at (0.87, −0.84, 0) m and (0, 0, −0.5) m, respectively. The target space a cuboid

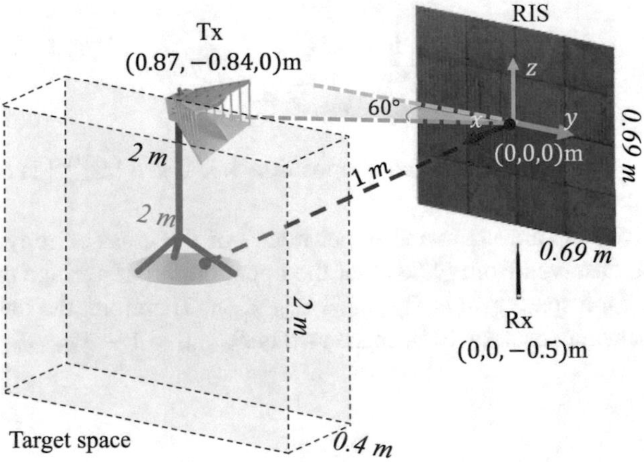

Fig. 4.19 Simulation layout

Table 4.3 Simulation parameters

Parameter	Value
Tx antenna gain (gT)	15.0 dBi
Rx antenna gain (gR)	6.0 dBi
Tx power (P)	100 mW
Number of reconfigurable elements per group (G)	144
Signal frequency (f_c)	3.198 GHz
Number of available states (N_S)	4
Number of frames (K)	10
Probability of space grid being nonempty ($p_{m,1}$)	0.5
Size of reflection vector set ($\|\mathscr{V}\|$)	100
Number of space grids (M)	18
Size of random sampled subset (Q_{-m}^{sam})	1000
Power of noise (ϵ)	10^{-9} dBm
Size of space of interest (l_x, l_y, l_z)	(0.1, 0.1, 0.1) m
Initial learning rate (α_0)	0.001
Variance of reflection coefficient (ϵ_{ref})	1

region located at 1 m from the RIS, and is divided into M space blocks each with size $0.1 \times 0.1 \times 0.1\,\text{m}^3$. The simulation parameters are summarized in Table 4.3.

4.2.6.2 Results

In Fig. 4.20, we compare the training results for different algorithms. Specifically, the first algorithm in the legend is the proposed PRPG algorithm where a sensing network (SensNet) and a policy network (PolicyNet) are adopted. The second algorithm adopts a sensing network but adopt a random beamformer patternmat. The third algorithm adopts both a sensing network and a policy network, but the sensing network does not contain a model-aided decoder as in the proposed algorithm. The fourth algorithm only uses the model-aided decoder to map the received signals to the sensing results.

It can be observed that the proposed PRPG algorithm converges with high speed and it results in the lowest cross-entropy loss among all the considered algorithms. In particular, Fig. 4.21 shows a ground-truth object and the corresponding sensing results versus the number of training epochs. As the number of training epochs increases, the sensing result approaches the ground truth and becomes approximately the same after 10^4 training epochs.

In Fig. 4.22, it shows the ground-truths and the sensing results for different algorithms and the target objects with different shapes. Comparing the sensing results with the ground truths, we can observe that the proposed algorithm outperforms other benchmark algorithms to a large extent. Besides, by comparing the sensing results of the proposed algorithm in the second column with the ground truths in

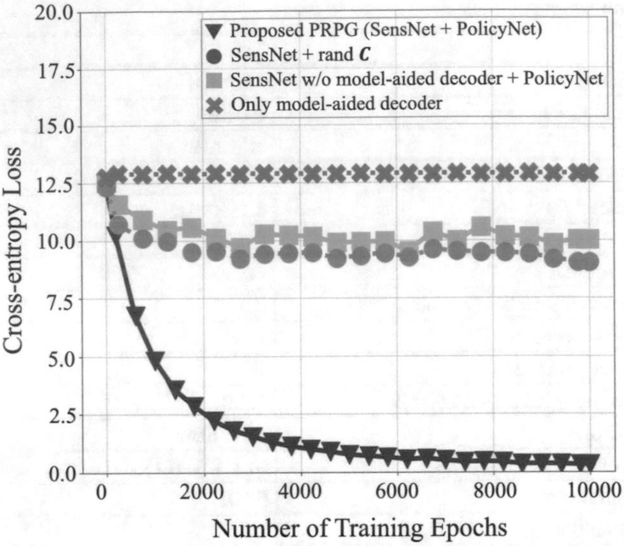

Fig. 4.20 Cross-entropy loss versus the number of training epochs for different algorithms

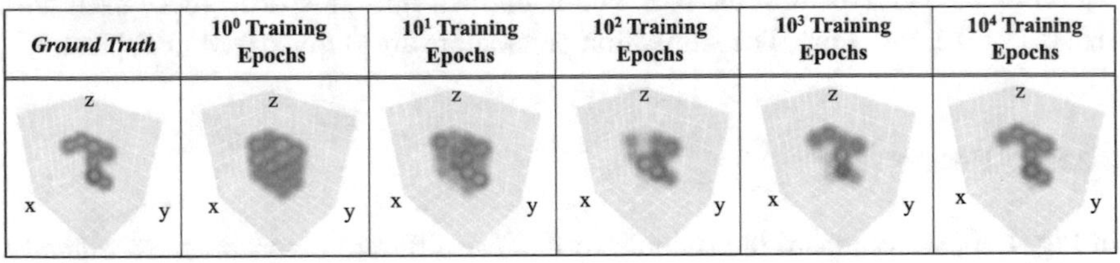

Fig. 4.21 Illustrations of ground-truth and the sensing results of different training epochs for a target object

the first column, we can observe that the proposed algorithm obtains the accurate sensing results despite the different shapes of the target objects.

In Fig. 4.23, the training results versus the number of training epochs for the PRPG algorithm are given and compared for different learning rates are compared. The initial learning rates in each case are set to be $\alpha_0 = 10^{-1}, 10^{-3}, 10^{-5}$, which then decrease inversely as the number of training epochs increases. It can be observed that large values of the learning rates prevent the algorithm to converge, while low values of the learning rates result in a slow decrease of the cross entropy loss. The setup $\alpha_0 = 10^{-3}$ outperforms the others, which verifies our learning rate selection.

In Fig. 4.24, it can be observed that as the size of the RIS, i.e., N_G, increases, the result cross-entropy loss after training decreases. This is because the received energy can be improved with more reconfigurable elements to reflect transmitted signals, as indicated by (4.54). Besides, more reconfigurable elements create a larger

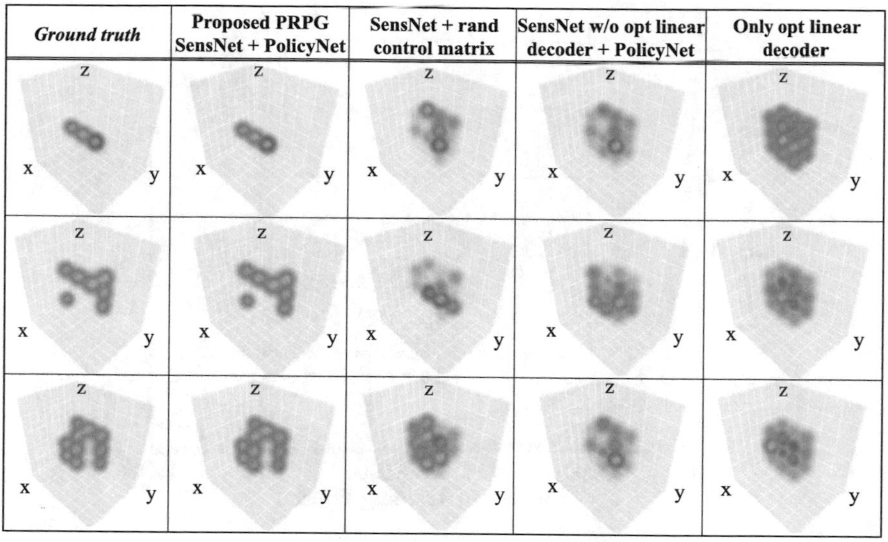

Fig. 4.22 Illustrations of the ground-truths and the sensing results of objects with different shapes for different algorithms

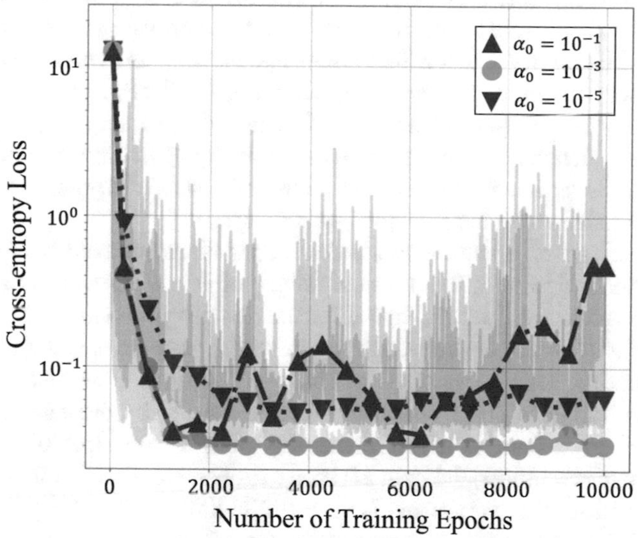

Fig. 4.23 Cross-entropy loss of the mapping of the received signals in high, normal, and low learning rate cases

design freedom and higher controllability of the beamforming, which makes gains of these reflection paths via different space grids more distinguishable. Therefore, objects at different space grids can be sensed with a higher precision. However, the cross-entropy cannot be reduced infinitely. When N_G is sufficiently large, the cross-

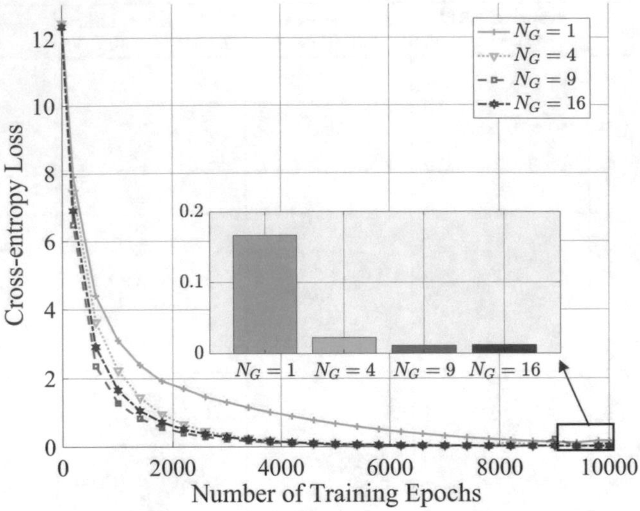

Fig. 4.24 Cross-entropy loss versus the number of training epochs for different sizes of the RIS

entropy will remains stable. As shown in Fig. 4.24, the cross-entropy loss results for $N_G = 9$ and $N_G = 16$ are almost the same. Besides, comparing the curves for $N_G = 9$ and $N_G = 16$ within the first 2000 training epochs, we can observe that increasing the number of reconfigurable elements when $N_G \geq 9$ has a negative impact on the training speed and convergence rate. This is because increasing the number of reconfigurable elements leads to a higher complexity of finding the optimal policy for the RIS to determine its beamformer patternmat, since the policy network of the RIS needs to handle a higher-dimensional state space.

In Fig. 4.25, we compare the theoretical upper-bound derived in (4.98) and the proposed PRPG algorithm for different values of M in 2D and 3D scenarios. It can be observed that, in both 2D and 3D scenarios, the probability of sensing error increases with M. Also, the cross-entropy loss in 3D scenarios is higher than those for 2D scenarios. This is because the space grids in the 3D scenarios are more closely spaced to each other, which make them hard to be distinguished. Finally, it can be observed that, as M increases, the cross-entropy loss of the proposed algorithm increases more quickly in 3D scenarios compared to that in 2D scenarios. This which verifies that 3D sensing is more difficult than 2D sensing.

In Fig. 4.26, we show the comparison between the proposed RIS assisted scenario and the benchmark, which is the MIMO RF sensing scenarios with no RIS. Both the RIS assisted scenario and the MIMO scenarios adopted a similar layout described in Sect. 4.2.6.1, and the result cross-entropy loss is obtained by Algorithm 14. Nevertheless, in the MIMO sensing scenarios, a static reflection surface takes the place of the RIS, which cannot change the beamformer pattern for the reflection signals. When the size of the MIMO array in Fig. 4.26 is $n \times n$ ($n = 1, 2, \ldots, 5$), it indicates that n Tx antennas and n Rx antennas are adopted in the scenario.

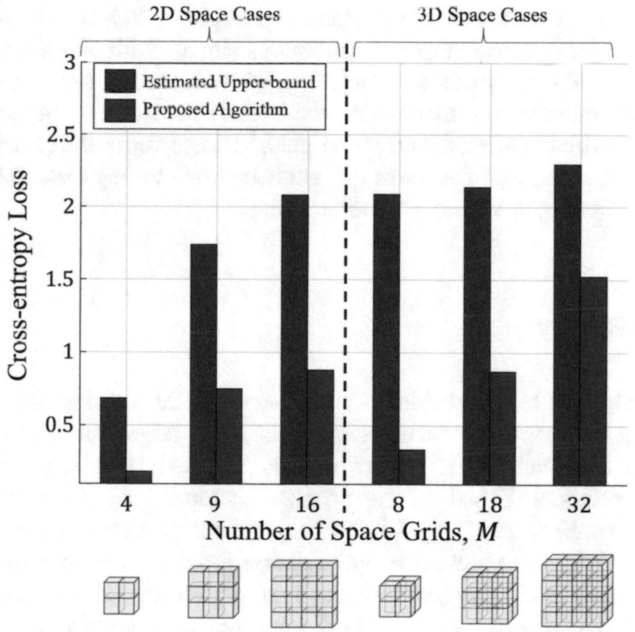

Fig. 4.25 Estimated upper-bound and the results of the proposed algorithm for the cross-entropy loss versus different numbers of space grids in 2D and 3D scenarios. The drawings at the bottom indicate the arrangement of the space grids

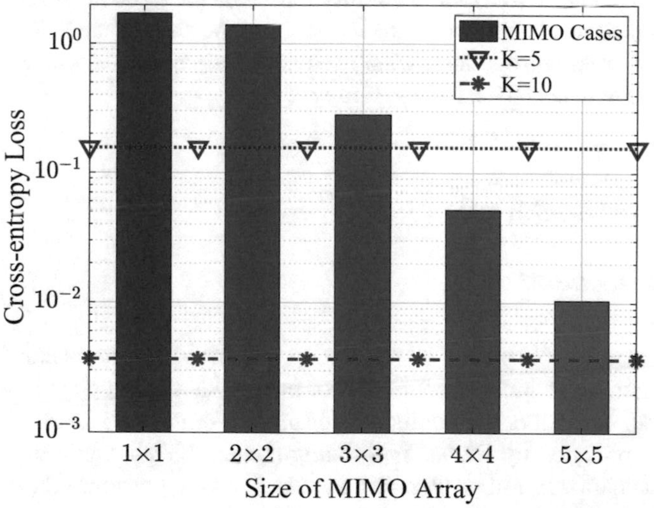

Fig. 4.26 Comparison between the RIS assisted scenario and the MIMO scenarios with different numbers of Tx/Rx antennas. The bars illustrate the results for different MIMO scenarios, and the dash lines depict the results of the RIS assisted scenario with different numbers of frames

Specifically, the Tx/Rx antennas are arranged along the y-axis with a space interval of 0.1 m. The Tx antennas transmit continuous signals with phase interval $2\pi/n$, and the n received signals of the Rx antennas with suppressed LoS signals are used as the measurement vector. Comparing the MIMO benchmarks and the proposed method, we can observe that the proposed method outperforms the $(1 \times 1) \sim (5 \times 5)$ MIMO sensing scenarios in terms of the resulted cross-entropy loss. This shows the significance of using RIS to assisted RF sensing.

4.2.7 Summary

In this section, we have considered a RIS assisted RF sensing scenario, where the existence and locations of objects within a 3D target space can be sensed. To facilitate the beamformer pattern design, we have proposed a frame-based RF sensing protocol. Based on the proposed protocol, we have formulated an optimization problem to design of the beamformer pattern and the mapping of the received signals. To solve the optimization problem, we have formulated an MDP framework and have proposed a deep reinforcement learning algorithm named PRPG to solve it. Also, we have analyzed the computational complexity and the convergence of the proposed algorithm and have computed a theoretical upper-bound for the cross-entropy loss given the beamformer patterns of the RIS. Simulation results have verified that the proposed algorithm outperforms other benchmark algorithms in terms of training speed and the resulted cross-entropy loss. Besides, they have also illustrated the influence of the sizes of the RIS and the target space on the cross-entropy loss, which provides insights on the implementation of practical RIS assisted RF sensing systems.

4.3 Indoor Localization

4.3.1 Motivations

Location-based services, such as navigation and mobile recommendation system, are considered as an indispensable part of many emerging applications in Mobile Internet [64]. In outdoor environments, location information with acceptable accuracy can be provided by Global Positioning System (GPS), while the localization in indoor scenarios is still challenging mainly due to the uncontrollable multi-path effects [65].

To enable indoor location based services, various localization techniques are proposed, among which the technique based on received signal strength (RSS) in wireless local area network (WLAN) has gained much attention [66]. Leveraging the widespread WLAN infrastructure in indoor environments, the RSS-based technique can provide location information for user devices equipped with existing Wi-Fi modules, thus avoiding the cost of installing any extra localization hardware. Besides, comparing with channel state information (CSI) based technique, RSS information can be easily obtained without the need of using some advanced WiFi network interface cards [67].

However, the performance of the RSS-based localization depends on the collected RSS values. Specifically, a RSS-based system usually contains two phases: the *offline* and the *online* phases. In the first phase, the system collects a specific RSS value for each sampling location, and these values are all stored into a database, i.e., the radio map. Then, in the online phase, the system estimates the user's location by comparing the RSS value measured by the user and the stored values in the radio map [68]. In the uncontrollable radio environment, the radio map is passively measured and cannot be customized, and the existence of adjacent locations whose RSS values in the radio map are similar to each other inevitably degrades the performance of the localization system [69].

Recently, RISs have been used as a potential tool to actively configure the radio environment in the wireless communication systems [2, 70]. By changing the reflectivities of RIS units independently, the RIS is capable of changing the reflected RF signals in a desired way, and therefore the surrounding radio environment can be customized [20]. This paves a new way to actively alter the radio maps and reduce the similarity of the RSS values corresponding to adjacent locations, which improves the localization accuracy.

In this section, we propose MetaRadar, an RIS assisted system for multi-user localization. With RISs, MetaRadar can generate favourable radio environment to achieve fine-grained localization. However, the integration of the RIS will significantly affect the RSS-based system and bring challenges to both the offline and the online phase.

- In the first phase, MetaRadar needs to generate radio maps for all the possible radio environments created by RIS in order to leverage the reconfiguration ability of the RIS. Since the number of possible radio environments is large, it is challenging to build radio maps for all of them.
- In the second phase, MetaRadar has to select favorable radio maps from the vast number of available radio maps and combine the information collected under different radio maps to provide high precision location results, which will complicate the localization process.

To tackle these challenges, we carefully design the two phases of MetaRadar: the radio map preparation phase and the fine-grained localization phase. For the first phase, we propose a compressive technique which eliminates the all measurements containing unnecessary information. By using the received signals recorded in several critical radio environments, we can construct all the potential radio maps. Then, the MetaRadar uses an iterative approach for multi-user localization in the second phase. We propose a configuration optimization algorithm to decide the most suitable radio map and the corresponding RIS reflectivity in each iteration. The RSS values measured by multiple users in this iteration, the corresponding radio map, and the information in previous iterations are all utilized to iteratively improve the accuracy of the estimated locations.

MetaRadar is implemented using an RIS made of electrically tunable metamaterial units, and commercial universal software radio peripheral (USRP) devices as the transceiver and the receiver, in an indoor setting. The system performance is evaluated under different scenarios with different number of users. The experimental results show that the RIS can largely improve the performance of the RSS based technique. For single users and multiple users in some scenarios, the localization accuracy can be up to centimeters. Specifically, for a single user with the distance of 1 m to the RIS, MetaRadar can provide the location information with average localization error of 1 cm within 2 s.

The rest of the section is organized as follows. In Sect. 4.3.2, we review the related work about existing indoor localization and RIS techniques. In Sect. 4.3.3, we describe the system architecture including the radio map preparation and the fine-grained localization phases. Details of the two phases are discussed in Sects. 4.3.4 and 4.3.5, respectively. We present the system implementation in Sect. 4.3.6 and show the evaluation results in Sect. 4.3.7. Finally, we discuss the extra challenges in Sect. 4.3.8 and conclude the work in Sect. 4.3.9.

4.3.2 Related Work

Recent years have witnessed much interest in indoor localization systems. According to the enabling technologies, these systems can be categorized into several types: Wi-Fi [71], RFID [72, 73], visible light [74, 75], and other technologies based systems [76, 77]. Compared with other systems, Wi-Fi based system can locate every Wi-Fi compatible device without installing extra hardware, thus becoming one of the most widespread indoor localization approaches [78].

Various techniques are adopted in Wi-Fi based systems such as RSS, CSI, angle-of-arrival (AoA), and time-of-arrival (ToA) techniques [79]. Among these techniques, the RSS technique is widely used because of the simplicity of measuring RSS and the minimum hardware requirements. Different localization systems have been designed by exploiting the RSS information. For example, the Radar system [71] uses the deterministic method to estimate the user location, where the information of the nearest neighborhood is utilized to infer the location. The authors in [80] store information about the RSS distributions from the access points and use the Bayesian Network approach to estimate the user location. In [81], a joint clustering and probabilistic determination technique is proposed to tackle the noisy wireless channel and manage the computation cost. Besides, the authors in [68, 82] explore methods to reduce the cost for radio map construction by integrating movement information from inertial sensors.

In the aforementioned works, the radio environment is passively adopted, and the RSS values at different locations cannot be reconfigured, which limits the localization accuracy. In comparison, the RIS-aided system can actively customize the radio environment by changing the reflectivities of metamaterial units [2]. Utilizing the customized radio environment, we can obtain a radio map with favorable RSS values at different locations, which helps increase the localization accuracy. Besides, RISs can also benefit the localization accuracy in multi-user systems, since it provides the flexibility of generating various radio maps which potentially suit for multi-user coexisting scenes.

4.3.3 System Overview

Hardware and software architectures of the MetaRadar are presented in Fig. 4.27a and b. As illustrated in Fig. 4.27a, MetaRadar is composed of a metasurface and an AP in hardware and a localization process in software, which are combined to provide location information for mobile users in indoor scenarios [83]. Specifically, the indoor space in MetaRadar, where the mobile users with hand-held devices to locate are, is referred to as the *space of interest* (SOI). Without loss of generality, we assume SOI to be a cubic region which is discretized into 3D blocks. As shown in Fig. 4.27b, the localization process consists of two phases: the radio map preparation phase and the fine-grained localization phase.

Radio Map Preparation Phase To locate multiple users, the 3D radio map for the SOI need to be obtained in this phase. Besides, to improve the localization capability, MetaRadar needs to select configurations of the metasurface to reconfigure the radio environment and provide favorable radio maps. Therefore, in prior to the localization phase, MetaRadar needs to first collect RSS values in SOI given each configuration.

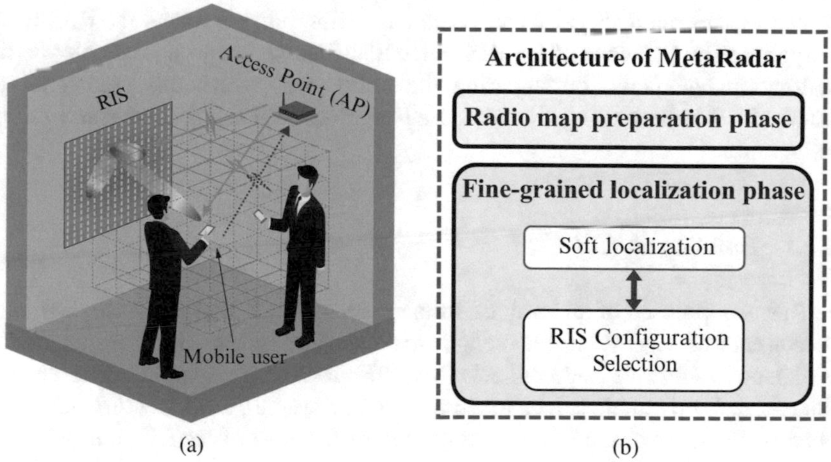

Fig. 4.27 Illustration on (**a**) hardware and (**b**) software architectures of MetaRadar

However, as the number of available configurations can be very large, it is costly to measure the RSS values in the SOI for all the configurations. To address this issue, we model the received signals in the reconfigurable radio environment of MetaRadar, and then propose a compressive construction technique for MetaRadar to obtain the radio maps given each configuration, which will be discussed in Sect. 4.3.4.

Fine-Grained Localization Phase As shown in Fig. 4.27a, the user requiring location information sends a localization request with the current measured RSS value from her/his mobile device to the AP, which initiates the fine-grained localization. As shown in Fig. 4.27b, the fine-grained localization phase is composed of a soft localization and an RIS configuration selection processes, which work iteratively. The soft localization process takes the measured RSS value and calculates the probability for the user to be at each location. Then, the RIS configuration selection process optimizes the configuration of the RIS in order to obtain a better radio map to locate the user based on the location probabilities. The user will measure the RSS of localization signals given the newly optimized configuration of RIS and send the RSS to the AP. The location probabilities can be utilized to estimate the locations of users, and its accuracy will be improved iteratively. The localization phase will terminate when a satisfied localization accuracy is obtained. In our experiment, each iteration takes 100 ms, and results with acceptable accuracy can be obtained within several seconds. More details will be introduced in Sect. 4.3.5.

4.3.4 Radio Map Preparation Phase

In this section, we first introduce the process of building the RIS which consists of an array of units, and then present an experimental example to show how the RIS customizes the RSS values at a location. After that, we model the RSS which incorporates the influence of the RIS reflection. Finally, we propose a compressive construction technique for preparing the radio map, which can predict all the possible RSS values using the measured received signal data for some critical configurations.

4.3.4.1 Building an RIS

The RIS is composed of an array of units organized on a planar surface [84]. Each unit consists of several subwavelength-scale metal patches which are printed on the dielectric substrate and connected by PIN diodes. By applying different bias voltages, each PIN diode can be tuned into two states, i.e., *ON* and *OFF* states. The ON and OFF states of the PIN diodes determine the *state* of the RIS unit [47].

The RIS unit is able to manipulate the phase and amplitude of the reflected signals. To characterize this capability, the concept of *reflectivity* is introduced, which is defined as the ratio of the reflected signals to the incident signals. Let r denote the reflectivity of an RIS unit, which is a complex number. The amplitude of r, i.e., $|r|$, denotes the ratio of the reflected signal amplitude to the incident signal amplitude, and the angle of r, i.e., $\angle r$, is the corresponding phase shift from the incident signal to the reflected signal. The reflectivity of an RIS unit is determined by its state, incident angle $\boldsymbol{\varphi}^I = (\varphi_1^I, \varphi_2^I)$ of the incident signals, and reflection angle $\boldsymbol{\varphi}^R = (\varphi_1^R, \varphi_2^R)$ of the reflected signals. By configuring these units into different states, the reflectivities of RIS units can be changed, and thus the radio environment can be modified.

Besides, in most cases, the reflectivity is also related to the frequency of the incident signals, and only for a narrow bandwidth the reflectivities under different states are desirable [84]. Therefore, we only use a single frequency sine-wave signal for localization. We denote the frequency of the signal as f_c, and thus the reflectivities of the RIS in different states are defined on f_c.

To reduce the mutual coupling among RIS units as well as the regulation burden [36], we group adjacent RIS units together, which is referred to as an element. The element is the minimal unit of the RIS which can be controlled independently. All the RIS units of one element are in the same state, which is referred to as the state of the element. The set of elements in the RIS is denoted by $\mathcal{M}$. We define the *configuration* of the RIS as the states of all the elements.

4.3.4.2 Changing the RSS Value at a Location

The radio environment reconfiguration capability of the RIS can be described as follows. Consider an RIN with M elements. Given an incident signal on the RIS with frequency f_c and incident angle $\boldsymbol{\varphi}^I$, based on [22], the reflected signals at a location can be calculated by

$$y(\boldsymbol{c}) = \sum_{m \in [1,M]} \frac{\lambda_c \exp(-j2\pi d_m/\lambda_c)}{4\pi d_m} \cdot r(\boldsymbol{\varphi}^I, \boldsymbol{\varphi}_m^R, c_m) \cdot \zeta_m, \tag{4.99}$$

where $\boldsymbol{c} = \{c_1, \cdots, c_M\}$ is the configuration of the RIS, c_m denotes the state of the m-th element. Here, $c_m \in \mathcal{C}_a$, $\forall m \in \mathcal{M}$, where $\mathcal{C}_a = \{c_1^a, \cdots, c_{N_a}^a\}$ denotes the set of available states of an element and $N_a = 4$. λ_c denotes the wavelength of the f_c signals, d_m is the distance from the m-th element to that location, and ζ_m denotes the incident signal on the m-th element. Therefore, by changing the configuration of the metasurface $\boldsymbol{c}$, the reflected signals at different locations can be modified and the radio environment can be customized.

To illustrate how the RIS configures the radio environment, we measure the 3D radio map on three planes 0.5, 0.7, and 0.9 m in front of the RIS given different configurations. The RIS elements in MetaRadar have four different states, and the incident signals are of frequency $f_c = 3.2$ GHz and incident angle $\boldsymbol{\varphi}^I = (60°, 90°)$. Figure 4.28 shows the configurations and the corresponding measured RSS, where

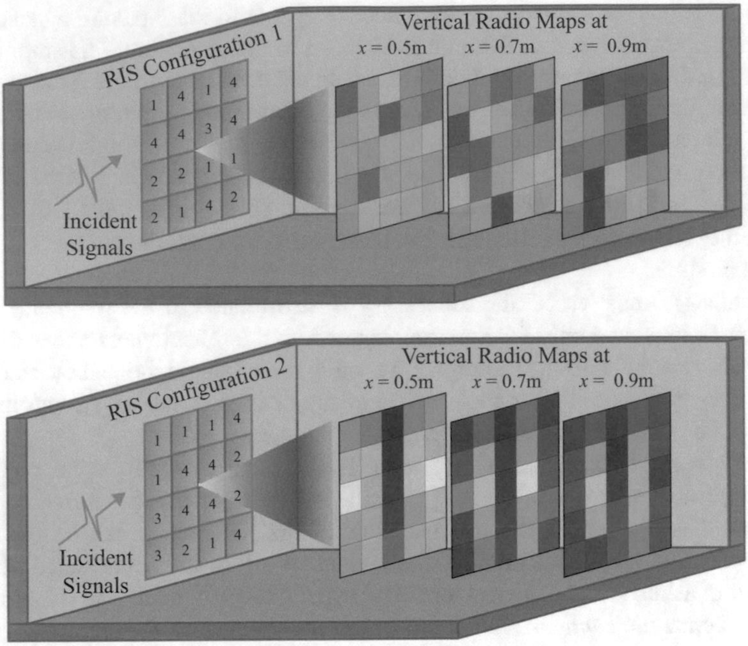

Fig. 4.28 Constructing a radio map in 3D space through the RIS's capability of changing radio environments. The numbers on the RIS denote the states of elements, and the radio maps are formed by the RSS values measured on the vertical plane in front of the RIS, where a light (e.g., light yellow), or dark (e.g., dark blue) color indicates a high, or low RSS value. The size of each block is 10 cm. The radio maps are measured at three different planes/locations, i.e., 0.5, 0.7, and 0.9 m, in front of the RIS. Note that the visualized radio maps measured at $x = 0.5$ m can be easily distinguished from those radio maps measured at $x = 0.7$ or 0.9 m

the number on the RIS element denotes the corresponding state. Besides, the color represents the value of the RSS at the center of each block, where a light (e.g., light yellow), or dark (e.g., dark blue) color indicates a high, or low RSS value. The length of the block edge is 10 cm.

Comparing the two visualized radio maps measured at $x = 0.5$ m (or at another location), the RSS values at the same block are different under the two different RIS configurations, indicating the ability of RIS to customize the radio environment. With this ability, we can obtain a sequence of RSS values at each block by setting different RIS configurations. As observed in Fig. 4.28, the RSS value sequence at each block is distinct from those at other blocks, which can be utilized to identify a certain location.

4.3.4.3 RSS Modeling

Since the reflectivity of the RIS is determined by the configuration, the incident angles and the reflection angles, when the location of the AP is fixed, the RSS can be described by the user's location and the RIS configuration. Suppose the user is located at one of the blocks in the SOI. The set of blocks in the SOI is denoted by $\mathcal{N} = \{1, \cdots, N\}$, and all the blocks have equal size with edge length e.

As shown in Fig. 4.27a, the wireless channel between the AP and the user contains a line-of-sight (LOS) channel, M reflected channels and multi-path channels. The m-th reflected channel is the reflection link through the m-th element. The multi-path channel accounts for the reflection and scattering in the indoor environment [85].

Therefore, given emitted signal ζ, user's location n, and RIS configuration $\boldsymbol{c}$, and assuming the superposition property of reflected channels [40], the signal received by the user can be expressed as

$$y(\boldsymbol{c}, n) = h_n^{LOS}\zeta + \sum_{m\in\mathcal{M}} h_{m,n}(c_m)\zeta + h_n^R\zeta + \xi, \tag{4.100}$$

where h_n^{LOS} is the gain of the LOS channel, $h_{m,n}(c_m)$ is the channel gain of the m-th reflected channel, h_n^R is the gain of the multi-path channel, and ξ denotes the noise signal. Therefore, the RSS for configuration $\boldsymbol{c}$ and block n can be expressed as $s(\boldsymbol{c}, n) = |y(\boldsymbol{c}, n)|^2$.

The distribution of the RSS can be approximated by the Gaussian distribution [81], which can be expressed as

$$\mathbb{P}(s(\boldsymbol{c}, n) = s) = \frac{1}{\sqrt{2\pi\sigma^2}} e^{-\frac{(s-\mu(\boldsymbol{c},n))^2}{2\sigma^2}}, \tag{4.101}$$

where σ is the standard deviation, and $\mu(\boldsymbol{c}, n)$ is the mean of RSS for configuration $\boldsymbol{c}$ at block n.

4.3.4.4 Compressive Construction Technique

According to (4.100), we can calculate the mean RSS at any block for any configuration if the channel gains h_n^{LOS}, $\{h_{m,n}(c_m)\}$, and h_n^R are known. However, it is difficult to measure the these channel gains directly. Instead, we measure the received signals for some critical configurations, which can be utilized to derive the RSS for any configuration.

Specifically, let $\boldsymbol{c}_{m,k}$ denote the critical configuration where the state of element m is c_k^a and the states of other elements are all c_1^a. After measuring the received signals under the configurations $\{\boldsymbol{c}_{m,k}\}$, $\forall m, k$, we can calculate the difference $\delta_{m,k}$, which is defined as

$$\delta_{m,n,k} = h_{m,n}(c_k^a) - h_{m,n}(c_1^a) = y(\boldsymbol{c}_{m,k}, n) - y(\boldsymbol{c}_{m,1}, n). \tag{4.102}$$

Based on (4.100), the received signal for $\boldsymbol{c}$ can be expressed as

$$y(\boldsymbol{c}, n) = y(\boldsymbol{c}_{1,1}, n) + \sum_{m \in \mathcal{M}} \delta_{m,n,k}. \tag{4.103}$$

In total, there are $N_a \times M - M + 1$ critical configurations (exclude $M - 1$ repeated configurations). Using the received values under these critical configurations, we can derive radio maps for all the possible N_a^M configurations.

However, in (4.100) we assume the superposition property of the reflected channels, and it is not satisfied unless the mutual coupling among elements can be ignored. To reduce the mutual coupling, we group neighboring units together, and design independent control and power supply circuits for each element, which is discussed in Sect. 4.3.6. The performance of the proposed compressive technique is also evaluated in Sect. 4.3.7.2.

4.3.5 Fine-Grained Localization Phase

In the fine-grained localization phase, MetaRadar locates multiple users with an acceptable localization accuracy by iteratively invoking two processes, i.e., *soft localization* and RIS configuration selection, as illustrated in Fig. 4.29.

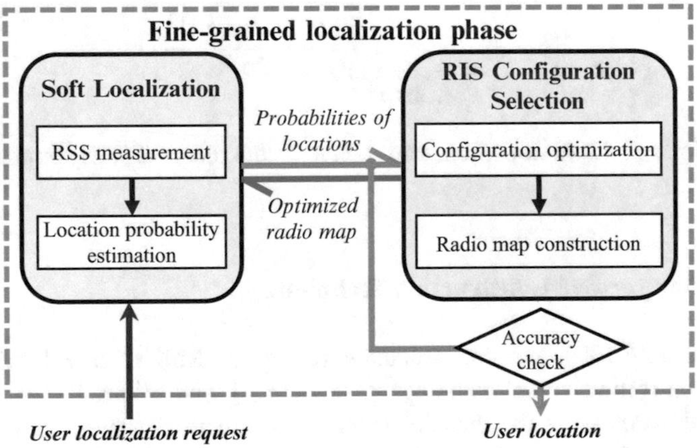

Fig. 4.29 Illustration on the fine-grained localization phase

4.3.5.1 Soft Localization

The soft localization process is shown in the left part of Fig. 4.29. As MetaRadar in the fine-grained localization phase receives a localization request from the user, it invokes the soft localization process first, which contains two tasks, i.e., the RSS measurement and location probability estimation.

RSS Measurement The RSS measurement takes place on the user device, which measures the RSS values of the received signals given the AP transmitting localization signals. The user sends the measured RSS values along with its localization request to the AP. Then, at the AP, it invokes the location probability estimation subprocess to calculate the probability of the user being at each block based on the user's measured RSS value and the current radio map.

Location Probability Estimation As the soft localization process is invoked iteratively, we denote the iteration index by k. Then, the estimated probability for user i to be at the n-th block can be expressed as $p_{i,n}^k$, the optimized radio map in the k-th iteration can be expressed as $\boldsymbol{\mu}_r^k$, and the measured RSS value of user i can be expressed as s_i^k. Based on the Bayes' theorem, the location probability estimation in the $(k+1)$-th iteration can be calculated as

$$p_{i,n}^{k+1} \approx \mathbb{P}(n|\boldsymbol{\mu}_r^k, s_i^k) = \frac{p_{i,n}^k \mathbb{P}(s_i^k|\boldsymbol{\mu}_r^k, n)}{\sum_{n \in \mathcal{N}} p_{i,n}^k \mathbb{P}(s_i^k|\boldsymbol{\mu}_r^k, n)}, \tag{4.104}$$

As the initial condition, we assume that $p_{i,n}^0$ is equal for every block, i.e., $p_{i,n}^0 = 1/N, \forall n \in \mathcal{N}$, indicating that the location probabilities estimated in the zeroth iteration are equal over all N blocks. After the location probabilities of users are obtained in each iteration, the AP decides whether the termination criterion is satisfied, which is discussed in Sect. 4.3.5.3.

4.3.5.2 RIS Configuration Selection

Configuration Optimization Based on the location probability of the users in current iteration, the RIS configuration selection process optimizes the configuration of the RIS to maximize the localization accuracy. To evaluate the accuracy of localization, we define the *localization loss* of the MetaRadar as follows. Let $\mathcal{I} = \{1, \cdots, I\}$ denote the set of all the users who participate in the fine-grained localization, and the localization loss can be formulated as

Algorithm 16: Configuration optimization algorithm

Input: Estimated Probabilities $\{p_{i,n}\}$.
Output: Configuration $\boldsymbol{c}$.
Initialize configuration $\boldsymbol{c}^1$ randomly, and calculate the corresponding RSS $\boldsymbol{\mu}^1$;
Set iteration index $z = 0$;
while $z \leq Z_u$ **do**
  Set $k = k + 1$, $\boldsymbol{\mu}^{z+1} = \boldsymbol{\mu}^z$, and $\boldsymbol{c}^{z+1} = \boldsymbol{c}^z$;
  Based on (4.107), generate the negative gradient $\boldsymbol{g}^z$ of the localization loss using $\boldsymbol{\mu}^z$ and $\{p_{i,n}\}$;
  for $m \in \mathcal{M}$ **do**
    Set $\boldsymbol{c}^* = \boldsymbol{c}^z$;
    Enumerate all the possible values $c \in \mathcal{C}_a$ for element m in $\boldsymbol{c}^*$;
    Calculate the RSS $\boldsymbol{\mu}^*$ for $\boldsymbol{c}^*$ and the normalized difference $\boldsymbol{d}^*$ between $\boldsymbol{\mu}^*$ and $\boldsymbol{\mu}^z$;
    If $|\boldsymbol{g}^z - \boldsymbol{d}^*| < |\boldsymbol{g}^z - \boldsymbol{d}^{z+1}|$ and the loss $l_u(\boldsymbol{c}^*) + \epsilon < l_u(\boldsymbol{c}^z)$, then replace $\boldsymbol{c}^{z+1}$ with $\boldsymbol{c}^*$, and replace $\boldsymbol{\mu}^{z+1}$ with $\boldsymbol{\mu}^*$;
  end
  If $\boldsymbol{c}^{z+1} = \boldsymbol{c}^z$, return $\boldsymbol{c}^z$;
end

$$l(\boldsymbol{c}) = \sum_{i \in \mathcal{I}} \sum_{n \in \mathcal{N}} p_{i,n} \sum_{n' \in \mathcal{N}} \gamma_{n,n'} \int_{\mathcal{R}_{n'}} \mathbb{P}(s_i|\boldsymbol{c}, n) \cdot ds_i, \tag{4.105}$$

where $p_{i,n}$ is the estimated probability that user i is located at the n-th block. $\gamma_{n,n'}$ is the error parameter when the user is at the n-th block while the estimated location is the n'-th block, which is defined as

$$\gamma_{n,n'} = |\boldsymbol{r}_n - \boldsymbol{r}_{n'}|, \tag{4.106}$$

where $\boldsymbol{r}_n$ is the location of the n-th block center. s_i is the RSS of user i, and $\mathcal{R}_{n'}$ is the decision region for block n'. That is, if $s_i \in \mathcal{R}_{n'}$, user i's location is estimated as n'. Since most of decision regions $\mathcal{R}_{n'}$ are irregular, it is difficult to compute the localization loss in a closed form. In the appendix, an upper bound of localization loss $l_u(\boldsymbol{c})$ is provided. In the following, we use $l_u(\boldsymbol{c})$ in replace of $l(\boldsymbol{c})$ to reduce the computational complexity. Besides, we also eliminate the blocks with insignificant estimated probabilities, i.e., $\sum_{i \in \mathcal{I}} p_{i,n} \leq \alpha$, to accelerate the computational speed.

Therefore, the configuration needs to be optimized to minimize the localization loss in this iteration. Generally, this problem is hard to solve due to the enormous number of available configurations and the complicated relationship between the configuration and the RSS. To solve this problem efficiently, we propose the configuration optimization algorithm. The basic idea is to use a modified gradient descent method which considers the limited states of RIS elements to minimize the localization loss.

Specifically, the configuration $\boldsymbol{c}^1$ is initialized randomly at first, and we calculate the corresponding vector of RSS values $\boldsymbol{\mu}^1 = \{\mu(\boldsymbol{c}^1, 1), \cdots, \mu(\boldsymbol{c}^1, N^e)\}$. Here, the

RSS vector only contains the values for blocks with estimated probabilities greater that α, and the number of remaining blocks is N^e. In the z-th iteration, we first treat all the elements in $\boldsymbol{\mu}^z$ as continuous variables, and the minimization of localization loss can be viewed as an unconstrained minimization problem. Based on the idea of gradient descent method, we use the negative gradient $\boldsymbol{g}^z$ of the localization loss with respect to $\boldsymbol{\mu}$ as the search direction to the optimal result. According to Appendix A, the n-th element in $\boldsymbol{g}^z$ can be expressed as

$$\frac{\partial l_u}{\partial \mu_n} = \sum_{i,n'} (p_{i,n} + p_{i,n'}) \frac{\gamma_{n,n'}}{2} \left(-\frac{\mu_n - \mu_{n'}}{4\sigma^2} \right) e^{-\frac{(\mu_n - \mu_{n'})^2}{8\sigma^2}}. \tag{4.107}$$

Next, we adjust $\boldsymbol{\mu}^z$ in the direction of $\boldsymbol{g}^z$ by altering configuration $\boldsymbol{c}$. To be specific, we set $\boldsymbol{c}^* = \boldsymbol{c}^z$, and all the possible element states in $\boldsymbol{c}^*$ are enumerated successively. The corresponding RSS vector is denoted by $\boldsymbol{\mu}^*$, and the normalized difference $\boldsymbol{d}^* = |\boldsymbol{\mu}^* - \boldsymbol{\mu}^z|/|\boldsymbol{\mu}^*||\boldsymbol{\mu}^z|$. If $|\boldsymbol{g}^z - \boldsymbol{d}^*| < |\boldsymbol{g}^z - \boldsymbol{d}^{z+1}|$, which means the direction of $\boldsymbol{\mu}^* - \boldsymbol{\mu}^z$ is closer to the direction of $\boldsymbol{g}^z$ comparing to that of $\boldsymbol{\mu}^{z+1} - \boldsymbol{\mu}^z$, we will replace $\boldsymbol{c}^{z+1}$ and $\boldsymbol{\mu}^{z+1}$ with $\boldsymbol{c}^*$ and $\boldsymbol{\mu}^*$, respectively. Besides, we require $l_u(\boldsymbol{c}^*) < l_u(\boldsymbol{c}^z)$ in order to assure the descent of localization loss. The iteration will end when no configuration element is changed or the iteration number z exceeds Z_u.

The convergence of Algorithm 16 can be analyzed as follows. Since the localization loss has lower bound 0 and decreases at least ϵ with each iteration, the iteration cannot go on indefinitely, and the algorithm will converge. In addition, an upper bound Z_u of iteration number is also set in order to terminate the algorithm within a limited period of time.

Radio Map Construction After the optimal configuration is obtained by Algorithm 16, the AP constructs the corresponding radio map based on the method described in Sect. 4.3.4.4, which is referred to as the optimized radio map $\boldsymbol{\mu}_r^k$. Different from the RSS vector, the optimized radio map contains the RSS values for all the blocks. As shown in Fig. 4.29, the optimized radio map will be utilized in the location probability estimation process in the next iteration.

4.3.5.3 Termination of the Localization Phase

Based on (4.105), if $l_u(\boldsymbol{c}_k) < \beta_1$ or $k > \beta_2$, the fine-grained localization phase will be terminated, where β_1 and β_2 denote the loss threshold and the maximal iteration number, respectively. In this case, the index of the block where the location probability of the user is maximized is output as the user location. The AP will estimate the locations of the users, and then send an ending signal to all the users together with their locations. Otherwise, the MetaRadar enters the next iteration and invokes the RIS configuration selection process.

If the user is moving during the localization process, the localization phase needs to terminate before the user moves to another block in order to assure the localization accuracy. Let l_b denote the edge length of the block, δ_C denote the time period of each iteration, and v denote the speed of the user. Thus, the maximum number of iterations β_2 has to be less than $l_b/v\delta_C$.

4.3.6 Implementation

In this section, we will provide detailed information on the RIS, the Access Point (AP) and the user's module.

4.3.6.1 RIS Module

The structure of the RIS module is illustrated in Fig. 4.30. It contains an RIS layer, a control layer and a power supply layer. Specifically, the RIS layer is utilized to reflect the RF signals. The control layer can modify the configuration of the RIS to obtain the desired reflected waves. Finally, the power supply layer is used to provide stable electricity supply for the above layers.

As shown in Fig. 4.31, the RIS is a square sheet with the size $69 \times 69 \times 0.52\,\text{cm}^3$. It contains 16 elements, and each element is composed of 12×12 units. Each unit has three sublayers [36]. On the top sublayer there are three copper patches which are connected by PIN diodes (BAR 65-02L). The substrate layer in the middle is made of Rogers 3010 with dielectric constant 10.2. A metallic layer is utilized as the ground at the bottom. The operation states of PIN diodes are controlled by the voltages applied on the via holes. When the applied bias voltage is 3.3 V(or 0 V), the corresponding PIN diode is at the ON (or OFF) state. The choke inductors (30 nH)

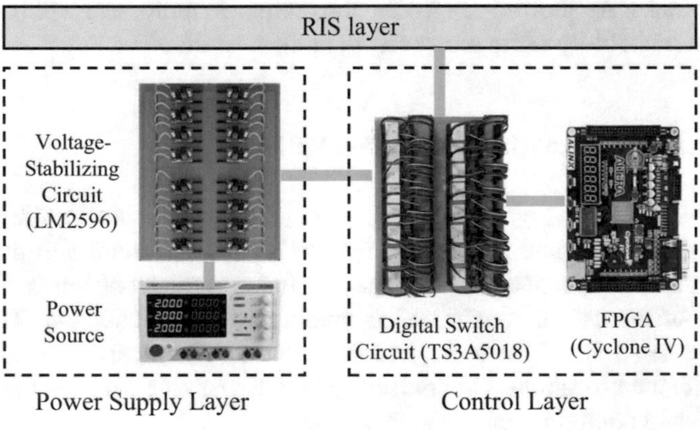

Fig. 4.30 Three layers in the RIS

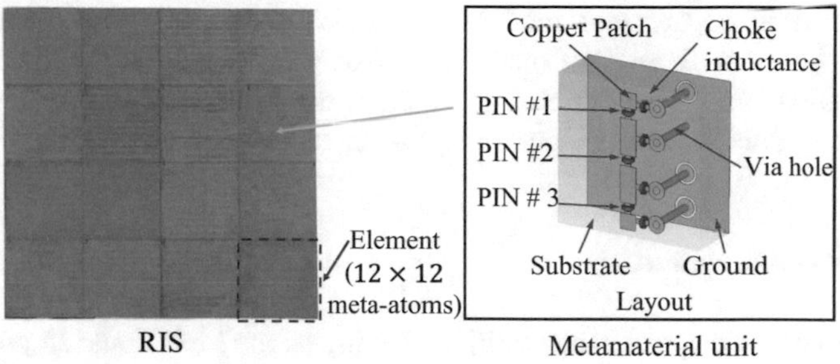

Fig. 4.31 The formation of the RIS with metamaterial units

Table 4.4 Reflectivity of an unit in different states

Reflectivity	c_1^a	c_2^a	c_3^a	c_4^a
Amplitude	0.95	0.97	0.93	0.88
Phase Shift	−33°	60°	134°	−136°

between the via hole and the via holes and the copper patches are utilized to separate the DC port and the RF signals.

The working frequency of the RIS is at 3.2 GHz. Using CST Microwave Studio, Transient Simulation Package, the reflectivity of an unit in different states are calculated, and four states are selected, which are denoted by $\{c_1^a, c_2^a, c_3^a, c_4^a\}$. As shown in Table 4.4, these states have high amplitude of reflectivity, and the phase shifts between two adjacent states are close to 90°, which demonstrates the effectiveness of the metamaterial unit design.

To reduce the mutual coupling in the control layer and the power supply layer, we use a specific digital switch (TS3A5018) to control the bias voltages of all the units in one element, and use the voltage-stabilizing circuit to supply the bias voltages. We use the FPGA (Cyclone IV) in the control layer to manipulate the digital switches and communicate with the AP.

4.3.6.2 Access Point and User Modules

The AP module consists of a laptop, a USRP, a low-noise amplifier and a horn antenna. The baseband signal generation, signal processing, configuration optimization and communication with FPGA are all performed on the laptop using python programs with GNU radio packet [39]. The USRP (LW-N210) connected to the laptop can convert baseband signals to the RF signals with frequency 3.2GHz and vise versa. Since we only utilize the signal strength of the received signals, the accurate synchronization among USRPs are not necessary. The RF signals generated by the USRP will first be amplified by the low-noise amplifier (ZX60-43-S+) and then sent to the directional double-ridged horn antenna (LB-800) which emits high gain RF signals to the RIS [86].

The structure of the user module is similar to that of the AP module. The difference between these two modules is that we use a small polymer antenna FXUWB10 with the size $3.5 \times 2.45 \times 0.02\,\text{cm}^3$ in replace of the horn antenna in the user module in order to receive signals in all directions.

4.3.6.3 Workflow Setting

In the following we specify the workflow setting of the MetaRadar in practice. The 3D radio map preparation phase using the compressive construction technique is carried out first. In this phase, the receiver antenna is placed at the center of each block in the SOI sequentially. When the receiver antenna is in the n-th block, the RIS changes to each of the $N_a \times M - M + 1$ critical configurations with an changing interval equaling to 0.5 s. By this means, necessary measurement is obtained to derive the radio maps for all the possible N_a^M configurations.

Then, MetaRadar enters the fine-grained localization phase and waits for the localization requests from users. The timeline is divided into cycles with duration 100 ms, and AP emits signals with frequency f_c from 30 to 80 ms in every cycle. The fine-grained localization will start if any user sends request to the AP in the last 20 ms in a cycle. The request contains the information of the average RSS value $\{s_i^0\}$ in this cycle. The time division multiplex (TDM) technique is adopted to separate the requests of different users. Specifically, the last 20 ms in every cycle is divided into I_s time slots, where I_s is the number of all the users, and each user sends its request signal during the assigned time slot.

After the localization phase starts, the MetaRadar will estimate the location probabilities, optimize the RIS configuration, construct the radio map, and send the starting signals to the RIS in the first 30 ms of the next cycle. The starting signal contains the optimized configuration $\boldsymbol{c}_1$. After receiving the starting signals, the RIS will change its configuration to $\boldsymbol{c}_1$. During the next 50 ms, the AP continuously emits sine wave signals with frequency f_c, and the users will record the value of the corresponding RSS and calculate the mean value. The mean value of RSS recorded by user i is denoted by s_i^1. During the last 20 ms in this cycle, the users will send their values to the AP in a TDM manner.

At the end of each iteration, the AP will decide whether the termination criterion is satisfied. Specifically, the loss threshold and the maximal iteration number are set to $\beta_1 = 0.1$ and $\beta_2 = 500$. The iteration will end if $l_u \leq \beta_1$ or the iteration $k > \beta_2$, and users' location result will be output.

4.3.7 *Evaluation*

In this section, we first introduce the setup of our test environment, and then present the evaluation results.

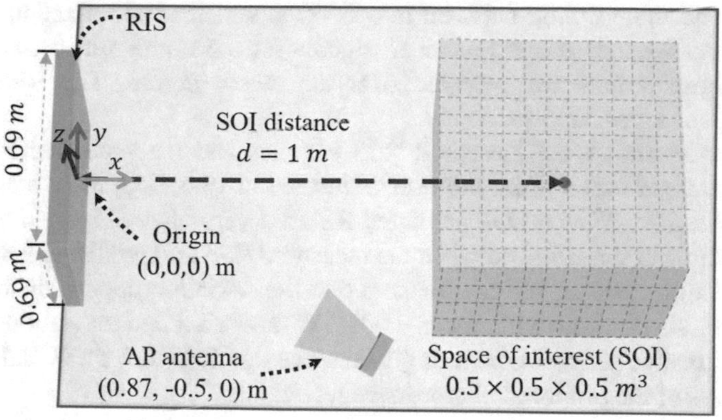

Fig. 4.32 Experimental layout

4.3.7.1 Experimental Setup

We perform the experiments in a classroom with size $25\,\mathrm{m}^2$, and the walls of the classroom are made of bricks and concrete. As illustrated in Fig. 4.32, the RIS is on the plane $x = 0$, and the center of the RIS is located at $(0, 0, 0)$ Besides, the AP antenna is 1 m away from the RIS and pointed at it. Moreover, the SOI is cuboid region with size $0.5 \times 0.5 \times 0.5\,\mathrm{m}^3$. The center of the SOI is located at $(d, 0, 0)$, where d denotes the distance between the SOI center to the RIS. To evaluate the system performance for different distances, we choose three SOIs with distance $d = 1$, 2, and 3 m, respectively. When building the radio map for each SOI, we discretize the SOI into blocks with size $5 \times 5 \times 5\,\mathrm{cm}^3$ and record the signals for each block.[4] There are no objects between the users and the RIS.

To evaluate the performance of the localization system, the localization error is introduced. The localization error of each user is defined as the distance between the actual and the estimated location. For the whole system with multiple users, we use the average localization error of all the users as the localization error of the system. Here, we choose the center of the estimated block as the estimated location.

4.3.7.2 Results for Radio Map Construction

Since the residual mutual coupling among RIS elements will influence the RSS model and the proposed method, in this subsection, we perform experiments to evaluate the performance of the proposed method. Specifically, we measure the amplitude and the phase of the received signals for the critical configurations and the RSS for 1000 random configurations. When measuring the received signals, the USRPs of the AP and the receiver USRSs are connected by the MIMO cable to

[4]The signals are obtained by the user antenna stuck on a tripod. If human is holding the antenna, the localization accuracy will degrade due to the obstruction of the signals from the RIS by human body or the interference of signals reflected from human body.

realize the accurate timing between two USRPs, which is necessary to obtain the signal phase. We measure the received signals for 0.5 s under each configuration, and the signal values are averaged over the whole period. The RSS for each configuration is measured for 50 ms.

Figure 4.33 shows the measured RSS samples and the predicted RSS for 25 different configurations when the user is located at (0.5, 0, 0) m. We can observe that the measured RSS and the predicted RSS are very close. Figure 4.34 presents the distribution of deviation between the measured RSS and predicted RSS for 1000 different configurations. We can observe that the deviation approximately follows the normal distribution with mean 0.0811W and standard deviation 0.1717W, which are relatively small comparing to the average RSS 1.7753W, indicating the effectiveness of the proposed compressive technique.

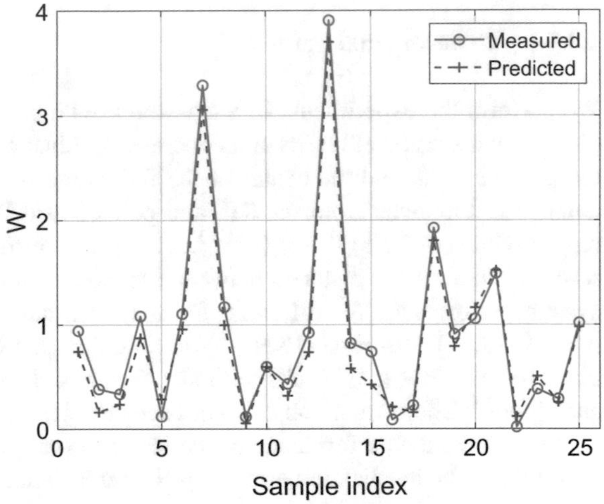

Fig. 4.33 Measured RSS samples and the predicted RSS for different configurations

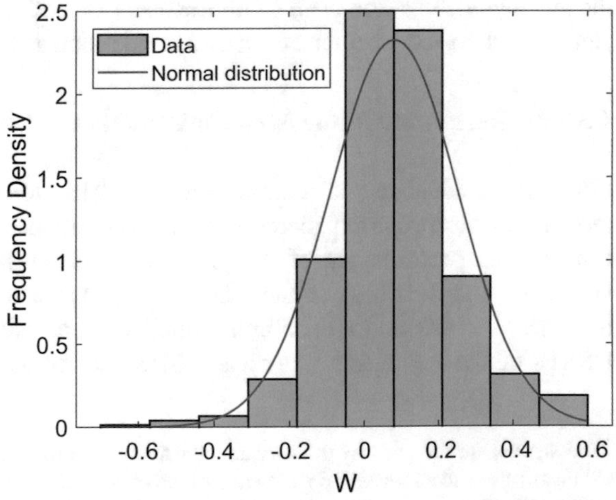

Fig. 4.34 The distribution of deviation between measured RSS and predicted RSS and the normal distribution fit for the deviation

4.3.7.3 Results for Single User Localization

In this subsection, we present the experimental results for single user localization.

Figure 4.35a illustrates the performance of fine-grained localization for single user when the distance from the SOI center to the RIS $d = 1$ m. To evaluate the performance of the MetaRadar, in Fig. 4.35a, we also give the performance obtained by another three schemes: the fixed configuration scheme, the random configuration scheme, and the simulated annealing (SA) scheme. In the fixed configuration scheme, the states of all the RIS elements are c_1^a. In the random configuration scheme, random configurations are generated in different iterations. And in the SA scheme, the simulated annealing method is utilized to optimize the RIS configurations [87]. We can observe that the localization error l_e of the fixed algorithm fluctuates between 0.30 and 0.32 m, while l_e of the other three schemes decrease when the number of iterations n_c increases. Besides, the localization error of the MetaRadar scheme is between those of the random configuration and the SA schemes. The execution time of these schemes is shown in Table 4.5. The execution time of randomly selecting configurations approximates 0 s, and the execution time of the SA method is much larger than that of the configuration optimization algorithm in MetaRadar. Based on the algorithm execution time, we modify the duration in each iteration (the third row), and derive the total execution time to reach the minimal localization error (the fifth row). Among the three schemes, the MetaRadar scheme has the lowest execution time. Although the minimum localization error of the SA scheme is slightly lower than that of the MetaRadar scheme, the execution time of the SA scheme is much longer, indicating that the MetaRadar scheme is more practical for real applications.

Figure 4.35b presents the minimum localization error l_e^m in different directions versus the distance from SOI center to the RIS d. We can observe that the minimum localization error increases with the distance, and the proposed system can achieve a centimeter error when the distance between the RIS and the user is smaller than 2 m. We can also observe that the localization error in the x axis is clearly larger than those in the y and z axes. Since the x axis is perpendicular to the RIS, the correlation of signals in the x direction is higher than those in the y and z directions, and therefore it is more difficult to distinguish different blocks in the x direction.

Figure 4.35c shows the minimum localization error l_e^m versus the distance from the RIS to the user. We choose a planar area (size $5 \times 0.5\,\text{m}^2$) on the plane $z = 0$ as the SOI with SOI distance $d = 3.25$ m. It can be observed that l_e^m increases with the distance from the RIS to the user (similar to the results in Fig. 4.35b), and the localization error obtained using the omni-directional antenna is higher than that obtained using the horn antenna. Since the beam emitted by the horn antenna is pointed to the RIS, the RIS can control more reflected signals, and thus the localization accuracy using the horn antenna is higher.

Figure 4.35d depicts the localization loss l_u versus the iteration number n_c. We can observe that the localization loss first decreases and then remains constant when the iteration number increases, which implies that the localization loss has a lower bound. We can also observe that for the same iteration number, the localization

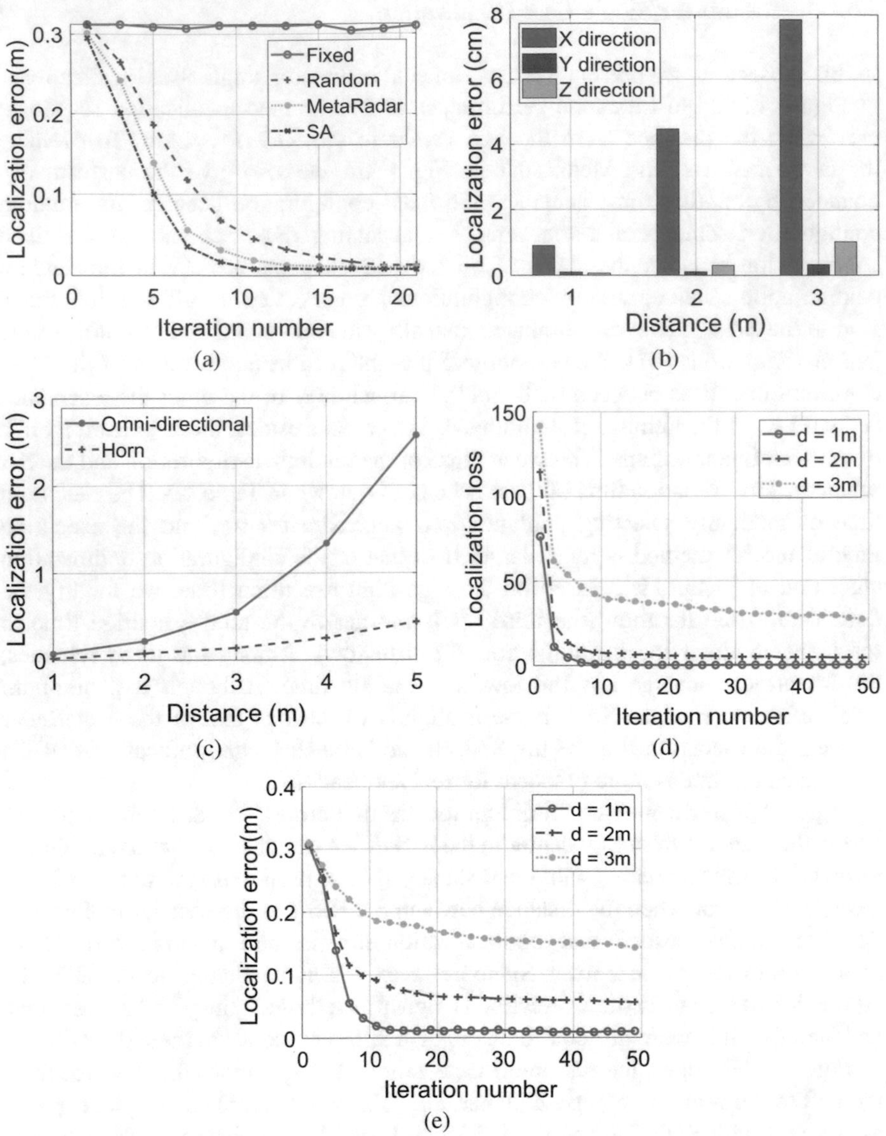

Fig. 4.35 The performance of fine-grained localization for single user: (**a**) Three configuration algorithms; (**b**) Minimum localization errors in different directions versus distance; (**c**) Minimum localization errors versus the distance using different antennas; (**d**) Localization loss versus iteration number; (**e**) Localization error versus iteration number

Table 4.5 Execution time of different localization schemes

Scheme	Random	MetaRadar	SA
Algorithm execution time (s)	0	0.0103	1.5311
Iteration duration (s)	0.085	0.1	1.62
Iteration number	19	13	11
Total execution time (s)	1.615	1.3	17.82

loss increases with the distance between the RIS and the SOI center. Besides, in Fig. 4.35e, we present the localization error l_e versus the iteration number n_c. Similar to the results in Fig. 4.35d, the localization error decreases when the iteration number increases, and increases with the distance. This implies that the localization error is positively correlated to the localization loss, which verifies the effectiveness of choosing localization loss to evaluate the accuracy of the localization algorithm of MetaRadar.

4.3.7.4 Results for Multiple User Localization Without Obstruction

In this subsection, we evaluate the system performance for multiple user localization without obstruction.

Figure 4.36 illustrates the process of multi-user localization without obstruction. For display simplicity, in this figure, we choose a planar SOI which is on the plane $z = 0$ with size $0.5 \times 0.5\,\text{m}^2$ and SOI distance $d = 1$ m. The ground truth of users' locations are labeled in the figures by the red triangle (the first user), yellow circle (the second user), and the black star (the third user), respectively. We can observe the RSS varies for different iterations. The probabilities are approximately uniformly distributed in each location in the first iteration (first row of subfigures in Fig. 4.36), while after several iterations, the probabilities of locations near the ground truth are obviously higher than those in the other locations, which implies the effectiveness of the multiple user localization scheme. We can also observe that the locations in the x direction near the ground truth have higher probabilities than the locations in the y direction, indicating that it is more likely to misjudge the x coordinate than the y coordinate of the user's location (similar to Fig. 4.35b, the x axis is perpendicular to the RIS, and the correlation of signals in this direction is higher than those in the y and z directions.). Besides, we can observe that for two user localization, the probability at location $(0.975, 0.125, 0)$ when $n_c = 9$ is smaller than that when $n_c = 5$. The decline in the probability is mainly due to the disturbance from the noise.

Figure 4.37 shows the fine-grained localization for multiple users without obstruction. In order to avoid the obstruction among users, all the users are all located at the plane parallel to the RIS. Since the LOS channel between the RIS and each user is not obstructed by other users, the interference caused by the existence of other users can be ignored.

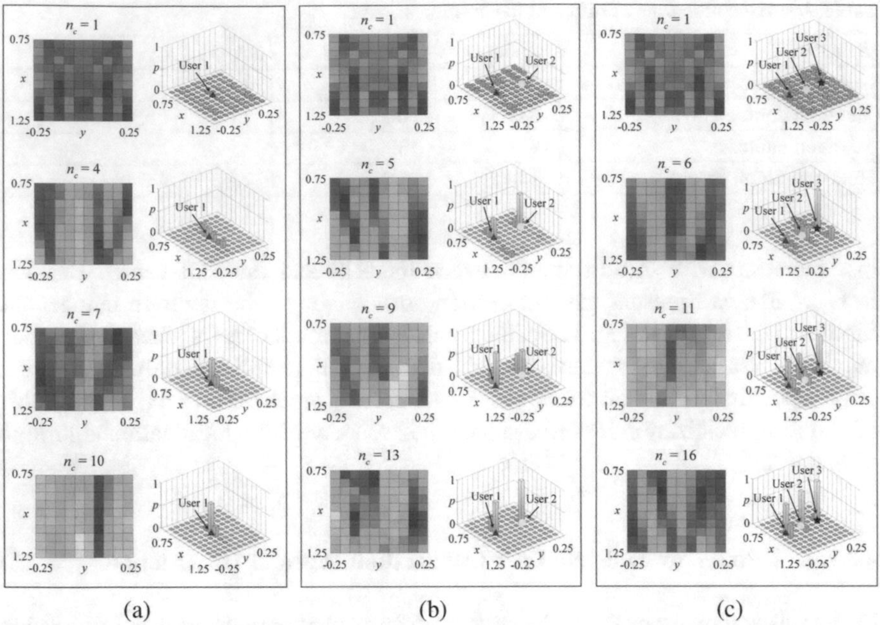

Fig. 4.36 Illustrations of the three-user localization process without obstruction: (**a**) One user; (**b**) Two users; (**c**) Three users. The first column in each subfigure shows the radio maps after different number of algorithm iterations, and the second column shows the corresponding probability distribution for different locations. Here, the probability is the sum of the probabilities of all the users. The ground truth of three users' locations are denoted by the red triangle, yellow circle, and the black star, respectively

Figure 4.37a presents the minimum localization error of the system l_e^m versus the distance d between the users' locations and the RIS when the number of users $I = 1, 2,$ and 3. We can observe that the localization error increases with the distance to the RIS. Besides, the localization error also increases with the number of users. Figure 4.37b shows the CDF of the localization error for different number of users when the distance $d = 2\,\text{m}$. It can also be observed that the localization accuracy decreases when the number of users increases. This is because the RIS configuration needs to be optimized for multiple users simultaneously, and the average signal variance introduced by the RIS for each user drops, which degrades the performance of the localization system.

Figure 4.37c depicts the localization loss l_u versus the iteration number n_c when the distance $d = 1\,\text{m}$, and Fig. 4.37d illustrates the localization error l_e versus iteration number n_c with the same distance between the RIS and the users. Similar to the results in Fig. 4.35d and d, the localization loss and the localization error for multiple user localization are also positively correlated. Besides, when the number of users increase, the localization loss and the error have slower decline speed, which implies a longer time for convergence.

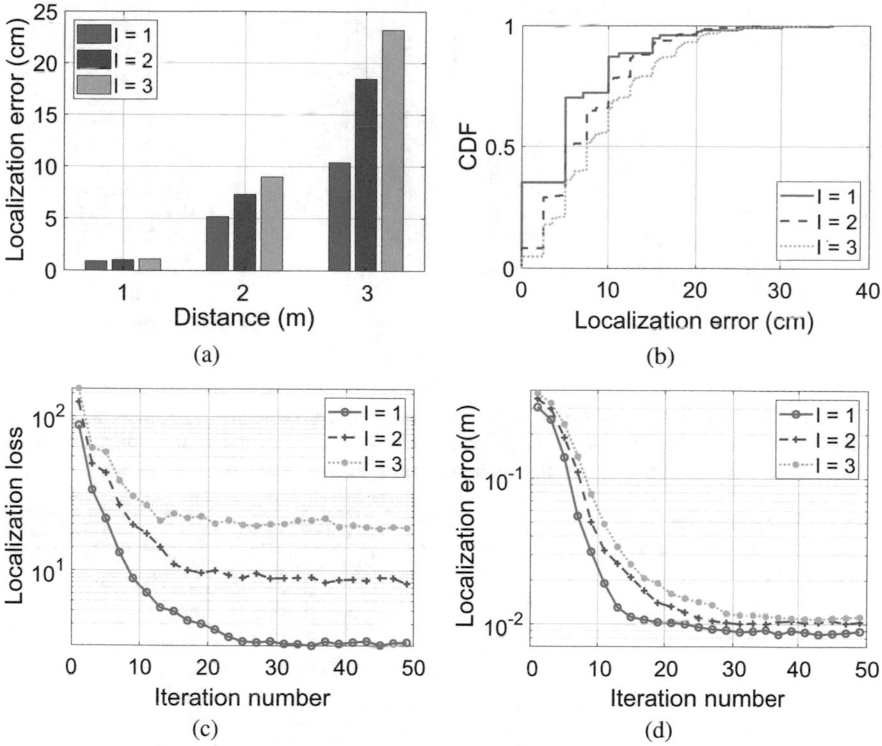

Fig. 4.37 When the number of users $I = 1, 2$, and 3, the performance of fine-grained localization for multiple users without obstruction: (**a**) Localization error versus distance; (**b**) The CDF of localization error; (**c**) Localization loss versus iteration number; (**d**) Localization error versus iteration number

4.3.7.5 Results for Multiple User Localization with Obstruction

We present the experimental results in Fig. 4.38 and Table 4.4 for multiple user localization with obstruction in this subsection. In Fig. 4.38a, we show the localization error l_e versus the distance d in this circumstance. Two users are placed in a line vertical to the RIS, and the distance between the two users is 0.5 m. The user closer to the RIS is denoted by i_1, and the other user is denoted by i_2. We can observe that at the same distance, the localization error of user i_2 is visibly larger than that of user i_1. This is because the existence of user i_1 will disturb the RF waves, and the received signals of user i_2 deviate from the results stored in the radio map.

Figure 4.38b presents the localization error l_e of user i_2 versus the iteration number n_c. The location of user i_2 is fixed at $(3, 0, 0)$. We can observe that for the same iteration number, the localization error of user i_2 decreases when the distance in the y direction between user i_1 and i_2 increases. This indicates that the localization error of a user is affected by the locations of other users.

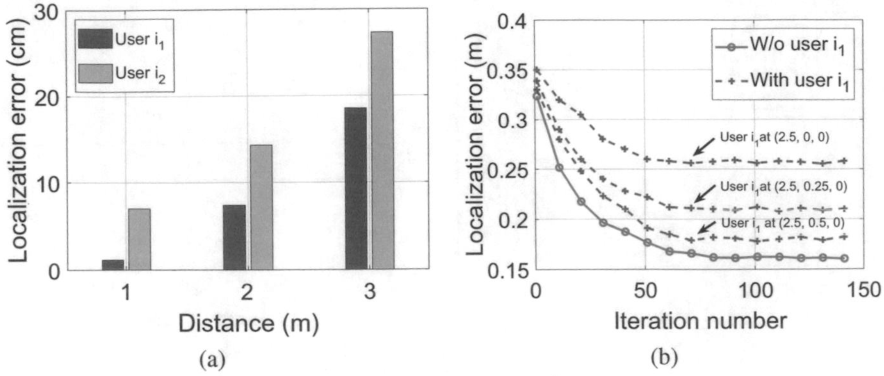

Fig. 4.38 The localization performance for two users with obstruction. The user closer to the RIS is denoted by i_1, and the other user is denoted by i_2: (**a**) Localization error versus distance; (**b**) Localization error of user i_2 versus iteration number with different users' locations

Table 4.6 The minimum localization error when $I = 3$ and $d = 1$ m

Case	Locations of users (m)	Error (cm)
1: same vertical plane $x = 1.375$ m	User 1: $(1.375, 0.025, 0.025)$ User 2: $(1.375, -0.125, -0.125)$ User 3: $(1.375, 0.175, 0.175)$	1.3
2: same horizontal plane $z = 0.025$ m	User 1: $(0.875, 0.025, 0.025)$ User 2: $(1.375, -0.125, 0.025)$ User 3: $(1.375, 0.175, 0.025)$	11.7
3: different horizontal planes	User 1: $(0.875, 0.025, -0.125)$ User 2: $(1.375, -0.125, 0.025)$ User 3: $(1.375, 0.175, 0.175)$	7.1

Table 4.6 presents the minimum localization error with three users in the same cubic SOI and $d = 1$ m. We can find that the minimum localization error varies in different cases where users have different positions. Specifically, case 1 has the lowest minimum localization error, and the minimum localization error of case 2 is higher than that of case 3. The reason is that the obstruction of the received signals is different in these cases. In case 1, users are in the same vertical plane, and thus there is no obstruction between the RIS and the users. However, the signals received by user 2 and 3 are partially obstructed by user 1 in case 2 and 3. Moreover, the obstruction is severer in case 3 because all the users are in the same horizontal plane, leading to lower localization accuracy in case 2.

4.3.8 Discussion

In this section, we investigate the indoor localization in a reconfigurable radio environment with RISs. However, to achieve the ubiquitous localization assisted by the RIS, some additional challenges need to be addressed.

Interference Among Multiple Users The fine-grained localization using the RSS based technique requires accurate radio maps. However, the existence of multiple users will introduce new multi-path effect, and the corresponding received signals are different from the values stored in the radio map which is constructed by a single user. As shown in Sect. 4.3.7.5, the localization accuracy will degrade if the LOS channel between the user and the RIS is obstructed by other users. In future designs, some heuristic methods can be developed to adjust the radio map to improve the localization accuracy. Based on the estimated locations of multiple users, the effect of interference among multiple users can be incorporated.

Support of a Larger Localization Area According to the experimental results, the localization error increases with the distance between the RIS and users. Therefore, the localization area is restricted comparing to conventional RSS based systems. There are two ways to increase the size of the localization area. The first way is to increase the size of the RIS. The RIS with larger reflection surface is able to control more reflected waves, and therefore a larger radio environment can be reconfigured for localization. The second way is to introduce more APs and RISs to enhance the localization accuracy for different areas. Specifically, we can deploy APs and RISs for every room in a building. Since the area of a room is comparable to the localization area of the RISs, in this way the localization accuracy of the whole building can be improved.

Uncontrollable Variance of Radio Environment Except the variance of the RIS reflection, the radio environment will also change with respect to time due to some uncontrollable factor, such as the movement of furniture, doors or any objects in the room. The system is unaware of the movements of objects, which is different from the interference among multiple users where the system knows the coarse locations of users. A research direction is to simultaneously locate the users and sense the variance of radio environment with the assist of the RIS. The sensing ability can be realized by imaging the indoor environment, which remains an active research topic of RISs applications [88]. Using the information of the sensed movements, the corresponding radio environment variance can be predicted to update the radio maps and improve the localization accuracy.

Narrow Bandwidth of RISs Since the reflectivity of the RIS is sensitive to the frequency of the wireless signals, the reflectivity of a metamaterial unit for different states can be distinguished only for a very narrow bandwidth. Therefore, we use a single frequency for localization. The signals in narrow bandwidth contain less information of the radio environment than the wideband signals, where the latter can further improve the localization performance. However, the design of a wideband RIS is still a challenging research topic.

Mobility of Users As shown in Table 4.2, the execution time of MetaRadar is 1.3 s, and thus the speed of the user needs to be less than 0.04 m/s according to the discussion in Section 5.3. If the user is with a higher speed, it will move to other blocks before the localization phase ends, resulting in the degradation of the localization performance. To support the high mobility scenario where users may be at different blocks in different iterations, a new location probability estimation function considering the correlation of these blocks over the time domain needs to be developed.

4.3.9 Summary

In this section, we have proposed the RIS assisted indoor localization system, MetaRadar. Unlike traditional RSS method, MetaRadar can control the RSS values in the radio environment by changing the configuration of the RIS, which contributes to the improvement of localization accuracy. We have proposed the compressive construction method to build the radio map for all the possible radio environment, and designed the configuration optimization algorithm to select the favorable radio environment for fine-grained localization. MetaRadar has been implemented using USRPs and the experiments have been conducted in a classroom with different distances to the RIS. The evaluation results have shown considerable improvement of the localization accuracy comparing to traditional RSS based systems with decimeter accuracy. Specifically, the proposed system equipped with a $0.48\,\mathrm{m}^2$ RIS can achieve a centimeter localization accuracy with up to 2 m localization range for single user and multiple users without obstruction.

4.3.10 Appendix

We first utilize the union bound method to derive an upper bound for the integral in (4.105), which is quite tight for high signal-to-noise ratios (SNRs) [89]. Specifically, we define the region $\mathscr{R}_{n',n}$ as

$$\mathscr{R}_{n',n} = \{s_i : (s_i - \mu_{n'})^2 \leq (s_i - \mu_n)^2\}. \tag{4.108}$$

Since $\mathscr{R}_{n'} \subseteq \mathscr{R}_{n',n}$, we have

$$\int_{\mathscr{R}_{n'}} \mathbb{P}(s_i|\boldsymbol{c}, n) \cdot ds_i \leq \int_{\mathscr{R}_{n',n}} \mathbb{P}(s_i|\boldsymbol{c}, n) \cdot ds_i. \tag{4.109}$$

Next, a closed form expression is provided for the upper bound. The region $\mathscr{R}_{n',n}$ can be expressed as

$$\begin{aligned}\mathscr{R}_{n',n} &= \{s_i : \mu_{n'}^2 - 2s_i\mu_{n'} \leq \mu_n^2 - 2s_i\mu_n\} \\ &= \left\{s_i : s_i\mu_{n'}^2 - s_i\mu_n^2 - \mu_{n'}\mu_n + \mu_n^2 \geq \frac{(\mu_{n'} - \mu_n)^2}{2}\right\} \\ &= \{s_i : (s_i - \mu_n)(\mu_{n'} - \mu_n) \geq 2d_{n,n'}^2\},\end{aligned} \tag{4.110}$$

where parameter $d_{n,n'}$ is

$$d_{n,n'} = \frac{|\mu_{n'} - \mu_n|}{2}. \tag{4.111}$$

Since $(s_i - \mu_n)(\mu_{n'} - \mu_n)$ follows Gaussian distribution with mean being 0 and variance being $\sigma^2(\mu_{n'} - \mu_n)^2$, we have

$$\int_{\mathscr{R}_{n',n}} \mathbb{P}(s_i|\boldsymbol{c}, n) \cdot ds_i = Q\left(\frac{2d_{n,n'}^2}{\sigma|\mu_{n'} - \mu_n|}\right) \leq \frac{1}{2}e^{-\frac{d_{n,n'}^2}{2\sigma^2}}, \tag{4.112}$$

where the relationship $Q(x) \leq \frac{1}{2}e^{-x^2/2}$ is utilized in the last step. Therefore, the upper bound of the localization error can be expressed as

$$l(\mathbf{C}) \leq l_u(\boldsymbol{C}) = \sum_{i\in\mathscr{I}} \sum_{n\in\mathscr{N}} p_{i,n} \sum_{n'\in\mathscr{N}} \frac{\gamma_{n,n'}}{2} e^{-\frac{d_{n,n'}^2}{2\sigma^2}}. \tag{4.113}$$

References

1. A. Bourdoux, A.N. Barreto, B. Liempd, C. Lima, D. Dardari, D. Belot, E.-S. Lohan, G. Seco-Grandos, H. Sarieddeen, H. Wymeersch, J. Suutala, J. Saloranta, M. Guillaud, M. Isomursu, M. Valkama, M.R.K. Aziz, R. Berkvens, T. Sanguanpuak, T. Svensson, Y. Miao, 6G white paper on localization and sensing (2020). arxiv: https://arxiv.org/abs/2006.01779
2. M. Di Renzo, M. Debbah, D.-T. Phan-Huy, A. Zappone, M.-S. Alouini, C. Yuen, V. Sciancalepore, G.C. Alexandropoulos, J. Hoydis, H. Gacanin, J.D. Rosny, A. Bounceu, G. Lerosey, M. Fink, Smart radio environments empowered by AI reconfigurable meta-surfaces: An idea whose time has come. EURASIP J. Wireless Commun. Netw. **2019**(1), 120 (2019)
3. S. Kianoush, S. Savazzi, F. Vicentini, V. Rampa, M. Giussani, Device-free RF human body fall detection and localization in industrial workplaces. IEEE Internet Things J. 4(2), 351–362 (2017)
4. P.W.Q. Lee, W.K.G. Seah, H. Tan, Z. Yao, Wireless sensing without sensors - An experimental approach, in *Proc. IEEE Int. Symp. Pers. Indoor Mobile Radio Commun., Tokyo* (2009)
5. T. He, S. Krishnamurthy, L. Luo, T. Yan, L. Gu, R. Stoleru, G. Zhou, Q. Cao, P. Vicaire, J.A. Stankovic, T.F. Abdeizaher, J. Hui, B. Krogh, Vigilnet: An integrated sensor network system for energy efficient surveillance. ACM Trans. Sensor Netw. **2**(1), 1–38 (2006)
6. D.J. Cook, M. Schmitter-Edgecombe, Assessing the quality of activities in a smart environment. Methods Inform. Med. **48**(5), 480–485 (2009)
7. M.G. Amin, Y.D. Zhang, F. Ahmad, K.D. Ho, Radar signal processing for elderly fall detection: The future for in-home monitoring. IEEE Signal Process. Mag. **33**(2), 7180 (2016)
8. T. Le, M. Nguyen, T. Nguyen, Human posture recognition using human skeleton provided by Kinect, in *Proc. IEEE ComManTel, Ho Chi Minh City* (2013)
9. B. Kellogg, V. Talla, S. Gollakota, Bringing gesture recognition to all devices, in *Proc. USENIX Conf. Netw. Syst. Des. Implementation, Seattle, WA* (2014)

10. L. Yao, Q. Sheng, W. Ruan, T. Gu, X. Li, N. Falkner, Z. Yang, RFcare: Device-free posture recognition for elderly people using a passive RFID tag array, in *Proc. Int. Conf. Mobile Ubiquitous Syst. Comput. Netw. Services, Coimbra* (2015)
11. H. Wang, D. Zhang, Y. Wang, J. Ma, Y. Wang, S. Li, RT-Fall: A real-time and contactless fall detection system with commodity WiFi devices. IEEE Trans. Mobile Comput. **16**(2), 511–526 (2016)
12. F. Adib, C.-Y. Hsu, H. Mao, D. Katabi, F. Durand, Capturing the human figure through a wall. ACM Trans. Graphics **34**(6), 219 (2015)
13. Q. Xu, Y. Chen, B. Wang, K.R. Liu, Radio biometrics: Human recognition through a wall. IEEE Trans. Inf. Forens. Secur. **12**(5), 11411155 (2017)
14. D. Sasakawa, N. Honma, T. Nakayama, S. Iizuka, Human posture identification using a MIMO array. Electronics **7**(3), 37 (2018)
15. N. Honma, D. Sasakawa, N. Shiraki, T. Nakayama, S. Iizuka, Human monitoring using MIMO radar, in *Proc. IEEE Int. Workshop Electromagn.: Appl. Student Innovation Competition, Nagoya* (2018)
16. N. Kaina, M. Dupre, G. Lerosey, M. Fink, Shaping complex microwave fields in reverberating media with binary tunable metasurfaces. Sci. Rep. **4**(1), 18 (2014)
17. T. Zhou, H. Li, D. Ye, J. Huangfu, S. Qiao, Y. Sun, W. Zhu, C. Li, L. Ran, Short-range wireless localization based on meta-aperture assisted compressed sensing. IEEE Trans. Microw. Theory Technol. **65**(7), 25162524 (2017)
18. J. Hu, H. Zhang, B. Di, L. Li, L. Song, Y. Li, Z. Han, H.V. Poor, Reconfigurable intelligent surfaces based radio-frequency sensing: Design, optimization, and implementation. IEEE J. Sel. Areas Commun. **38**(11), 2700–2716 (2020)
19. B. Di, H. Zhang, L. Li, L. Song, Y. Li, Z. Han, Practical hybrid beamforming with limited-resolution phase shifters for reconfigurable intelligent surface based multi-user communications. IEEE Trans. Veh. Technol. **69**(4), 45654570 (2020)
20. H. Zhang, B. Di, L. Song, Z. Han, Reconfigurable intelligent surfaces assisted communications with limited phase shifts: How many phase shifts are enough? IEEE Trans. Veh. Technol. **69**(4), 44984502 (2020)
21. Y. Huang, A. Charbonneau, L. Talbi, T.A. Denidni, Effect of human body upon line-of-sight indoor radio propagation, in *Proc. Canadian Conf. Elect. Comput. Eng., Ottawa, Ont.* (2006)
22. W. Tang, M. Chen, X. Chen, J. Dai, Y. Han, M. Di Renzo, Y. Zeng, S. Jin, Q. Cheng, T. Cui, Wireless communications with reconfigurable intelligent surface: Path loss modeling and experimental measurement. IEEE Trans. Wireless Commun. **20**(1), 421–439 (2021)
23. B. Di, H. Zhang, L. Li, L. Song, Z. Han, H.V. Poor, Hybrid beamforming for reconfigurable intelligent surface based multi-user communications: Achievable rates with limited discrete phase shifts. IEEE J. Sel. Areas Commun. **38**(8), 18091822 (2020)
24. Y.C. Eldar, G. Kutyniok, *Compressed Sensing: Theory and Applications* (Cambridge University Press, New York, 2012)
25. Z. Han, H. Li, W. Yin, *Compressive Sensing for Wireless Networks* (Cambridge University Press, New York, 2013)
26. M. Elad, Optimized projections for compressed sensing. IEEE Trans. Signal Process. **55**(12), 56955702 (2007)
27. M.D. Migliore, D. Pinchera, Compressed sensing in electromagnetics: Theory, applications and perspectives, in *Proc. European Conf. Antennas Propag., Rome* (2011)
28. R.M. Lewis, A. Shepherd, V. Torczon, Implementing generating set search methods for linearly constrained minimization. SIAM J. Sci. Comput. **29**(6), 25072530 (2007)
29. I. Goodfellow, Y. Bengio, A. Courvile, Y. Bengio, *Deep Learning* (MIT Press, Cambridge, 2016)
30. C.M. Bishop, *Pattern Recognition and Machine Learning* (Springer, New York, 2006)
31. F.J. Pineda, Generalization of back-propagation to recurrent neural networks. Physical Rev. Lett. **59**(19), 2229 (1987)

32. Z. Ugray, L. Lasdon, J. Plummer, F. Glover, J. Kelly, R. Martí, Scatter search and local nlp solvers: A multistart framework for global optimization. INFORMS J. Comput. **19**(3), 328340 (2007)
33. S. Ledesma, M. Torres, D. Hernández, G. Aviña, G. García, Temperature cycling on simulated annealing for neural network learning, in *Mexican Int. Conf. Artificial Intell., Aguascalientes* (2007)
34. H.V. Poor, *An Introduction to Signal Detection and Estimation*, 2nd ed. (Springer, New York, 1994)
35. G.B. Arfken, H.J. Weber, *Mathematical Methods for Physicists* (Academic, San Diego, 1995)
36. L. Li, H. Ruan, C. Liu, Y. Li, Y. Shuang, A. Alu, C.-W. Qiu, T.J. Cui, Machine-learning reprogrammable metasurface imager. Nat. Commun. **10**(1), 1082 (2019)
37. F. Hirtenfelder, Effective antenna simulations using CST MICROWAVE STUDIO®, in *Proc. Int. ITG Conf. Antennas, Munich* (2007)
38. O. Holland, H. Bogucka, A. Medeisis, The universal software radio peripheral (USRP) family of low-cost SDRs, in *Opportunistic Spectr. Sharing White Space Access: The Practical Reality*, vol. 323 (2015)
39. E. Blossom, GNU radio: Tools for exploring the radio frequency spectrum. J. Linux **2004**(122), 4 (2004)
40. A. Goldsmith, *Wireless Communications* (Cambridge University Press, Cambridge, 2005)
41. R. Obermeier, J.A. Martinez-Lorenzo, Sensing matrix design via capacity maximization for block compressive sensing applications. IEEE Trans. Comput. Imag. **5**(1), 2736 (2019)
42. S. Boyd, L. Vandenberghe, *Convex Optimization* (Cambridge University Press, Cambridge, 2004)
43. R. Obermeier, J.A. Martinez-Lorenzo, Sensing matrix design via mutual coherence minimization for electromagnetic compressive imaging applications. IEEE Trans. Comput. Imaging **3**(2), 217229 (2017)
44. Z. Li, Y. Xie, L. Shangguan, R.I. Zelaya, J. Gummeson, W. Hu, K. Jamieson, Programmable radio environments with large arrays of inexpensive antennas. GetMobile: Mobile Comput. Commun. **23**(3), 2327 (2019)
45. H. Gacanin, M.D. Renzo, Wireless 2.0: Towards an intelligent radio environment empowered by reconfigurable meta-surfaces and artificial intelligence. IEEE Veh. Technol. Mag. **15**(4), 74–82 (2020)
46. S. Zhang, H. Zhang, B. Di, Y. Tan, Z. Han, L. Song, Beyond intelligent reflecting surfaces: Reflective-transmissive metasurface aided communications for full-dimensional coverage extension. IEEE Trans. Veh. Technol. **69**(11), 13905–13909 (2020)
47. M.A. ElMossallamy, H. Zhang, L. Song, K.G. Seddik, Z. Han, G.Y. Li, Reconfigurable intelligent surfaces for wireless communications: Principles, challenges, and opportunities. IEEE Trans. Cognitive Commun. Netw. **6**(3), 9901002 (2020)
48. H. Hashida, Y. Kawamoto, N. Kato, Intelligent reflecting surface placement optimization in air-ground communication networks toward 6G. IEEE Wireless Commun. **27**(6), 146–151 (2020)
49. Y. Chen, B. Ai, H. Zhang, Y. Niu, L. Song, Z. Han, H.V. Poor, Reconfigurable intelligent surface assisted device-to-device communications. IEEE Trans. Wireless Commun. **20**(5), 2792–2804 (2021)
50. R. Sutton, A. Barto, *Reinforcement Learning: An Introduction* (MIT Press, Cambridge, 2018)
51. J. Hu, H. Zhang, K. Bian, M.D. Renzo, Z. Han, L. Song, MetaSensing: Intelligent metasurface assisted RF 3D sensing by deep reinforcement learning. IEEE J. Sel. Areas Commun. arxiv: https://arxiv.org/pdf/2011.12515.pdf
52. L. Dai, B. Wang, M. Wang, X. Yang, J. Tan, S. Bi, S. Xu, F. Yang, Z. Chen, M.D. Renzo, C.B. Chae, L. Hanzo, Reconfigurable intelligent surface-based wireless communications: Antenna design, prototyping, and experimental results. IEEE Access **8**, 45913–45923 (2020)
53. R. McDonough, A. Whalen, *Detection of Signals in Noise* (Academic Press, San Diego, 2004)
54. J. Bezdek, R. Hathaway, Convergence of alternating optimization. Neural Parallel Sci. Comput. **11**(4), 351368 (2003)

55. M. Volodymyr, K. Koray, S. David, A.A. Rusu, V. Joel, M.G. Bellemare, G. Alex, R. Martin, A.K. Fidjeland, O. Georg, Human level control through deep reinforcement learning. Nature **518**(7540), 529533 (2015)
56. J. Hu, H. Zhang, L. Song, Z. Han, H.V. Poor, Reinforcement learning for a cellular Internet of UAVs: Protocol design, trajectory control, and resource management. IEEE Wirel. Commun. **27**(1), 116–123 (2020)
57. A. Zappone, M. Di Renzo, M. Debbah, T.T. Lam, X. Qian, Model aided wireless artificial intelligence: Embedding expert knowledge in deep neural networks for wireless system optimization. IEEE Veh. Technol. Mag. **14**(3), 6069 (2019)
58. R.Y. Rubinstein, D.P. Kroese, *Simulation and the Monte Carlo Method* (Wiley, Berlin, 2008)
59. J.F. Bailyn, Generalized inversion. Nat. Lang. Linguist. Theory **22**(1), 150 (2004)
60. S.S. Skiena, *Sorting and Searching* (Springer, London, 2012)
61. M. Sipper, A serial complexity measure of neural networks, in *Proc. IEEE ICNN, San Francisco, CA* (1993)
62. M.D. Petkovi, P.S. Stanimirovi, Generalized matrix inversion is not harder than matrix multiplication. J. Comput. Appl. Math. **230**(1), 270–282 (2009)
63. Y. Xu, W. Yin, Block stochastic gradient iteration for convex and nonconvex optimization. SIAM J. Optim. **25**(3), 16861716 (2015)
64. F. Zafari, A. Gkelias, K.K. Leung, A survey of indoor localization systems and technologies. IEEE Commun. Surv. Tutorials **21**(3), 2568–2599 (2019)
65. L. Mainetti, L. Patrono, I. Sergi, A survey on indoor positioning systems, in *Proc. Int. Conf. Software, Telecommun. Computer Networks, Split* (2014)
66. A. Yassin, Y. Nasser, M. Awad, A. Al-Dubai, R. Liu, C. Yuen, R. Raulefs, E. Aboutanios, Recent advances in indoor localization: A survey on theoretical approaches and applications. IEEE Commun. Surv. Tutorials **19**(2), 1327–1346 (2016)
67. M. Ibrahim, M. Torki, M. ElNainay, CNN based indoor localization using RSS time-series, in *Proc. IEEE Symp. Computers Commun., Hague* (2018)
68. Z. Yang, C. Wu, Y. Liu, Locating in fingerprint space: wireless indoor localization with little human intervention, in *Proc. ACM Mobicom, Istanbul* (2012)
69. H. Zhang, H. Zhang, B. Di, K. Bian, Z. Han, L. Song, Towards ubiquitous positioning by leveraging reconfigurable intelligent surface. IEEE Commun. Lett. **25**(1), 284–288 (2021)
70. S. Zeng, H. Zhang, B. Di, Z. Han, L. Song, Reconfigurable intelligent surface (RIS) assisted wireless coverage extension: RIS orientation and localization optimization. IEEE Commun. Lett. **25**(1), 269–273 (2021)
71. P. Bahl, V. Padmanabhan, Radar: an in-building RF-based user location and tracking system, in *Proc. IEEE INFOCOM, Tel Aviv* (2000)
72. P. Yang, W. Wu, M. Moniri, C.C. Chibelushi, Efficient object localization using sparsely distributed passive RFID tags. IEEE Trans. Ind. Electron. **60**(12), 5914–5924 (2012)
73. Y. Ma, N. Selby, F. Adib, Minding the billions: Ultra-wide band localization for deployed rfid tags, in *Proc. ACM MobiCom, Snowbird* (2017)
74. L. Li, P. Xie, J. Wang, Rainbowlight: Towards low cost ambient light positioning with mobile phones, in *Proc. ACM MobiCom, New Delhi* (2018)
75. C. Zhang, X. Zhang, Pulsar: Towards ubiquitous visible light localization, in *Proc. ACM MobiCom, Snowbird* (2017)
76. L.-X. Chuo, Z. Luo, D. Sylvester, D. Blaauw, H.-S. Kim, Rf-echo: A non-line-of-sight indoor localization system using a low-power active RF reflector ASIC tag, in *Proc. 23rd Annu. Int. Conf. Mobile Computing and Netw., Snowbird* (2017)
77. C.X. Lu, Y. Li, P. Zhao, C. Chen, L. Xie, H. Wen, R. Tan, N. Trigoni, Simultaneous localization and mapping with power network electromagnetic field, in *Proc. ACM MobiCom, New Delhi* (2018)
78. Z. Farid, R. Nordin, M. Ismail, Recent advances in wireless indoor localization techniques and system. J. Comput. Netw. Commun. **2013**, 1–12 (2013)
79. S. Liu, M. Rao, Y. Tao, L. Liu, P. Zhang, A virtual TDOA localization scheme of Chinese DTMB signal in radio monitoring networks. China Commun. **12**(11), 1–13 (2015)

80. P. Castro, P. Chiu, T. Kremenek, R. Muntz, A probabilistic room location service for wireless networked environments, in *Proc. Int. Conf. Ubiquitous Computing, Berlin, Heidelberg* (2001)
81. M. Youssef, A. Agrawala, The Horus WLAN location determination system, in *Proc. ACM MobiCom., Seattle, WA* (2005)
82. A. Rai, K.K. Chintalapudi, V.N. Padmanabhan, R. Sen, Zee: Zero-effort crowdsourcing for indoor localization, in *Proc. ACM MobiCom., Istanbul* (2012)
83. H. Zhang, J. Hu, H. Zhang, B. Di, K. Bian, Z. Han, L. Song, MetaRadar: Indoor localization by reconfigurable metamaterials. IEEE Trans. Mobile Comput. arxiv: https://arxiv.org/pdf/2008.02459.pdf
84. A.B. Li, S. Singh, D. Sievenpiper, Metasurfaces and their applications. Nanophotonics **7**(6), 989–1011 (2018)
85. H. Hashemi, The indoor radio propagation channel. Proc. IEEE **81**(7), 943–968 (1993)
86. C.A. Balanis, *Antenna Theory: Analysis and Design* (Wiley, Hoboken, 2016)
87. P.J.M. Laarhoven, E.H.L. Aarts, *Simulated Annealing: Theory and Applications* (Springer, New York, 1987)
88. J. Hunt, J. Gollub, T. Driscoll, G. Lipworth, A. Mrozack, M.S. Reynolds, D.J. Brady, D.R. Smith, Metamaterial microwave holographic imaging system. J. Opt. Soc. Am. A Opt. Image Sci. Vis. **31**(10), 2109–2119 (2014)
89. J.G. Proakis, M. Salehi, *Digital Communications*. (McGraw-Hill Companies, New York, 2007)

Printed in the United States
by Baker & Taylor Publisher Services